Deleuze|Guattari & Ecology

Deleuze|Guattari & Ecology

Edited by

Bernd Herzogenrath
University of Frankfurt

palgrave
macmillan

First published 2009 by
PALGRAVE MACMILLAN

Palgrave Macmillan in the UK is an imprint of Macmillan Publishers Limited,
registered in England, company number 785998, of Houndmills, Basingstoke,
Hampshire RG21 6XS.

Palgrave Macmillan in the US is a division of St Martin's Press LLC,
175 Fifth Avenue, New York, NY 10010.

Palgrave Macmillan is the global academic imprint of the above companies
and has companies and representatives throughout the world.

Palgrave® and Macmillan® are registered trademarks in the United States,
the United Kingdom, Europe and other countries.

ISBN-13: 978–0–230–52744–7 hardback

This book is printed on paper suitable for recycling and made from fully
managed and sustained forest sources. Logging, pulping and manufacturing
processes are expected to conform to the environmental regulations of the
country of origin.

A catalogue record for this book is available from the British Library.

A catalog record for this book is available from the Library of Congress.

10 9 8 7 6 5 4 3 2 1
18 17 16 15 14 13 12 11 10 09

Transferred to Digital Printing in 2010.

Contents

Acknowledgements

I would like to thank Dan Bunyard and all the good people at Palgrave for giving me this opportunity...

Thanx to Hanjo Berressem for kick-starting my interest in Deleuze and adding fuel to the fire ever since...

Thanx to Janna "Best-Jamni-Ever!" Wanagas, without whose help this would not have been possible...

Contributors

Yves Abrioux is professor of English literature at the University of Paris VIII. He is the author of *Ian Hamilton Finlay—A Visual Primer* (Reaktion Books, 1994), and has published extensively on literary theory, science and literature from a theoretical angle, English and American literature, garden and landscape theory, and contemporary art. He has edited several issues of the journal *TLE* (Presses universitaires de Vincennes), with contributions from European, American, etc., scholars, including an issue on Gilles Deleuze ("Deleuze-chantier," 2001: coeditor) and directed a research project on public policies and landscaping for the French Ministry of the Environment (2004). He is a practicing artist and has exhibited internationally collaborative projects on questions of landscape and gardens.

Hanjo Berressem teaches American Literature and Culture at the University of Cologne. His interests include literary theory, contemporary American fiction, media studies and the interfaces of art and science. Apart from articles situated mainly in these fields, he has published books on Thomas Pynchon (*Pynchon's Poetics: Interfacing Theory and Text*. University of Illinois Press, 1992) and on Witold Gombrowicz (*Lines of Desire: Reading Gombrowicz's Fiction with Lacan*. Northwestern University Press, 1999). He is the editor, together with D. Buchwald and H. Volkening, of the collection *Grenzüberschreibungen: Feminismus und Cultural Studies*, Aisthesis, Bielefeld, 2001, and, together with D. Buchwald, of *Chaos-Control|Complexity: Chaos Theory and the Human Sciences* (Special Issue of *American Studies* 1:2000). At the moment, he is completing a book on Deleuze: *Radical Deleuze: Radical Constructivism and the Deleuzian Event*.

Ronald Bogue is a Distinguished Research Professor in the Comparative Literature Department at the University of Georgia, United States. He is the author of *Deleuze and Guattari* (1989), *Deleuze on Literature* (2003), *Deleuze on Cinema* (2003), *Deleuze on Music, Painting, and the Arts* (2003), and *Deleuze's Wake: Tributes and Tributaries* (2004).

Verena Andermatt Conley is Visiting Professor of Literature and of Romance Languages and Literatures at Harvard University. Publications include *Ecopolitics: The Environment in Poststructuralist Thought* (Routledge, 1997), *The War with the Beavers: Learning to be Wild in the North Woods* (Minnesota, 2003; 2005), and *Fictional Spaces: Space, Globalization and Subjectivity in Contemporary French Thought* (in progress).

Manuel DeLanda is the author of four philosophy books, War in the Age of Intelligent Machines (1991), A Thousand Years of Nonlinear History (1997),

Intensive Science and Virtual Philosophy (2002) and A New Philosophy of Society (2006), as well as of many philosophical essays published in various journals and collections. He teaches two seminars at University of Pennsylvania, Department of Architecture: Philosophy of History: Theories of Self-Organization and Urban Dynamics, and Philosophy of Science: Thinking about Structures and Materials.

Matthew Fuller is David Gee Reader in Digital Media at Goldsmiths College, University of London. Prior to this he was Lector in Media Design at the Piet Zwart Institute in Rotterdam. He is author of a number of books including *Behind the Blip: Essays on the Culture of Software* (Autonomedia) and *Media Ecologies: Materialist Energies in Art and Technoculture* (MIT Press). His edited titles include the recent *Software Studies, a Lexicon* (MIT Press) a book which develops efforts to discuss software from the perspectives of cultural theory and art methodologies. He is involved in a number of projects in art, experimental software and media. Research towards this essay was supported by the Fonds voor Beeldende Kunst, Vormgeving en Bouwkunst of the Netherlands.

Gary Genosko is Canada Research Chair in Technoculture at Lakehead University, Canada. He is editor of *The Guattari Reader* and the three-volume collection *Deleuze and Guattari: Critical Assessments*. He is author of *Felix Guattari: An Aberrant Introduction* and *The Party Without Bosses: Lessons on Anti-Capitalism from Felix Guattari and Lula da Silva*.

Mark Halsey teaches in the School of Law at Flinders University. He is the author of *Deleuze and Environmental Damage* (Ashgate), and his work on Deleuze and ecology has appeared in *Angelaki, Ethics & the Environment, Rhizomes*, and in the edited collection *Deleuzian Encounters* (Palgrave).

Bernd Herzogenrath teaches American Literature and Culture at the University of Frankfurt. He is the author of *An Art of Desire: Reading Paul Auster* (Rodopi, 1999), and the editor of *From Virgin Land to Disney World: Nature and Its Discontents in the USA of Yesterday and Today* (Rodopi, 2001), and *The Cinema of Tod Browning* (Black Dog, 2006). His fields of interest are nineteenth and twentieth-century American literature, critical theory, and cultural|media studies. He has just finished a project on a "Deleuzian History of the American Body|Politic" – future publications include an anthology of essays on the "King of the Bs," Edgar G. Ulmer, and a collection of essays on *Intermedia[lity]* (both forthcoming 2008).

Jonathan Maskit is Assistant Professor of Philosophy at Denison University (Ohio, United States). He has published in *Philosophy & Geography* (now *Ethics, Place, and Environment*) and *Research in Philosophy and Technology*, and has contributed to volumes on deep ecology, continental philosophy, and others. His research focuses on the intersection of continental philosophy, esthetics, and environmental philosophy. He is at work on a book entitled *On Nature and Culture in the Discourse of Modernity*.

Dorothea Olkowski is Professor and Chair of Philosophy, and former Director of Women's Studies at the University of Colorado at Colorado Springs. She has recently completed three books. *The Universal (In the Realm of the Sensible)*, a co-publication of Edinburgh University Press Columbia University Press (2007); *Feminist Interpretations of Maurice Merleau-Ponty*, coedited with Gail Weiss (Penn State University Press, 2006); and *The Other—Feminist Reflections in Ethics*, Helen Fielding, Gabrielle Hiltman, Dorothea Olkowski, Anne Reichold, eds. (Palgrave Publishers, 2007). She is also the author of *Gilles Deleuze and The Ruin of Representation* (University of California Press, 1999), and edited *Resistance, Flight, Creation, Feminist Enactments of French Philosophy* (Cornell University Press, 2000), co-edited *Re-Reading Merleau-Ponty, Essays Beyond the Continental-Analytic Divide* (with Lawrence Hass), (Humanity Books, 2000); *Merleau-Ponty, Interiority and Exteriority, Psychic Life and the World* (with James Morley), (SUNY Press, 1999); and *Gilles Deleuze and the Theater of Philosophy* (with Constantin V. Boundas), (Routledge Press, 1994). She is currently working on a book, *Nature, Ethics, Love*, for Columbia and Edinburgh University Presses.

Luciana Parisi conveys the MA Interactive Media at the Centre for Cultural Studies, Goldsmiths University of London. She has published various articles in *Tekhnema, Parallax, Ctheory, Social Text*, and *Theory, Culture & Society* on the relation between science, technology and the ontogenetic dimensions of evolution in nature, culture and capitalism. Her research has also focused on the impact of biotechnologies on the concepts of the body, sex, femininity and desire. In 2004, she published *Abstract Sex. Philosophy, Biotechnology and the Mutations of Desire* (Continuum Press). Most recently, her interest in interactive media technologies has led her new research towards the study of cybernetic memory, generative or soft architecture in relation to perceptive and affective space. She is currently working on digital architecture.

John Protevi is Associate Professor of French Studies at Louisiana State University. Among other works, he is the coauthor of *Deleuze and Geophilosophy* (Edinburgh UP, 2004) and editor of *The Edinburgh Dictionary of Continental Philosophy* (Edinburgh UP, 2005).

Stephen Zepke teaches Philosophy at the University of Vienna. His work concerns modern and contemporary intersections of art, philosophy and politics. He is the author of "Art as Abstract Machine: Aesthetics and Ontology in Deleuze and Guattari" and the editor (with Simon O'Sullivan) of *Deleuze, Guattari and the Production of the New* (Continuum, forthcoming 2008) and *Deleuze and Contemporary Art* (Edinburgh University Press, forthcoming 2009).

1
Nature|Geophilosophy| Machinics|Ecosophy

Bernd Herzogenrath

> It is probable that at a certain level nature and industry are two separate and distinct things.
>
> (*Anti-Oedipus* 3–4)

> There is no such thing as either man or nature now, only a process that produces the one within the other and couples the machines together.
>
> (*Anti-Oedipus* 2)

The last two decades have witnessed an immense interest in ecological questions. Book titles in Philosophy and Cultural Studies with reference to (deep) ecological issues abound, and the environment is on the public agenda—and no wonder, facing a world on the brink (as Al Gore and others state) of a climate collapse.

Association Test One: If I said "Postmodernism," what would you say? Maybe something like "deconstruction" or "death of the author." I guess the response "Nature" would be quite unlikely. In fact, "nature" has been having a hard time in postmodernism. "Nature" in postmodern thought—and I am aware of the danger of generalizing here—is by default culturized, semiotic nature, the nature of social or|and linguistic constructivism, connected mainly with the theories of Michel Foucault, Jacques Lacan, Jacques Derrida, and Judith Butler. Postmodernism writes a history of nature as a retro-effect of culture, of the different modes of the construction of nature, but it is not concerned with nature as materiality, or with nature|culture feedback loops other than one-sided cultural constructions|representations of nature. Such a position implies a move away from essentialism, however defined— something postmodernism fears most. There is no natural and originary nature—at least there is no access to such a chimera, since it is *always already* outside (and constructed by) culture and language. Although the alternative cannot be to go back behind the findings and analyses of postmodernism, it should be nevertheless noted that the dominant version of poststructuralism

1

in the guise of cultural|linguistic constructivism has ultimately dismissed the category of nature—the materiality of nature—by aiming to translate it without remainder (or only as negativity, as the impossible real) into the realm of representation. Much of postmodern ecocriticism has followed this route—Cheryl Glotfelty, in her introduction to the influential anthology *The Ecocriticism Reader* (1996), answers the question "What then *is* ecocriticsm?" in the following way:

> Simply defined, ecocriticism is the study of the relationship between literature and the physical environment. Just as feminist criticism examines language and literature from a gender-conscious perspective, and Marxist criticism brings an awareness of modes of production and economic class to its reading of texts, ecocriticism takes an earth-centered approach to literary studies.
>
> Ecocritics and theorists ask questions like the following: How is nature represented in this sonnet? What role does the physical setting play in the plot of this novel? Are the values expressed in this play consistent with ecological wisdom? How do our metaphors of the land influence the way we treat it? How can we characterize nature writing as a genre? (xix)

Ecocriticism thus wants to evaluate texts and ideas using "the environment" as a secure and transcendent measuring rod—the very "nature" that the texts in question are said to "construct" in the first place!

According to Lacan, one can only know about the impossible real (which denotes the strategic place of the unconscious, the body, materiality and the referent) through representation. The speaking human subject *qua* signifier is *always already* inscribed in the symbolic, so that "[n]ature provides...signifiers, and these signifiers organize human relations in a creative way, providing them with structures and shaping them" (*Fundamental Concepts* 20). In fact, for Lacan, "the symbol manifests itself first of all as the murder of the thing" (*Écrits* 104).

In a similar fashion, Derrida claims that "*[t]here is nothing outside of the text* [there is no outside-text; *il n'y a pas de hors-texte*]....there has never been anything but writing;...that what opens meaning and language is writing as the disappearance of natural presence" (158–159). He indicates that even if outside the text there are material conditions, these "outside conditions" are *always already* represented, materialities turned into mere con*text*. These versions of poststructuralism have exorcised nature and materiality out of representation and have thus closed representation in on itself. "Nature" in much of postmodern theory is "a product of discourse or intersecting textualities, as the world becomes a ceaseless play of interlocking and conflicting texts, spoken from different locations and negotiated across different perspectives" (Zita 89).

Even Judith Butler, whose forays into "bodies that matter" are among the most sophisticated attempts to "incorporate" matter into discourse and representation, ultimately fails to escape the exclusionary logic of belatedness, according to which materiality is *always already* a function of discourse. Butler's theory of gender performativity is ultimately an updated version of cultural|linguistic constructivism, construction being understood here as a process of materialization constituting types of bodies by means of the repetition of gender norms. Butler goes so far as to acknowledge a dynamism to matter, but this dynamism is a product of the discursive powers that matter is subjected to and that impose (symbolically constructed) forms from the outside, "a regulatory practice that produces the bodies it governs" (*Bodies That Matter* 1). Butler's materiality is ultimately one suspended in quotation marks:

> It must be possible to concede and affirm an array of 'materialities' that pertain to the body, that which is signified by the domains of biology, anatomy, physiology, hormonal and chemical composition, illness, age, weight, metabolism, life and death. None of this can be denied. But the undeniability of these 'materialities' in no way implies what it means to affirm them, indeed, what interpretative matrices condition, enable and limit that necessary affirmation. (67)

For Butler, then, "material" amounts to "factual," to a materiality that matters, that is always already cited: it is a result of the one way influence of discourse (in)forming materiality. This concept does not allow for the reverse operation of materiality itself affecting discourse. As she herself admits, "I am not a very good materialist. Every time I try to write about the body, the writing ends up being about language" (*Undoing Gender* 198). In this self-generating circularity, matter for Butler is ultimately what *"we call matter"* (*Bodies That Matter* 9), what we perceive as matter, *"a process of materialization that stabilizes over time to produce the effect of boundary, fixity, and surface"* (9). Thus, for Butler, matter is its own cultural script.

This intricate and complex connection between the materiality and the realm of representation might be argued to be the "blind spot" of a constructivism that, as Gilles Deleuze has observed, is "directed at rendering…representation infinite (orgiastic)" (*Difference and Repetition* 262). Thus, a new perspective that allows for the incorporation of the workings of the "repressed" of representation (namely of the "real," of "nature," of "matter") is needed. But how can materiality be thought differently? Or, to put the question in Butler's words: "How can there be an activity, a constructing, without presupposing an agent who precedes and performs that activity" (*Bodies That Matter* 7)?

Association Test Two: If I said "Deleuze," what would you say? Maybe the buzz-word here would be "machine," though another likely choice, the word "rhizome," would already hint at a kind of "ecological register" in Deleuze's (and Guattari's, and Deleuze and Guattari's) work; a concept that they themselves do not call "ecology," but rather "geophilosophy," in which the earth constitutes a fundamental concept since it "is not one element among others but rather brings together all the elements within a single embrace, using one or another of them to deterritorialize territory" (*What is Philosophy?* 85). As Mark Bonta and John Protevi state, "if you open *A Thousand Plateaus* at any page you'll find terms such as 'plateaus,' 'deterritorialization,' 'rhizome,' 'cartography,' and so on. If you would then say to yourself that this is a 'geophilosophy,' you'd be right" (vii).

But already the notion of the machine has more in common with ecological concerns—or, in fact, can be used to tweak the common sense notion of "ecology"—than meets the eye. In its conceptual reconfiguration of questions that are of central relevance in ecology and environmental ethics, Deleuze|Guattari's concept of machines, as developed in *Anti-Oedipus*, provides an enlightening model for understanding the notion of relations. Their model also affords a single mode of articulating developmental, environmental, and evolutionary relations within ecological systems, and makes room for a conceptualization of a general, non-anthropomorphic affectivity within dynamic systems. Categories such as "nature" and "man," "human" and "nonhuman," cannot anymore be grounded in an essentialist and clear-cut separation, as Deleuze states, "now that any distinction between nature and artifice is becoming blurred" (*Negotiations* 155):

> we make no distinction between man and nature: the human essence of nature and the natural essence of man become one within nature in the form of production or industry...man and nature are not like two opposite terms confronting each other—not even in the sense of bipolar opposites within a relationship of causation, ideation, or expression (cause and effect, subject and object, etc.); rather they are one and the same essential reality, the producer-product. (*Anti-Oedipus* 4–5)

The *inconvenient* truth of Al Gore, with its call to *re*-storing nature's balance before it's too late,[1] is thus only an illusion that ultimately conceals the "*real* truth of the matter—the *glaring, sober* truth that resides in delirium—is that there is no such thing as relatively independent spheres or circuits" (*Anti-Oedipus* 3–4, emphasis mine), which is why "the image of nature as a balanced circuit is nothing but a retroactive projection of man. Herein lies the lesson of recent theories of chaos: 'nature' is already, in itself, turbulent, imbalanced" (Žižek 38).

Thinking the environment with a Deleuze|Guattarian ecology|machinics is thus far removed from what might be termed "(intellectual) tree-hugging"—it is basically a call to think complexity, and to complex thinking, a way to think the environment as a negotiation of dynamic arrangements of human *and* nonhuman stressors, *both* of which are informed and "intelligent." It refers to a pragmatic and site-specific tracing of infinitely complex ecological arrangements, and as such cannot rely either on a theory of cultural|linguistic constructivism or on a natural|biological determinism. Deleuze|Guattari provide a useful toolbox for such a project—in his book *Chaosmosis*, Guattari, for example, has delineated what he calls an *eco*sophy, "a science of ecosystems" (91) and a "generalized ecology—or ecosophy" (91), which could also be read as a "generalized machinics": resonances, alliances and feedback loops between various regimes, signifying and non-signifying, human and non-human, natural and cultural, material and representational.

The term "ecology" refers to the interaction between human and non-human stressors and their environment[s], with the word *oikos*, the Greek word from which the word *eco*logy derives, meaning "house, domestic property, habitat, natural milieu" (Guattari, *Three Ecologies* 91 *n* 52), thus denoting the place where these interactions and encounters take place. If the eminent eco-socialist Barry Commoner's "First Law of Ecology: Everything is Connected to Everything Else" (33) meets with Deleuze|Guattari's idea "[t]hat what make a machine, to be precise, are connections" (*Kafka* 82), then it follows for the Deleuze|Guattarian "generalized machinics|ecology" that "[n]ature is like an immense Abstract Machine . . . its pieces are the various assemblages and individuals, each of which groups together an infinity of particles entering into an infinity of more or less interconnected relations" (*Thousand Plateaus* 512).

The "intelligent materialism"[2] of Deleuze does not fall into the "trap of essentialism" of which cultural|linguistic constructivism is so scared—simply because "essence" in itself does not exist in his account of things. What we see as essences are in fact machinic aggregations. In fact, Deleuze|Guattari categorically state that a machinics is "at work everywhere, functioning smoothly at times, at other times in fits and starts. It breathes, it eats. It shits and fucks. What a mistake to have ever said *the* id. Everywhere *it* is machines" (*Anti-Oedipus* 1). The concept of the machine neither proceeds from nor leads to an organic whole, a unity—an essence. Yet, the Deleuzian machine is not a machine in the sense of a "mechanical apparatus" or tool. It starts "in the middle" of things (like grass)—neither at the beginning, nor at the end—in order to think and describe an immanent production, without intention or end, with neither subjectivity nor any other outside controlling agency. The machine is nothing more—and nothing less—than the connections and assemblages it consists of, and its productions. Matter is machinic in the sense that the world is a multiplicity consisting of a variety of machines, such as self-organizing machines, ordered and static machines, dynamic machines, biological machines, and also the discursive and cultural machines of representation—this last machine, however, is only one among many, and not the "overriding machine" that cultural|-linguistic constructivism wants it to be. Lacan, for example, ultimately equates materiality with the materiality of the signifier, the body with the body of language, the machine with the symbolic: "The most complicated machines are made only with words.... The symbolic world is the world of the machine" (*Seminar II* 47). But, there is not only language and an "outside" to which it has no recourse. There are various machines, and there are the feedback loops between them. Accordingly, for Deleuze, "neither do... differences pass between the natural and the artificial since they both belong to the machine and interchange there. Nor between the spontaneous and the organized, since the only question is one of modes of organization" (*Dialogues* 143). For Deleuze|Guattari, matter is "molecular material" (*Thousand Plateaus* 342) equipped with the capacity for self-organization—matter is thus *alive, informed* rather than *informe* ('formless'): "matter... is not dead, brute, homogeneous matter, but a matter-movement bearing singularities or haecceities, qualities and even operations" (512) . True to the dictum of Chaos Physics and Complexity Theory, two disciplines that underlie much of Deleuze's thought, its autopoietic capacities reveal themselves at states "far from equilibrium," when matter crosses thresholds [e.g., phase states]. These capacities are hidden at a state of equilibrium, and yet it is exactly this state of equilibrium that in "traditional science" is regularly taken as *the* characteristic and essential feature of matter. Thus strategies of slowing down, stabilizing, and homogenization of matter result in an account of matter as passive, chaotic, and "stupid"—a mere "mass" or object to be "informed" by an outside spirit, force, subject, or God.

"Intelligent materialism" is designated thus not because it is supposed to be a more intelligent version of classical materialism, but because it is preoccupied with "intelligent matter" and supports a belief in the force and richness of matter itself: one that is not dominated by form, one that does not need form to be imposed on it to become alive but is in and of itself animate and informed. Matter engenders its own formations and differentiations because it carries them in itself, as potentialities, so that form|soul| mind is not something *external* to matter, but *coextensive* with it. Deleuze's "intelligent materialism" thus claims that matter is not (only) an effect of representation—matter is *productive*, and this productivity must be accounted for by its own, immanent criteria. His "transcendental empiricism" (as he himself called his position in an early phase of his thinking) is an empiricism that thinks of experience as having no foundation outside of itself, for example, in a subject, in a consciousness that is there *first* and *then* experiences, reflects, and categorizes the world. For Deleuze, it is not so much that the conscious subject explains the world—it is more a question of accounting for how a subject is formed from experience, from a singular affect or perception, from a "pre-individual" relation to materiality. It thus aligns itself with such various fields as realism, materialism, and pragmatism, but without the specter of "essentialism." All these "practices" simply denote a turn towards matter and materiality, and a move against the constructivist, impoverished concept of matter as passive and chaotic, where an "organizing" and transcendent agent is needed for making matter work, making it alive—if matter is passive, it cannot by itself account for the emergence of newness; if matter is chaotic, it cannot by itself account for order. Deleuze's "intelligent materialism," focusing on the autopoietic potential of matter as outlined in the research conducted by Chaos and Complexity Theory—see, for example, the work of Michel Serres, Prigogine|Stengers and Maturana|Varela[3]—*can* account for the world's order and creativity without taking recourse to essentialism or determinism, nor to any "transcendent vitalism," since *life* for Deleuze is the very property of matter itself.

Whereas cultural|linguistic constructivism is mostly (some would say *only*) concerned with representation, the symbolic, and ultimately psychic reality, an intelligent materialism widens the spectrum by being concerned with production, the real, lived reality. In the question concerning nature *or* nurture, such a position obviously claims that there is no *either|or*—all there is are feedback loops. Materiality—the unconscious, nature, ultimately life—is productive and autopoietic; the culturally|discursively constructed materiality|unconscious|nature is only one small part of the whole, and not even the most important one, more like the tip of the iceberg. Below the socially|linguistically constituted reality, there is the noise of the non-human, of the viral, chemical, biological, and so on, energy transformations. Deleuze refers, among others, to Lucretius, Spinoza, Hume and Leibniz in their respective work, uncovering a tradition of materiality and nature quite at odds with the Platonic model which Butler equates with Western Philosophy *per se*. Against the Platonist view, Deleuze claims, "[th]e Epicurean thesis is entirely different: Nature as the production of the diverse can only be an infinite sum, that is, a sum which does not totalize its own elements....Nature is not attributive, but rather conjunctive: it expresses itself through 'and,' and not through 'is'" (*Logic of Sense* 266–267). It is this immanent expressionism of nature that Deleuze also finds in Spinoza and that "forms the basis of a new naturalism" (*Expressionism* 232), a naturalism that re-establishes "the claims of a Nature endowed with forces or power" (228)[4]:

> if we are Spinozists we will not define a thing by its form, nor by its organs and functions, nor as a substance or a subject. Borrowing terms from the Middle Ages, or from geography, we will define it by *longitude* and *latitude*. A body can be anything: it can be an animal, a body of sounds, a mind or idea; it can be a linguistic corpus, a social body, a collectivity. We call longitude of a body the set of relations of speed and slowness, of motion and rest, between particles that compose it from this point of view, that is between *unformed elements*. We call latitude the set of affects that occupy a body at each moment, that is, the intensive states of an *anonymous force* (for existing, capacity for being affected). In this way we construct the map of a body. The longitude and latitude together constitute Nature, the plane of immanence or consistency, which is

always variable and is constantly being altered, composed and recomposed, by individuals and collectives. (*Practical Philosophy* 127–128)

In this materialist tradition, natural sciences and politics are closely connected and related not to an ethics derived from any presupposed transcendent model of morality, but to an "ethics of immanence." In their development of complex "machinic interactions" these "anomalous" philosophers point in a direction that Deleuze's own concept of the machinic is clearly indebted to.

The machine in the Deleuzian sense encompasses both culture *and* nature—both are parts of the same continuum. Thus, it would seem somewhat one-sided to concentrate on culture, psychic reality, and representation only. If Lacan, Derrida and Butler deal with nature, materiality, and the "real" body at all, they do so as a belated effect of language, the symbolic. The very resistance to seriously engaging with the "outside" of language is revealed in their distrust of and disinterest in the natural sciences. Although Butler concedes the importance of "the domains of biology, anatomy, physiology, hormonal and chemical composition" (*Bodies That Matter* 67), she ultimately shies away from discussing the "real" workings of matter— her insistence on the discursive formations of matter does not allow her to consider a working on a non- or prediscursive level.

While cultural|linguistic constructivism *is* immensely important, in particular in its questioning of the "grand narratives" and in showing the constructedness of presences and essences, it does not go far enough in so far as it by default blends out the field of materiality and nature, and the sciences that most prominently deal with these issues.[5] Against Derrida's (and cultural|linguistic constructivism's) agenda of deconstructing the *metaphysics of presence*, Deleuze decidedly poses an *ontology of difference*—instead of pointing out the impossibility of grounding *Being* in a transcendent or unitary entity or structure (God, or the Signifier), Deleuze develops a *differential* metaphysics, focusing on *becoming* and multiplicities.[6] In an interview with Raymond Bellour and François Ewald, Deleuze stated, "I've never been worried about going beyond metaphysics or any death of philosophy. The function of philosophy, still thoroughly relevant, is to create concepts" (*Negotiations* 136). This affirmative function of philosophy is also a call to transdisciplinarity, so that even when Deleuze was working on "painting and cinema: images, on the face of it...[he] was writing philosophy books" (*Negotiations* 137). The philosophical practice of creating concepts, as a creation of newness as well, necessitates, according to Deleuze, that philosophy enters into manifold relations with arts and sciences, since philosophy "creates and expounds its concepts only in relation to what it can grasp of scientific functions and artistic constructions.... Philosophy cannot be undertaken independently of science or art" (*Difference and Repetition* xvi). It is these resonances and exchanges between (or ecologies of) philosophy, science, and art that make philosophy *creative*, not *reflective*.

From within a Deleuzian context, the interesting and important question is how to develop a way of folding nature and the physical into culture and the psychic (and vice versa) other than culture metaphorically *representing* nature, which is seen as either a fixed substance or as passive and uninformed (or, more often than not, both at the same time). In fact, Deleuze|Guattari's intelligent materialism aims at "abolishing of all metaphor; all that consists is Real" (*Thousand Plateaus* 69). Another way of putting it amounts not only to a re-thinking of the concept of nature, but also to a re-thinking of the concept of writing—in a more radical way than Derrida's opening-up of the concept of "text." To extend the field of writing into the fields of materiality and the body, "writing now functions on the same level as the real, and the real materiality writes" (141). Adding material differences to a Derridean *difference* might also explain why Deleuze (in contrast to Derrida and Butler) is so interested in the natural sciences, most notably in Chaos and Complexity Theory and the New Physics.[7] Deleuze thus accommodates the paradigm shift that has taken place in the sciences. Traditional metaphysics takes from traditional physics generalizations and abstractions and then turns them into immutable "givens"—transcendence, it can be argued, is in fact produced from material immanence and then posited as an overcoding system of Truth. A "new metaphysics" (or ontology) in the Deleuzian sense is inextricably linked to the Material Sciences, the Natural Sciences, and the Life Sciences. In contrast to the positivistic approach of the traditional sciences, the "modern sciences" call for a different ontology—an ontology of difference. A New Metaphysics for (in line with) the New Physics—this is part of the Deleuzian project: "I consider myself a Bergsonian, because Bergson says that modern science has not found its own metaphysics, the kind of metaphysics that it would need. It is this metaphysics I am interested in....I consider myself a pure metaphysician."[8] The New Physics—in particular Chaos Theory—also yields a specific relation to the field of possibilities and multiplicities, to the field of the *virtual* that philosophy is so engaged with. In fact, philosophy, (the New) science(s), and art, according to Deleuze|Guattari, are involved in a *struggle against chaos* (*What is Philosophy?* 203)—or *chaotic virtuality* (155)—that "does not take place without an affinity with the enemy" (203), that is not aimed at reducing the dynamic differences to a conceptual identity. And it is this engagement with the *virtual*, with *chaos*, that attracts Deleuze|Guattari to Chaos|Complexity Theory, to a science that

> is inspired less by the concern for unification in an ordered actual system than by a desire not to distance itself too much from chaos, to seek out potentials in order to seize and carry off a part of that which haunts it, the secret of the chaos behind it, the pressure of the virtual. (156)[9]

Deleuze's metaphysics puts the stronger focus on immanence (*versus* transcendence), on production (*versus* representation), on materiality (*versus* language).

If Deleuze's work in general is fuelled by the quest for what he often calls "a superior empiricism" (*Nietzsche* 46; *Difference and Repetition* 57)—superior because of its focus on *immanence*—then this might also include the search for a "superior eco-thought," a way of thinking that ultimately reconfigures central tenets of ecology. Thus, Manuel DeLanda kick-starts this anthology by delineating the ontology underlying this reconfiguration—approaching ecological questions from a philosophical point of view involves first of all defining the ontological commitments of a philosophy. With idealist ontological commitments it would seem impossible to discuss even the simplest problems that ecology poses: if one believes that words shape human experience and that only the contents of that experience (phenomena) actually exist, then most processes of interest that have only lately been identified and named would be taken to be recent social constructions, not real ancient processes that have shaped human evolution for millennia. Similarly, with positivist ontological commitments (only what is directly observable has a mind-independent existence) all ecological processes that are either too fast or too slow, too small or too large, to be experienced directly would not count as real. However, most ecosystems contain many processes that escape direct observation. But old realist ontological commitments have their own problems: we may accept that unobservables, such as oxygen, nitrogen, or carbon, have the same reality as observables such as large plants and animals, but then postulate nonexistent entities like essences to explain their mind-independent identity, such as possession of a given number of protons in a nucleus for chemical species or possession of a certain DNA for biological species. Thus, realism must be renewed if it is to serve as a basis for a philosophical approach to ecological questions. DeLanda's essay attempts to do just that, taking as a point of departure Deleuze's new materialist philosophy, but going well beyond his own writings.

Using terms proposed by Arne Naess, it might be argued that Deleuze and Guattari are practitioners of "ecophilosophy," in that they engage philosophical issues of importance to the conceptualization of ecology, and that they promote an "ecosophy," that is, an ecological philosophy with a clear ethical orientation. That orientation, however, as Ronald Bogue's essay shows, is not fully consonant with the ethics of Naess and many other deep ecologists. Deleuze and Guattari make no fundamental distinction between human and nonhuman systems, and hence their ecosophy argues neither for a wholesale rejection of technology nor a blanket valuation of nonhuman wilderness. Their "machinism" avoids both technophilia and technophobia, its guiding principle being that of the invention of possibilities of life. For Deleuze and Guattari values are perspectival, and hence unavoidably allied to what deep ecologists might deem "speciesism." However, Deleuze and Guattari's problematization of the concept of the human ensures that their perspectivism is not anthropocentric, at least in the conventional sense of the term. The invention of possibilities of life summons us to a process of transforming the human by "becoming other," and especially by "becoming animal." The possibilities for "becoming animal" are enhanced by the number of species in the world, and hence Deleuze and Guattari's ecosophy commits them to the promotion and sustenance of biodiversity.

Hanjo Berressem, in his essay "structural couplings: radical constructivism and a deleuzian eco*logics*" argues that, while Guattari develops the politics|sciences for a "radical ecology," Deleuze offers a "radical philosophy" to the ecological project. The essay relates this "radical philosophy for a radical ecology" to the notion of "radical constructivism," creating a conceptual ecology made up of Guattari's sociopolitical subject (the emergent subject), Maturana|Varela's biological subject (the autopoietic subject), the subject of systems theory (the "eigen" organized subject) and Deleuze's philosophical subject (the habitual subject). In analogy to the radical split introduced into the system of life by radical constructivism, Berressem argues, Deleuze conceptualizes the world as a multiplicity of recursively nested machinic aggregates that are radically immanent to a plane of intensities|energetics although they are, simultaneously, radically separated from that plane in terms of their operational|informational registers. Deleuze negotiates this radical logic through his alignment of the "plane of immanence|consistence," and the "plane of transcendence| organization," which address philosophically the radical relation between self-organizing systems (eigenorganizations) and energetics. Against the background of the radically constructivist mediologies of Vilém Flusser and Niklas Luhmann, Berressem argues that Deleuze's radical philosophy allows for a fundamental realignment of the ecological notions of the environment and the medium. While the former is the result of the perceptual|cognitive operations of eigenorganizations, the latter consists of the field of their intensive|energetic operations|propagations. In cybernetic terms, the threshold between these two realms concerns the transfer of analog intensities into digital data. The radical addition Deleuze provides to ecology and to radical constructivism is the notion of a "philosophical projective plane," from within whose topo*logics* one can develop a "flat eco*logics*." The topo*logics* of this plane allow for a "radical" reconceptualization of this ecological threshold, which Berressem's essay illustrates by a reading of Robert Smithson's radically ecological "flow works."

While deep ecology subjectifies and shallow ecologies objectify nature, Deleuze's flat eco*logics* intensify it, by opening up the "philosophical subject" to the realm of nonhuman machines, affects, *haecceities* and to what Maturana|Varela call "structural couplings." As "a philosophy of structural couplings," in fact, Deleuzian philosophy becomes a "radical haecology." In acknowledging its emergence from the immanent level of nonhuman, anonymous machines, this haecology is what second-order cybernetics calls an "eigentheory." According to Berressem, the functional parameters of a "radical haecology" are as follows. As a didactics, it aims at making the operations of "structural couplings" conscious (Guattari). As a science, it aims to generate knowledge about structural couplings and to develop routines that allow for regulated responses to structural disequilibria (von Foerster, Maturana|Varela). As a practice, it aims at implementing parameters within which the equilibrium between the autopoietic systems and their media can be kept, both locally and globally (Bateson). As an art form, it provides "blocks of sensation" (Smithson). As a philosophy, it creates concepts that radically link immaterial concepts and material movements (Deleuze).

In his book *vampyrotheutis infernalis* Flusser describes the life of a species of squid that is in every way the human's intensive other. Deleuze's radical haecology asks us to embrace the "intensive" life of the *vampyroteuthis infernalis* as "our" own, and it provides the radical parameters for this embrace.

Gary Genosko's essay "Subjectivity and Art in Guattari's *The Three Ecologies*" is the first of a number in this collection that shift the emphasis to the work of Félix Guattari. His essay takes up key ideas presented in Guattari's bold tri-ecological vision involving biospherical, social, and mental ecologies. In a detailed consideration of Guattari's thought, Genosko explains that his real innovation was to develop the relationship between art and ecology through the question of the formation of subjectivity. Guattari believed that ecology, generalized as ecosophy, could help summon forth new, productive traits of subjectivity. His positive models were drawn from a range of arts, and in his view esthetics had to be supported by an ethical responsibility that traversed micro- and macro-levels of concern. Genosko explores the prospects of transdisciplinarity as a nontranscendent ecology of knowledge through examples of ecosophic esthetic and ethical practices.

Both Gilles Deleuze and Felix Guattari argue for an *oecumenon*, a world of diverse forms nourishing and sustaining and sustaining themselves. It is Guattari, however, as Verena Andermatt Conley argues in her essay "Artists or 'Little Soldiers?' Felix Guattari's Ecological Paradigms," who explicitly addresses the question of the state of the world in *The Three Ecologies* and *Chaosmosis*, two enduring testaments to his ecological commitment. Conley's essay reviews several tenets of his argument—the necessity to create new paradigms, to turn technologies toward humans, to reconstruct singular and collective processes of *subjectivation*—before testing the validity today of his pronouncements (in the guise of the practicing psychoanalyst that he had been) on the relation between creative resistance and militantism, that is, between artists and "little soldiers," for the sake of encouraging active engagement with pressing environmental issues. Conley revisits Guattari's essays almost 20 years after their publication to inquire whether an alliance with the author for ecological purposes is still possible. Although the world is in a more fragile condition than it was when Guattari wrote his two pieces, their hypotheses are, hardly paradoxically, of greater urgency today than they were then.

For starters, the French philosopher's debt to Gregory Bateson will outline Guattari's debt to the anthropologist who informs his main arguments concerning ecology. Guattari does not focus on the environment but on other ways of exchanging (of economy, indeed of relations with the world or *oikos*) between humans and between humans and nature. He emphasizes aesthetics and ethics that, when allied, show how creativity can be a form of resistance. In this way ethics becomes a necessary responsibility of all those who are in a position, like the psychoanalyst, to mould or shape the psyches of others in an era where Marxist distinctions of infrastructure and superstructure have collapsed, and where regimes of signs are determined by economic policy (or the logic of cost-effectiveness). Different ways of exchanging and of domesticating the earth are a precondition to making changes in natural ecology. To produce openings through which the possibility of change can be countenanced and to reconstruct processes of subjectivation currently under the sway of the media, Guattari develops an ecosophy. It is, in view of his writings of the later 1980s, an updated version of what he called schizoanalysis.

Conley argues that Guattari writes in a post-68 vein that is predicated on the belief of a generalized resistance to capitalism that may no longer be ours today. Guattari's emphasis on the creation of new forms of intelligence, desire and sensibilities away from the media continue to be most important for addressing questions of the environment that cannot be tackled solely by technocrats. However, when social experimentation has disappeared and "development" falls under the sign of profit in a world where nature has become quasi-invisible, creativity has to be complemented by stronger forms of militantism. Without falling back into old forms of associative commitment, one has to be "analytically militant" in Guattari's terms but also go *through* institutions that, in the passage, find themselves transformed or even reinvented. As the psychoanalyst says himself, theory—like science— has to be continually updated. In a world where the urgency of ecological problems has increased at a logarithmic progression, we can add a new twist to Guattari's theories of aesthetics and ethics. To the necessity of analyzing and producing subjectivities, of linking theory and practice, we have to add an appeal to adventitious or grassroots organizations that engage with, and bring promise of, ongoing institutional change.

While almost all environmental problems can be traced back to either the size of the human population or the level of material consumption, environmental philosophers have written little about consumerism. When they have done so, Jonathan Maskit argues in his "Subjectivity, Desire, and the Problem of Consumption," they have tended to treat consumerism merely as a problem of what we do rather than how or who we are. They have proposed addressing the ill effects of consumerism by consuming as much but otherwise (substitution), by making production more efficient, or by means of governmental policies to reduce consumption (e.g., through carbon taxes). Maskit argues that consumption must also be addressed by way of subjectivity, that is, by taking into account who we are and who we want to be. Environmentalism might then be an opportunity to rethink who we are as subjects and how we want to live rather than a set of practices often perceived to be ascetic.

Following Guattari, Maskit shows that addressing subjectivity also requires that we address what he terms social ecology (and to a lesser degree his third ecology, that of nature). A robust environmental position requires that we rethink who we are and who we want to be both as individuals and as a society as well as what nature is and what we would like nature to be (although Maskit's essay does not address this latter issue here). Addressing socioeconomic organization and subjectivity together means calling into question the form of subjectivation which makes a high-consumption life-style in a capitalist economy synonymous with human flourishing. Rejecting this form of subjectivity and social organization does not mean returning to some earlier or "authentic" way of being. Instead we must be open to exploring alternative forms of subjectivity and social organization. To that end we must think differently in order to act (and thus be) different. In closing, Maskit suggests some alternative practices that may lead to alternative ways of being, although, like Guattari, he is hesitant to prescribe too much, preferring to leave such deliberations to agents.

In her essay "Political *Science* and the Culture of Extinction," Dorothea Olkowski claims that Bruno Latour argues that virtually no intellectual effort has been directed to the opposition between modern science and modern politics, but that the two may be defined by a common text, that of the *constitution*. Modern politicians drafted a constitution for states and modern scientists drafted one for nature, but each constitution was drafted with the aim of excluding the power of the other, giving moderns unprecedented power in society but none in nature. To remedy this, Latour become premodern again by forming hybrids out of politics and science. Yet, as Isabelle Stengers and Ilya Prigogine argue, even hybrids belong to the classical scientific model that suffers from the inability to give a coherent account of the relationship between humans and nature as it constitutes them independently of one another. Olkowski argues that for Deleuze (and Guattari) the same terms of analysis operate. They too utilize a version of the classical model, updated to take probability into account. But if the structure of hybrids accurately describes our current state, then it appears that capitalism axiomatizes with one hand what it decodes with the other, that nature can be characterized as nothing but processes of production, and that we humans are utterly captive to capitalism and the endless proliferation of its axioms.

Foucault often refers to Kant's "What is Enlightenment?" essay as the beginning of a new form of philosophy, as the event of philosophy posing to itself a new duty: to critically examine the events that constitute the lives of people at a certain time in history. This new duty is not meant to replace the traditional philosophical objective of determining universal conditions of being, knowledge, beauty, justice, and so on, but it has, Foucault says, become more and more important to us to perform this self-analysis, this "history of the present" or "critical ontology of ourselves."

Although Deleuze and Guattari do not always thematize this Kantian–Foucaultian demand, John Protevi argues, they do adhere to it, as can be seen from the many analyses of the contemporary world that pepper *Anti-Oedipus* and *A Thousand Plateaus*. In his essay "Katrina," Protevi takes up the Kantian–Foucaultian challenge and, using concepts from Deleuze and Guattari's encounter with Complexity Theory, develops a naturalizing account of Hurricane Katrina that can contribute to such a "critical ontology of ourselves." By a "naturalizing" account, Protevi means that he uses the same conceptual field—that of complex material systems—to understand both the natural and social aspects of the event of Katrina. Furthermore, he uses the term "event" not simply in the sense of something that happens, but also as a possible turning point for our social system, one that starkly reveals the poverty of neoliberalism in conceiving spontaneous social solidarity. This poverty of neoliberalism is revealed by discussing how the spontaneous rescue efforts of the people of Louisiana were stopped by government order, provoked by a racialized panic which also delayed government rescue efforts until sufficient military force was ready to begin the "combat operations" thought to be necessary given the presence of thousands of black people with "insufficient" police power.

Luciana Parisi, in her essay "Technoecologies of Sensation," argues that changes in technical machines are inseparable from transformations in the material, cognitive, and affective capacities of a body to feel. It suggests that current modifications in cybernetic and bioinformatic machines of communication are leading to the formation of a technoecology of informa-tion sensing, implying a new level of relatedness between organic and inorganic milieus of transmission. With the ubiquitous distribution of digital media architecture, the rewiring of modes of feeling operates no longer through perfectly integrated circuits of communication, but with the anticipation of feeling (i.e., the feeling of feeling) or the sensation of preemption. Within such digital architecture, media have ceased to be instruments of communication and have become part of an atmospheric grid of connection where distinct milieus adapt together as microclimates in complex weather systems. Whether you are at the airport, the shopping centre or the underground, a mediatic ecology immerses modes of audio-visual and video-telephonic mobile connections ready to envelop a techno-culture addicted to constant feeling. Borrowing Deleuze and Guattari's notion of machinic arrangements of mentalities, affectivities, desires, Parisi suggests that such technoculture leads to a new understanding of ecological systems. Hence, it may no longer be a question of exploring how the organism and its environment exist in strict feedback copulation, but rather a matter of metaphysical enquiry into an artificial ecology, constituted by processes of unnatural combinations, dispensing from the naturalness of the natural object. From this standpoint, the article investigates new levels of permutations of such processes involving the advancing of symbiosensation, irreducible either to direct sensory feeling or to cognitive emotions. Symbiosensation, whilst being the product of a biodigital technoculture, is also the result of the ecotechnics of change in machinic processes at once implicated into a neostratum of control that ceaselessly rubs up against metaphysics.

According to Stephen Zepke's essay "Eco-Aesthetics: Beyond Structure in the work of Robert Smithson, Gilles Deleuze and Felix Guattari," there is a break that traverses both the art of Robert Smithson and the philosophy of Deleuze and Guattari. It is the break between Smithson's Site-Nonsite works (1966–1969) and his Earthworks (1970–1973), which separates the structure of "Nature's" signs from Nature as a process of production, and divides conceptual art's immaterial structures from the vitality of the earth. Deleuze's work of the same time also manifests this break. The transition from *Difference and Repetition* (1968) to *Anti-Oedipus* (1972) takes us from a conceptual mapping of immaterial structures to a material machinery of production. On the one hand, Smithson finds entropy erupting in the midst of his structural oppositions, opening up the gallery space to eternal drifts and inexorable deformations. Deleuze, on the other hand, struggles to avoid idealizing genesis in the strange structuralism he creates, and ends up abandoning this for a renewed and vitalist materialism. Zepke's essay traverses this break, and charts the echoes between the artist and the philosophers as laying out a concise cartography of the time. But, more importantly, it shows how Smithson and Deleuze|Guattari establish this break as the condition to the exploration of an eco-esthetics that comes after, and affirms the artistic construction and expression of Nature itself, and that finally defines the terms of our very contemporaneity.

In his recent project *The Place Where You Go To Listen* (2006), which, in a way, translates Land Art into Music, the Alaskan composer John Luther Adams connects weather stations and seismic stations and creates a "sound machine" where climatic and geological data are transposed into tonal sculptures. In 1852, Henry David Thoreau described in his book *Walden* the "accidental music" created by the wind blowing through telegraph wires. In both cases, as Bernd Herzogenrath, in his essay "The 'Weather of Music': Sounding Nature in the 20th and 21st Century," shows, nature itself creates a kind of "ambient music," sounds based on a nature|culture-continuum—an ecology of music. Based on Thoreau's sonic esthetics, a genuine American tradition of avant-garde composers from Charles Ives, via John Cage, to John Luther Adams, make "nature" the focus of compositorial reflection|contemplation, not so much in the way of program music, but as a physical and acoustic phenomenon—the "musicalization of nature" is complemented by the "naturalization of music."

Mark Halsey's essay "Deleuze and *Deliverance*: Body, Wildness, Ethics" explores the environment in popular culture. Specifically, it uses the work of Deleuze and Guattari to examine constructions of nature, law and identity in the literary and cinematic representations of *Deliverance*. The book and film are treated primarily as meditations on the molecular transformation of bodies in the context of negotiating the plane of consistency—the place where science, law and morality (and the comfort these provide) prove absent and|or senseless. With regard to *Deliverance*, Halsey's essay poses and responds to two key questions. First, what happens to "nature" during the descent into and return from the Cahulawassee River? And second, how does nature impress upon and offer up alternative renderings of the body, of urbanity, and of law within the text(s) which deliver *Deliverance* to its audience(s)? Following a summary of the plot and an outline of central characters, Halsey uses key concepts from Deleuze and Guattari to chart the tension between body and identity, plane of consistency and nature, as well as ethics and law, at key points in the literary and cinematic narratives. The body is shown to be necessarily something of unknown potential which is nonetheless rendered predictable and ascribed limits by its immersion in urbano-industrialized rhythms and routines. The practice of forgetting (or rendering invisible) the city's ongoing violence (its original destruction of wildness and of the body) is shown, via reference to the novel and film, to be dependent on repeated (if occasional) traumatic experiences of wild (violent) and distant (uncivilized) places. But, more than this, Halsey discusses *Deliverance* in terms of the richness of the encounter between law and ethics—between prescribed action (image of thought) and deliberate experimentation (plane of consistency). He argues that what ultimately descends into the Cahulawassee is rationality (the Cartesian subject), law (right), and the trappings and habits of urbanity (sedentary life, the rise of the *Urstaat*). What returns—what emerges—are bodies (molecular), wildness and the singular experience of an ethics. Halsey concludes by positioning the book and film as seminal devices for thinking through the implications of (post)industrial ecological impacts and the personal–social costs associated with the demise of "wildness."

Where geography opposes land as terrain to landscape as the infrastructure imposed upon this, the ideological approach more typically indexes landscape, considered as a way of seeing, on the relationships between a given society and the land, while aesthetics opposes landscape, again as a cultural formation, to mere unmediated land and ponders the possibility of societies devoid of a notion of landscape. The implicit model of human behavior underlying all such conceptions casts the subject confronted with land or territory as an implicitly male actor faced with matter which veers towards "female" passiveness, in a struggle for control over the external world. The suspicion remains that this is a historically specific and ethnocentric view of landscape, stemming from its historical association with European imperialism. It ignores the existence of ecologies which, regulating relations between humans and nonhumans on lines other than those of subject and object, suggest the possibility of a quite different approach to landscape.

Yves Abrioux, in his essay on "Intensive Landscaping," explores the hypothesis that a truly operative definition of landscape can be achieved through a careful consideration of the process that Deleuze and Guattari name territorialization and which they associate with a simultaneous process of *de*territorialization. A close reading of the definition of the *ritournelle* expounded in the eleventh plateau of *A Thousand Plateaus* allows (de) territorializing to be seen as a matter of tempo, whose principal operator is the *ritournelle* or ditty (a term Abrioux proposes to substitute for the standard translation of *ritournelle* by "refrain"). Detailing the three phases of the *ritournelle*, he shows how this considers art as an ethological, rather than merely anthropological, phenomenon. Importantly, it is one which produces difference through repetition. Deleuze and Guattari's analysis shows the invention of home as a *milieu* or medium, in the jump out of chaos that defines phase 1. In phase 2, home stabilizes into territory by interiorizing forces *as* earth. In phase 3, the pressure of the territorialized forces of creation opens a deterritorializing crack in the circle of earth, creating another region which Deleuze and Guattari name the World. Whereas the forces of chaos belong to phase 1 of the *ritournelle*, terrestrial forces belong to phase 2 and cosmic forces to phase 3. Each of these three phases inheres in the others: chaos, earth and World are forces or tempos confronting one another. They converge within the *ritournelle* as, simultaneously, territorialization and deterritorialization.

The concepts of rhythm, motif, melody, expression, style and signature, which contribute to the emergence and the interplay of the three phases of the *ritournelle*, flesh out a notion of landscape and gardens as artistic practices which may participate in the repeated inventions of home as a territory always already in the process of passing into something else. Abrioux briefly approaches in these terms the land art of the 1960s and 1970s, suggesting that different styles of land art make it possible to identify specific lines of becoming in a body of work which is still commonly regarded as homogenous. He also analyzes a modest landscaping project in France, which demonstrated the creative power of intensive landscaping (i.e., landscaping which is expressive in the terms developed by Deleuze and Guattari). Abrioux shows that this project did not so much stage a power struggle over some physical or institutional territory as provoke unpredictable processes of resonance among the perceptive and affective qualities of its site. By suggesting that forces which could thus be made to fluctuate might jump into a different order of being, the garden liberated the creative potential of landscape. In the spirit of Deleuze and Guattari, landscaping as style offers the promise of a social spacing yet to come.

In the last essay of this collection, Matthew Fuller examines "Art For Animals." Whilst art that depicts animals and animality has a long history, art *for* animals, that is, with animals intended as its key users or audience has a relatively short history. It is one, however, that is particularly capable of revealing insights into the changing nature of art, of thought and of "meta-ecological" conditions.

Paul Perry, a Canadian|Dutch artist, has installed small robotic devices to spray pheromones high on trees to attract the engagement of particular animals. These animals form the main audience for these almost invisible, or rather, "insensible" pieces. Vito Acconci has constructed architectural spaces from ladders, buildings and bird cages aimed at use by birds and humans. Thomas Schütte installs a "Hotel For Birds" on a plinth in London's Trafalgar Square. Made of brightly colored layers of perspex, this is a sculpture designed to catch light, and to act as a "public space" for urban rock doves displaced by a cleansing policy established by a different branch of the body commissioning the work. The Bureau of Inverse Technology makes a robotic goose, the aim of which is to interact with a small group of Brent geese, and Marcus Coates stages a series of actions with animal materials and behaviors with interaction with other species as the prime goal. Some of this work is absurdist, whimsical, self-trivializing. But all of these works set up actual, nonrepresentational and imaginal relations with animals, and there are other examples, artists that make scenarios that animals live in, work on and complete or render definitively unfinished. Deleuze and Guattari, who have, following von Uexküll and Kafka amongst others, placed animal subjectivity at the core of their reinvigoration of thought, provide some dynamic formulations of conceptual personae as animal-beings and of animals as engaged in reciprocal relations of life shaped by color, growth and habitat formation. In *What is Philosophy* art and nature are alike because they combine the interplay between House and Universe, the homely and the strange, and the specific articulation of the possible with the infinite plane of composition. As Fuller argues, "Art for Animals" takes up such work for the category of art.

In engaging animal cultures and sensoria, these projects also make art step outside itself, and make us imagine a nature in which nature itself must be imagined, sensed and thought through. At a time when human practices are rendering the earth definitively *unheimlich* for an increasing number of species, abandoning the human as the sole user or producer of art is one perverse step towards doing so. More widely, a core process of Guattari's writing, one which it amplifies from that of Deleuze, is the project of understanding ecology on multiple scales, from the social, to the medial, technical and aesthetic, to that of subjectification. Fuller's essay draws upon such processes to develop the question of animal–human subjectivation as a cultural and inventive process.

As Dianne Chisholm is right to point out, the *geo* in Deleuze|Guattari's geophilosophy "evokes no singular (geological, biological, hydrological, thermodynamical, etc.) activity but, instead, emits a multiplicity of interconnecting 'geos'—geology, geography, geophysics, and geopolitics, and emerging composites such as geophysiology, geomicrobiology, ad infinitum"[10]...similar to the *eco* in the "generalized ecology," which, according to Guattari, consists of the interplay of at least "three ecological registers (the environment, social relations and human subjectivity" (*Three Ecologies* 28). However, this observation has to be complemented by the fact that, as Hanjo Berressem has already pointed out in his "Emergent Eco:logics: Cultural and Natural Environments in Recent Theory and Literature," one should not think of *one* eco*logy*, but rather of eco*logics*—both nature|matter and culture|representation are dynamic, open and ultimately machinic aggregates that operate according to different but interrelated eco|logics. It is thus necessary to propagate

> a logic of open systems (with a minimum of structural stability) and morphogenesis that links various fields of research within a chaotic and eco:sophic reference. The project is to turn an eco:logics and thus an environmentics—which is by definition both local *and* global—into a "general project" that traverses philosophy, sociology, politics, art, [and] the sciences. (54–55)

This is exactly what the present anthology—the first book-length study to address the issue of a Deleuze|Guattarian Ecology—is aiming at.[11]

Notes

1. See Gore's earlier book, *Earth in the Balance: Ecology and the Human Spirit.* Boston: Houghton Mifflin, 1992.
2. I am borrowing the term "intelligent materialism" from Hanjo Berressem. In his essay "Matter that Bodies: Gender in the Age of a Complex Materialism," he develops an "intelligent materialism|realism" with Deleuze, and against Butler. *Genderforum* 2 (2002). See http://www.genderforum.uni-koeln.de/mediating/btm/btm.html. Last accessed August 29, 2008.
3. See, for example, (Prigogine and Stengers) and (Maturana and Varela).
4. For a thorough analysis of Deleuze's naturalism (and one of the earliest essays to address the issue of a Deleuzian ecology), see Patrick Hayden. 'Deleuze and Naturalism: A Convergence with Ecological Theory and Politics.' *Environmental Ethics* 19:2 (Summer 1997), 185–204.
5. Thus, Deleuze's work has close affinities with Radical Constructivism, in particular the type of Maturana|Varela, who include the whole range of "materiality" in

their concept of constructedness: not only culture constructs and is constructed—nature produces and is produced as well. See also Hanjo Berressem's essay in this anthology.
6. Compare Deleuze's *Difference and Repetition* in particular.
7. For a detailed assessment of Deleuze's engagement with contemporary science, see (DeLanda).
8. "Réponsers à une série de questions." Arnaud Villani. *La Guêpe et l'Orchidée. Essais sur Gilles Deleuze.* Éditions Paris: Belin, 1999, 129–130, 130: "Je me sens bergsonien, quand Bergson dit que la science moderne n'a pas trouvé sa métaphysique, la métaphysique don't elle aurait besoin. C'est cette métaphysique qui m'intéresse.... [J]e me sens pur metaphysician" (my translation).
9. In a footnote, which refers the reader to James Gleick's book *Chaos: Making a New Science*, they add: "Science feels the need not only to order chaos but to see it, touch it, and produce it" (*What is Philosophy?* 229:14).
10. Dianne Chisholm. "Rhizome, Ecology, Geophilosophy (A Map to this Issue)." April 1, 2008.
11. Although this is indeed the first *book*-length study, this issue has already been addressed by the online journal *Rhizome* 15 (Winter 2007), in their Special Issue on "Deleuze and Guattari's Ecophilosophy" (ed. Dianne Chisholm). See www.rhizomes.net/issue15, last accessed April 1, 2008.

Works cited

Berressem, Hanjo. "Emergent Eco:logics: Cultural and Natural Environments in Recent Theory and Literature" in *Space – Place – Environment*. Eds Lothar Hönnighausen, Marc Frey, James Peacock, and Niklaus Steiner. Tübingen: Stauffenberg Verlag, 42–61.

Bonta, Mark and John Protevi. *Deleuze and Geophilosophy*. Edinburgh: Edinburgh University Press, 2004.

Butler, Judith. *Bodies That Matter. On the Discursive Limits of "Sex"*. New York and London: Routledge, 1993.

——. *Undoing Gender*. New York and London: Routledge, 2004.

Commoner, Barry. *The Closing Circle: Nature, Man, and Technology*. New York: Knopf, 1971.

DeLanda, Manuel. *Intensive Science and Virtual Philosophy*. London and New York: Continuum, 2002.

Deleuze, Gilles. *Nietzsche and Philosophy*. Trans. J. Tomlison. New York: Columbia University Press, 1983.

——. *Spinoza: Practical Philosophy*. Trans. R. Hurley. San Francisco: City Lights, 1988.

——. *The Logic of Sense*. Trans. Mark Lester with Charles Stivale. New York: Columbia University Press, 1990.

——. *Expressionism in Philosophy: Spinoza*. Trans. M. Joughin. New York: Zone Books, 1992.

——. *Difference and Repetition*. New York: Columbia University Press, 1994.

——. *Negotiations 1972–1990*. Trans. M. Joughin. New York: Columbia University Press, 1995.

——. "Réponsers à une série de questions." Arnaud Villani. *La Guêpe et l'Orchidée. Essais sur Gilles Deleuze*. Éditions Belin, 1999, 129–130.

——. *Cinema 2. The Time-Image*. Trans. Hugh Tomlison and Robert Galeta. London: The Athlone Press, 2000.

Deleuze, Gilles and Claire Parnet. *Dialogues*. New York: Columbia University Press, 1987.

Deleuze, Gilles and Félix Guattari. *Anti-Oedipus: Capitalism and Schizophrenia*. Trans. R. Hurley, M. Seem, and H.R. Lane. Minneapolis: University of Minnesota Press, 1983.

——. *Kafka. Toward a Minor Literature*. Trans. Dana Polan. Minneapolis and London: University of Minnesota Press, 1986.

——. *A Thousand Plateaus: Capitalism and Schizophrenia*. Trans. B. Massumi. Minneapolis: University of Minnesota Press, 1987.

——. *What Is Philosophy?* Trans. H. Tomlinson and G. Burchell. New York: Columbia University Press, 1994.

Glotfelty, Cheryl and Harold Fromm. *The Ecocriticism Reader. Landmarks in Literary Ecology*. Athens, Georgia: University of Georgia Press, 1996.

Guattari, Félix. *Chaosmosis. An Ethico-Aesthetic Paradigm*. Trans. Paul Bains and Julian Pefanis. Bloomington and Indianapolis: Indiana University Press, 1995.

——. *The Three Ecologies*. Trans. Gary Genosko. London: Athlone Press, 2000.

Jacques Lacan. *Écrits: A Selection*. Trans. A. Sheridan. New York: Norton, 1977.

——. *The Seminar of Jacques Lacan Book II: The Ego in Freud's Theory and in the Technique of Psychoanalysis 1954–55*. Trans. S. Tomaselli. Cambridge: Cambridge University Press, 1988.

——. *The Four Fundamental Concepts of Psycho-Analysis*. Trans. A. Sheridan. Harmondsworth: Penguin, 1991.

Maturana, Humberto R. and Francisco J. Varela. *Autopoiesis and Cognition. The Realization of the Living*. Dordrecht: Kluwer Academic Publishers, 1980.

——. *The Tree of Knowledge. Biological Roots of Human Understanding*. Boston: Shambhala Publications, 1987.

Prigogine, Ilya and Isabelle Stengers. *Order Out of Chaos. Man's New Dialogue with Nature*. London: Flamingo, 1984.

Zita, Jacqueline N. *Body Talk – Philosophical Reflections on Sex and Gender*. New York: Columbia University Press, 1998.

Žižek, Slavoj. *Looking Awry: An Introduction to Jacques Lacan through Popular Culture*. Cambridge: MIT Press, 1991.

2
Ecology and Realist Ontology
Manuel DeLanda

Introduction

Ecosystems are one of the most complex entities in this planet, their study involving the contributions of many scientific fields. We need, for example, thermodynamics to study the circulation of energy in food webs, starting from their main source: solar energy. When the latter is transformed via photosynthesis into sugars, that is, when it is stored in chemical form, we need to apply not only organic chemistry but also quantum mechanics to decipher the mechanism of this transformation. The following links in a food web, herbivores and carnivores must, in turn, be understood in historical terms, so genetics and evolutionary biology must be brought to bear. And when we need to understand the effect of the extinction of a species on the fate of an ecosystem we must apply tools from mathematical fields, such as topology, to map the connectivity of the nodes of a web and track the cascading effects that the disappearance of a node may have on other nodes. In short, it would be foolish to approach the study of ecosystems using the tools from fields like literary criticism, semiotics, or hermeneutics. And yet, it is the latter fields that have dominated philosophy in the last few decades. Fortunately, a few late twentieth-century philosophers were materialists and it is the work of these authors that we can use as a resource in properly conceptualizing these complex entities.

One may ask, of course, why do we need philosophy in this task at all? The answer is that most scientists are too specialized to actually care about devising a proper framework for understanding not only ecosystems but also other complex assemblages of heterogeneous parts. Worse yet, most scientists subscribe to the philosophical school known as "positivism," according to which their only responsibility is to devise compact descriptions of phenomena useful for prediction and control. A typical positivist, for example, believes in the objective existence of directly observable entities, such as large plants and animals, but not energy, oxygen, microorganisms and many other entities that figure not only in descriptions of ecosystems but, much more importantly, in the explanation of ecological processes. To an old-fashioned positivist these unobservable entities are mere theoretical constructs whose legitimacy depends solely on their predictive power or their pragmatic effects on the increased control of variables in laboratory experiments. There are, of course, realist scientists, that is, scientists who believe that entities like hydrogen atoms or electrons are every bit as real as large animals and plants. But many of them would defend the mind-independent identity of these unobservables in a naive way, that is, by asserting that this identity depends on the possession of an essence. The essence of a hydrogen atom, for example, would be its possession of a single proton in its nucleus. Why? Because if we added another proton the atom would lose its identity as hydrogen and become helium. The essence of a given animal or plant, in turn, would lie in its DNA, since drastic enough changes in its genes would generate a different species. Essentialism, however, is not a viable position when it is examined in detail and it is this implausibility that has made so many scientists, as well as intellectuals from the humanities, turn against realism. So a crucial task for any materialist today is to offer a realist ontology that does not depend on essences. This task is what is attempted in the following essay.

One warning is necessary before we begin. What follows is an ontological discussion, not an epistemological one. I will, for example, take for granted that there is overwhelming evidence that the atomic species listed in the periodic table of the elements are real, but I do not give an account of how this evidence was produced or how different pieces of evidence were put together to create compelling arguments for their mind-independent existence. That is a task for a different essay. Nonrealists and, in particular, idealists do not separate ontology and epistemology. Since according to them all nonobservables are mere theoretical constructs, there can be no real account of these entities that does not include an explanation of the sources of this "construction." I do not offer such an account here, but one is forthcoming. The one thing I can say about that future project is that it will not attempt to explain what "science in general" is as a social and historical entity, because there is no such thing as science in general. To believe so is to fall into essentialism and to identify "science" with the possession of some necessary property. There is only a multiplicity of scientific fields, each one historically individuated as a separate entity, even if borrowings of methods, ideas, or technologies have produced some shared properties. Far from following a convergent path to a final absolute truth these fields are diverging more and more, with new fields constantly appearing at the boundaries (or in between) old ones: physical chemistry, biochemistry, biophysics and so on. The reason for this must also be ascribed to ontology: the world itself is divergent and scientific fields are being forced to diverge to track this reality.

Finally I must add that I do not follow Deleuze in every detail; in fact, I break with his position at many points. He and Guattari, for example, believed in the existence of "science in general" (its essence being some obscure entity they call "functives") and that by itself is enough to reject their account. Thus, here as elsewhere, I perform a *reconstruction* of his work. When I refer to Deleuze's ideas and positions in what follows I am merely giving him credit for having inspired a particular line of argument, not trying to derive legitimacy from his name. If someone can show that he did not, in fact, subscribe to these ideas or did not take these positions, then all the better for me: the ideas and positions are mine and I do not owe him anything. On the other hand, I believe that the best way of honoring his memory is not to stick to what he said in every detail but to push the line of flight that he rode in his life and work to its ultimate consequences.

Deleuze's ontology: A reconstruction

A philosophy's ontology is the set of entities it is committed to assert actually exist, or the types of entities that according to that philosophy populate reality. Although historically there have been a great variety of ontological commitments, we may classify them into three main groups. To begin with, there are philosophers for whom reality has no existence independently of the human mind that perceives it, so their ontology consists mostly of mental entities, whether these are thought of as transcendental concepts or, on the contrary, as linguistic representations or social conventions. This ontological stance is usually labeled "idealism." Then there are philosophers who may grant to the objects of everyday experience a mind-independent existence, but who remain skeptical that theoretical entities (both unobservable relations such as physical causes and unobservable entities such as electrons) possess such mind-independence. Pragmatists, positivists and instrumentalists of different stripes all subscribe to one version or another of this ontological stance.

Finally, there are philosophers who grant reality full autonomy from the human mind, arguing that to base an ontology on the distinction between the observable and the unobservable betrays a deep anthropocentrism. This ontological stance is referred to as "realism." Deleuze is such a realist philosopher. There is, on the other hand, plenty of room for realists to disagree when it comes to specifying the contents of this mind-independent reality. Naive realists, for example, believe in the existence of both *general* categories and their *particular* instantiations. The crucial relation here is one of class membership, a set of particulars belonging to a given class or category if they share a common core of properties (an essence). Deleuze rejects all forms of essentialism replacing the relation between the general and the particular with that between the *universal* and the *singular*, or, more exactly, with the relation between individual singularities and universal singularities. Let me give a brief example of each, one from biology, the other from physics.

For a long time philosophers thought of biological species as representing general categories, with individual organisms being particular instantiations of that category. But by the 1930s, as the theories of Darwin and Mendel were combined into modern evolutionary theory, this old model began to change. Today it is widely accepted, though still controversial in some quarters, that species are as singular, as unique, as historically contingent as organisms. Natural selection ensures that some homogeneity will be achieved in the gene pool of the species (a homogeneity that explains the resemblances among organisms) while reproductive isolation, the closure of the gene pool to external flows of genetic materials, ensures that a species will be capable of retaining its identity. Since reproductive isolation is a contingent achievement that may be reversed, nothing guarantees that this identity will in fact endure for ever. Also, since the anatomical, physiological and behavioral properties of a species are produced by contingent historical processes of selection, processes which cannot be exactly duplicated, driving a species to extinction is like killing a unique individual, one that will never return again. Thus, the new view is that species are individual entities, only *spatiotemporally larger* than organisms. It follows that the relation between organisms and species is not one of tokens belonging to types, but one of parts composing a larger whole. In other words, it is not a class membership relation but a machine-like relation between working components and the whole that emerges from the causal interactions between components. In this way, general categories defined by an essence are replaced by a nested set of *individual singularities* (individual cells, individual organisms, individual species) each existing at a different spatiotemporal scale.

On the other hand, general laws and the particular events or processes that obey them should be replaced by *universal singularities*. Here the best example comes from classical physics. While in its original formulation the basic ideas of this field were given the form of general laws (Newton's laws of motion, for example), in the eighteenth and nineteenth centuries it acquired an alternative form: most classical processes from optical to gravitational were seen to conform to a "least principle," that is, they were viewed as governed by a mathematical singularity in the form of a minimum point. This minimum had in a sense greater universality than the laws themselves, since the laws of optics, of motion and of gravitation could all be seen as regularities in processes governed by one and the same universal singularity. Physicists do not make an ontological distinction between the two versions. As positivists (that is, as nonrealists) they aim at producing compact descriptions useful for prediction and control, and if two versions of the same theory yield the same predictions then these are exactly equivalent as far as they are concerned. However, from the perspective of a realist ontology it makes all the difference in the world which version one takes to be real. One version commits philosophers to reified general entities (laws) while the other entails a commitment only to the abstract structure of these laws, a structure consisting of universal singularities.

In short, in place of the relation between the general and the particular, a relation that, when reified (that is, when treated realistically), implies the existence of essences, Deleuze puts the universal-singular and the individual-singular, a much more radical maneuver than the simple nominalist move of disregarding general classes and sticking to particulars. Similarly, his proposal is more radical than the conventionalist maneuver of simply declaring general categories to be "social constructions." No doubt, there are many categories which do not pick out a real larger-scale individual in the world (the category "schizophrenia," for example, may actually group together several different mental conditions) and to that extent these are mere social constructions. But it would be wrong to argue that every category is like this, or to argue that not to view general categories as mere conventions is to espouse a form of essentialism. In fact, the opposite is true: to simply replace essences with social conventions quickly degenerates into a form of "social essentialism." Essences and general categories (not to mention general laws) are very hard to get rid of and simple nominalist or conventionalist maneuvers do not achieve the desired goal. I will spend the rest of this essay sketching how Deleuze proposes to perform this difficult feat, but in a nutshell it boils down to this: the identity of each individual entity (as well as any resemblances among the individuals belonging to a given population) needs to be accounted for by the details of the *individuation process* that historically generated the entity in question; and any regularities in the processes themselves (especially regular or recurrent features in different processes) must be accounted for in terms of an *immanent* (nontranscendent) abstract structure. Deleuze uses the term "intensive" when discussing individuation processes, and the term "virtual" to refer to the ontological status of abstract structures, so I will start by defining these two terms.

But before I begin let me take care of a possible objection. All the theoretical resources which I will use to define processes of individuation come from the hard sciences: physics, chemistry, biology. Similarly, all the resources needed to define immanent process-structures come from mathematics: topology, group theory, dynamical systems theory. This immediately raises the following objection: how can one develop a realist ontology which is supposed to serve as a foundation for objective knowledge while from the start one presupposes that there is such thing as "objective knowledge?" If the point of a realist ontology was *foundational* this would indeed constitute a vicious circle. But one does not have to believe in rock-solid foundations at all. One may alternatively view the role of the philosopher as allowing the *bootstrapping* of an ontology. The term comes from computer science where it refers to the way the vicious circle between hardware and software is broken. When a computer is turned on (or "booted up") the software must be loaded into the hardware, but "loading" is a software function. This circularity is broken by hardwiring a little bit of software (a hardware mini-loader) which loads the software loader which, in turn, loads the rest of the software. Similarly, a realist ontology may be lifted by its own bootstraps by assuming a minimum of objective knowledge to get the process going and then accounting for the rest. After all, that is how the earliest versions of the field of classical physics bootstrapped themselves: Galileo toured the Venetian Arsenal, where the complex ships that Venice used to enforce its monopoly of trade in the Levant were built, and observed all kinds of causal knowledge (about materials, forces, mechanisms) applied in practice. This was not formal or mathematical knowledge, to be sure, but causal know-how, yet it was solid enough to serve as a bootstrapping platform.

The same point applies to an ontological account: the minimum of presupposed knowledge need not constitute a foundation at all but only a bootstrapping platform. Whether the choice of minimum to start with is correct or not can be checked by the overall coherence of the resulting ontology and by verifying that it does indeed avoid the postulation of general entities (ideal types, eternal laws). Clearly, an ontology where general laws are not among the contents of reality would radically break with standard scientific conceptions and in this sense it would not be dependent on physical science's own ontology. But why bootstrap via the physical or the biological sciences when one could begin with the social component of science (its institutions, its ideologies)? The short answer is that, for a realist whose goal is to create a mind-independent ontology, the starting point must be those areas of the world that may be thought of as having existed prior to the emergence of humanity on this planet. This does not mean that the institutional component of scientific fields is unimportant, but it does imply a rejection of the social constructivist (or network-actant) approach to the study of social factors in hard science as misguided. Having said this, let me discuss Deleuze's conception of individuation processes, that is, of the intensive part of reality.

The science of thermodynamics distinguishes between *extensive* and *intensive* physical properties. Extensive properties include basic spatial properties such as length, area, and volume, as well as quantities such as amount of energy or entropy. All of these are defined as *intrinsically divisible* quantities: dividing a given volume of matter into two equal halves yields two volumes, each half the extent of the original one. On the other hand, there are properties such as temperature, pressure, speed, or density which cannot be so divided, and are referred to as "intensive:" breaking up a volume of water at 90 degrees of temperature into two equal parts does not yield two half volumes at 45 degrees each, but two half volumes at the original temperature (see van Wylen 16). Deleuze takes this textbook definition in terms of indivisibility but argues that it would be more accurate to say that an intensive property *cannot be divided without involving a change in kind*.[1] In a sense, a given volume of water can in fact be divided in intensity if one heats up the container from underneath, creating a temperature difference between its top and bottom portions. But this operation changes the thermodynamic nature of the system: while prior to the heating the system was at equilibrium, once the temperature difference is created the system will be away from equilibrium.

Beyond this minor redefinition, there are two other crucial aspects of intensive properties which matter from the point of view of individuation. One is that *intensive differences drive processes.* At its simplest, a difference in intensity will tend to cancel itself out and in the process it will drive a system back to equilibrium. This tendency explains why temperature or pressure cannot be divided in extension: whatever differences are created during the division process will be objectively averaged out and the original equilibrium temperature or pressure will be restored. The second idea is that intensities are characterized by *critical thresholds* marking points at which a material spontaneously and abruptly changes in structure. The sudden change of liquid water into ice, or the equally abrupt change into steam, are familiar examples of these critical events occurring at very well defined intensive thresholds. The crucial ontological role played by intensive differences in Deleuze's philosophy is expressed in the following quote:

> Difference is not diversity. Diversity is given, but difference is that by which the given is given…Difference is not phenomenon but the noumenon closest to the phenomenon…Every phenomenon refers to an inequality by which it is conditioned…Everything which happens and everything which appears is correlated with orders of differences: differences of level, temperature, pressure, tension, potential, difference of intensity. (Deleuze 222)

This does not imply, however, that the treatment of intensive differences in nineteenth-century thermodynamics can provide the bootstrapping element we need, the reason being that this branch of physics focused exclusively on the final equilibrium state of a given system and basically ignored the difference-driven process giving rise to that final state. Given that intensive differences are supposed to define individuation processes, not individual products, to study systems where these differences are already canceled defeats the very purpose of the concept. What we need is a version of thermodynamics where the systems studied are away from equilibrium, that is, where the experimental setup is designed not to allow the intensive differences to disappear. Thus, for bootstrapping purposes we need far-from-equilibrium thermodynamics (see Nicolis and Prigogine). This change greatly enriches the repertoire of endogenous tendencies guiding the processes which produce individual entities. While in classical thermodynamics the only endogenous tendency explained by intensive differences is the tendency towards a *simple and unique* equilibrium, in far-from-equilibrium conditions a wider variety of endogenous tendencies appears: systems may still tend to a steady state but now these equilibria may come in bunches, and, more importantly, these *multiple equilibria* may not be steady state but cyclical or even turbulent (periodic and chaotic attractors).

In addition to this traditional meaning of the term "intensive," a meaning which, as I just said, relates to the endogenous *tendencies* of a process, Deleuze uses the term in a second but closely related sense, one referring to the *capacities* of final products to enter into further processes. In particular, in this second sense the term refers to the capacity of individual entities to enter as components of *heterogeneous assemblages*, that is, compositions in which the differences among the parts are not canceled through homogenization (reference to "differences that are not canceled" is what unites the two senses of the term "intensive" in Deleuze's work). This idea may be illustrated with examples from biology. One of the intensive processes which fascinate Deleuze is the process of embryogenesis, the process that converts a fertilized egg into an individual organism. This process is driven by intensive differences (for example, different densities of certain chemical substances) and as such is an example of intensive individuation in the first sense. The extensive properties of an actual organism (as well as the qualities which define its identity) are produced by spatiotemporal dynamisms driven by intensive differences. In other words, individual organisms are "actualized" via a difference-driven morphogenetic process. As Deleuze puts it:

> How does actualization occur in things themselves? . . . Beneath the actual qualities and extensities [of things themselves] there are spatio-temporal dynamisms. They must be surveyed in every domain, even though they are ordinarily hidden by the constituted qualities and extensities. Embryology shows that the division of the egg is secondary in relation to more significant morphogenetic movements: the augmentation of free surfaces, stretching of cellular layers, invagination by folding, regional displacement of groups. A whole kinematics of the egg appears which implies a dynamic. (214)

Once the individual organism is produced, however, its extensities and qualities will hide the original intensive process and its endogenous tendencies. In other words, the process will become hidden under the product. But this product, in turn, will possess in addition to a well defined set of properties (extensive and qualitative) an *open set* of capacities to interact with other such individuals, organic and nonorganic. In particular, biological organisms are capable of forming the heterogeneous assemblages we call "ecosystems" playing a given role in a complex food web and its constant flow of matter and energy. Deleuze refers to these capabilities as "affects," the capacity of an individual entity *to affect and be affected by* other individual entities. Given that all the potential interactions which an organism may have cannot be given in advance, its affects (as opposed to its qualities and extensities) do not form a closed set, and many will remain forever unexercised. Perhaps due to this fact, philosophers tend to view properties (which are always given and enumerable) as the sole object of knowledge when studying entities. But Deleuze disagrees:

> We know nothing about a body until we know what it can do, what its affects are, how they can or cannot enter into composition with other affects, with the affects of another body, either to destroy that body or to be destroyed by it, either to exchange actions and passions with it or to join with it in composing a more powerful body. (Deleuze and Guattari 257)

What ontological status do capacities have? Given that, as I said, most capacities may remain unexercised (if the right object which affects or is affected by a given entity is not present), their ontological status is similar to that of tendencies, which may also remain unactualized. Deleuze uses the term "virtual" to define this status, not in the sense of a "virtual reality" as in computer simulations, but as a *real virtuality*, every bit as real as the intensive processes it governs and the final products the latter produce. In other words, interpreted in a realist way, capacities and tendencies (or, as he refers to them, affects and singularities) would constitute the abstract structure of intensive processes. As he argues, intensive and virtual thinking imply a completely different conception of matter as well as constituting a major shift in Western ideas on the genesis of form. An essentialist ontology assumes not only that forms preexist their material realization, but also that matter is an *inert receptacle* for eternal forms imposed on it from the outside. Deleuze refers to this conception of morphogenesis as "the hylomorphic model." Intensive and virtual thinking, on the other hand, break with essentialism by endowing matter with morphogenetic capabilities of its own. Artisans and craftsmen, in his view, understand this other conception of matter and form, at least implicitly: they tease a form out of an active material, collaborating with it in the production of a final product rather than commanding it to obey and passively receive a previously defined form. As Deleuze writes, the hylomorphic model:

> assumes a fixed form and a matter deemed homogeneous. It is the idea of the law that assures the model's coherence, since laws are what submits matter to this or that form, and conversely, realize in matter a given property deduced from the form... [But the] *hylomorphic* model leaves many things, active and affective, by the way side. On the one hand, to the formed or formable matter we must add an entire energetic materiality in movement, carrying *singularities*... that are already like implicit forms that are topological, rather than geometrical, and that combine with processes of deformation: for example, the variable undulations and torsions of the fibers guiding the operations of splitting wood. On the other hand, to the essential properties of matter deriving from the formal essence we must add *variable intensive affects*, now resulting from the

operation, now on the contrary, making it possible: for example, wood that is more or less porous, more or less elastic and resistant. At any rate, it is a question of surrendering to the wood, then following where it leads by connecting operations to a materiality instead of imposing a form upon a matter.... (408, italics in the original)

Tendencies and capacities are both *modal* terms; that is, unlike properties which are always fully realized in an individual entity, tendencies and capacities may not ever be realized. This creates a fundamental ontological problem for Deleuze because modal terms are typically treated in terms of the concept of "possibility," and this concept has traditionally been associated with essentialism. Although I cannot go into a full discussion of modal logic and its notion of "possible worlds," the link between essences and possibilities can be easily grasped if we think that, like an essence, which represents an eternal archetype resembling the entities which realize it, a possible state or relation also resembles that which realizes it. In other words, the process of realization seems to add very little to a possibility other than giving it "reality," everything else being already given. Possible individuals, for example, are pictured as already possessing the extensities and qualities of their real counterparts, if only potentially. It is to deal with this problem that Deleuze created the notion of the virtual. In his words:

What difference can there be between the existent and the nonexistent if the nonexistent is already possible, already included in the concept and having all the characteristics that the concept confers upon it as a possibility?... The possible and the virtual are... distinguished by the fact that one refers to the form of identity in the concept, whereas the other designates a pure multiplicity... which radically excludes the identical as a prior condition... To the extent that the possible is open to "realization" it is understood as an image of the real, while the real is supposed to resemble the possible. That is why it is difficult to understand what existence adds to the concept when all it does is double like with like... Actualization breaks with resemblance as a process no less than it does with identity as a principle. In this sense, actualization or differenciation is always a genuine creation. Actual terms never resemble the singularities they incarnate... For a potential or virtual object to be actualized is to create divergent lines which correspond to—without resembling—a virtual multiplicity. (Deleuze 211–212)

Let me give a simple example of how mathematical singularities (as part of what defines a virtual multiplicity) lead to an entirely different way of viewing the genesis of physical forms. There are a large number of different physical structures which form spontaneously as their components try to meet certain energetic requirements. These components may be constrained, for example, to seek a point of minimal free energy, like a soap bubble, which acquires its spherical form by minimizing surface tension, or a common salt crystal, which adopts the form of a cube by minimizing bonding energy. One way of describing the situation would be to say that a *topological form* (a universal-singular point) guides a process which results in many different physical forms, including spheres and cubes, each one with different *geometric* properties. This is what Deleuze means when he says, in the quote above, that singularities are like "implicit forms that are topological rather than geometric" (Deleuze and Guattari 408). This may be contrasted with the essentialist approach in which the explanation for the spherical form of soap bubbles, for instance, would be framed in terms of the essence of sphericity, that is, of geometrically characterized essences acting as ideal forms. Unlike essences (or possibilities) which resemble that which realizes them, a universal singularity is always *divergently actualized*, that is, it guides intensive processes which differentiate it, resulting in a open set of individual entities which are not given in advance and which need not resemble one another (for example, a spherical bubble and a cubic crystal).

The concept of a "universal singularity" in the sense in which I am using it here is a mathematical concept, so care should be taken to endow it with ontological significance. In particular, mathematical singularities act as *attractors* for trajectories representing possible processes for a given system *within a given dynamical model*. They are supposed to explain the long-term tendencies in the processes represented by those trajectories. How to move from an entity informing the behavior of a mathematical model (phase space) to a real entity objectively governing intensive processes is a complex technical problem which Deleuze tackles but which cannot be described here. Elsewhere I have tried to give my own account of this ontological interpretation in a way that would satisfy scientists, or, at least, analytical philosophers of science (see DeLanda). In what follows I will assume that the technical difficulties can be surmounted and that a realist interpretation of some features of these models can be successfully given. But I should at least quote Deleuze on his ontological commitment to the real counterparts of these mathematical entities:

> The virtual is not opposed to the real but to the actual. *The virtual is fully real in so far as it is virtual*...Indeed, the virtual must be defined as strictly a part of the real object as though the object had one part of itself in the virtual into which it plunged as though into an objective dimension...The reality of the virtual consists of the differential elements and relations along with the singular points which correspond to them. The reality of the virtual is structure. We must avoid giving the elements and relations that form a structure an actuality which they do not have, and with-drawing from them a reality which they have. (Deleuze 208–209, italics in the original)

Two more details must be added before arriving at the definition of a *virtual multiplicity*. First, singularities are not always topological points but also closed loops of different kinds, defining not only processes tending towards a steady state, but also processes in which the final product displays endogenous oscillations (periodic attractors) as well as turbulent behavior (chaotic attractors). Second, besides attractors we need to include *bifurcations*. Bifurcations are critical events which convert one type of attractor into another. These events tend to come in *recurrent sequences* in which one distribution of universal singularities is transformed into another, then another and so on. For example, there is a well studied sequence which begins with a point attractor that at a critical value of a control parameter becomes unstable and bifurcates into a periodic attractor. This cyclic singularity, in turn, becomes unstable at another critical value and undergoes a sequence of instabilities (several period-doubling bifurcations) which transform it into a chaotic attractor.

If this were a purely formal result in mathematics its ontological significance might be doubted, but, as it turns out, this cascade of bifurcations can be related to actual recurring sequences in physical processes. A realization of the above cascade occurs, for example, in a well studied series of distinct hydrodynamic flow patterns (steady-state, cyclic, and turbulent flow). Each of these recurrent flow patterns appears one after the other at well defined critical thresholds of temperature or speed. The cascade that yields the sequence conduction–convection–turbulence is, indeed, more complicated and may be studied in detail through the use of a special machine called the Couette–Taylor apparatus. At least seven different flow patterns are revealed by this machine, each appearing at a specific critical point in speed, and, thanks to the simple cylindrical shape of the apparatus, each transition may be directly related to a *broken symmetry* in the group of transformations of the cylinder. Although the mathematical concept of "symmetry" cannot be explained here, one can roughly define it as the degree to which an object lacks detail: the more bland or less detailed the object the more symmetry it has.[2] Hence, a sequence of events in which this blandness is progressively lost (a symmetry-breaking cascade) represents a process of *progressive differentiation*, a process in which an originally undifferentiated object (e.g., a bland, uniform flow of water) progressively acquires more and more detail (displaying waves of different periodicities and, later on, complex arrangements of vortices) (see Stewart and Golubitsky, chapter 5). The purely formal entity (the symmetry-breaking cascade of bifurcations and the resulting attractors) is what Deleuze refers to as a "virtual multiplicity," while the physical sequence which produces actual patterns of flow would correspond to the intensive embodiment of that multiplicity.

This is, in a nutshell, the realist ontology of Gilles Deleuze: a world of actual individual entities (nested within one another at different spatio-temporal scales), produced by intensive individuation processes, themselves governed by virtual multiplicities. I left out many details, of course, including a discussion of the space formed by multiplicities (called "the plane of immanence" or "plane of consistency"), the form of temporality of this space (a crucial question if multiplicities are to be different from timeless archetypes), as well as the way in which this virtual spacetime is constantly formed and unformed (this involves introducing one more entity, half-virtual half-intensive, called a "line of flight"). I will not discuss these further issues here, vital as they are, given that the elements already introduced are sufficiently unfamiliar to raise questions of their own. Even without discussing planes of immanence and lines of flight one may legitimately ask whether we really need such an *inflationary ontology*, an ontology so heavily laden with unfamiliar entities. The answer is that this ontology eliminates a host of other entities (general types and laws) and that, in the final balance sheet, it turns out to be leaner not heavier than what it replaces.

Chemical and biological entities in ecology

If the study of ecosystems involved only large plants and animals, positivists and realists (as well as the least self-deluded of idealists) would have little to disagree about. But there are many entities, such as basic nutrients like oxygen and nitrogen, that are not large enough for us to perceive directly and that must be given a proper ontological status to be included in a materialist philosophy. To illustrate this point let me give an example of how the universal-singular and the individual-singular can replace the old relation between the general and the particular in the world of chemistry. Here as elsewhere static classifications must be replaced by a symmetry-breaking abstract structure (accounting for the regularities in the classified entities) as well as concrete intensive individuation processes (accounting for the production of the classified entities). How to perform this replacement needs to be worked out case by case, a fact that illustrates that the study of the intensive as well as that of the virtual is ultimately empirical.

Perhaps the best example of a successful general classification which can already be replaced by virtual and intensive entities is the famous Periodic Table of the Elements which categorizes *chemical species*. The table itself has a colorful history, given that many scientists had already discerned regularities in the properties of the chemical elements (when ordered by atomic weight) prior to Mendelev stamping his name on the classification in 1869. Several decades earlier one scientist had already discerned a simple arithmetical relation between triads of elements, and later on others noticed that certain properties (like chemical reactivity) recurred every seventh or eighth element. What constitutes Mendelev's great achievement is that he was the first one to have the courage to leave *open gaps* in the classification instead of trying to impose an artificial completeness on it. This matters because in the 1860s only around 60 elements had been isolated, so the holes in Mendelev's table were like daring predictions that yet undiscovered entities must exist. He himself predicted the existence of germanium on the basis of a gap near silicon. The Curies later on predicted the existence of radium on the basis of its neighbor barium (see Atkins Chapter 7). I take the rhythms of the table as being as real as anything that science has ever discovered. I realize that within the ranks of the sociology of science there are many who doubt this fact, arguing that, for example, had Priestley's phlogiston triumphed over Lavoisier's oxygen an entirely different chemistry would have evolved. I strongly disagree with this assertion, but I will not engage this argument here. As I said before, this conventionalist maneuver only pretends to get rid of eternal archetypes and succeeds only in replacing them with social essences: conventional forms imposed upon an amorphous world very close to the inert matter of classical essentialism.

The virtual multiplicity underlying this famous classification has been recently worked out. Given that the rhythms of the table emerge when one arranges the chemical species by the number of protons their atomic nuclei possess (their atomic number) and that the nature of the outer orbital of electrons is what gives these elements their chemical properties, it should come as no surprise that the multiplicity in question is a symmetry-breaking abstract structure that relates the shape of electron orbitals to the atomic number. I mentioned before the sequence of bifurcations in fluid flow dynamics that unfolds as one increases the intensity of speed or temperature to certain critical thresholds. Similarly, the sequence of broken symmetries behind the table may be seen to unfold as one injects more and more energy into a basic hydrogen atom. The single electron of this atom inhabits a shell with the form (and symmetry) of a sphere. Exciting this atom to the next level yields either a second larger spherical orbital or one of three possible orbitals with a two-lobed symmetry (two lobes with three different orientations). This new type of orbital has indeed the right mathematical form to be what a sphere would be if it had lost some of its symmetry. Injecting even more energy we reach a point at which the two-lobed orbital bifurcates into a four-lobed one (with five different possibilities), which in turn yields a six-lobed one as the excitation gets intense enough. In reality, this unfolding sequence does not occur to a hydrogen atom but to atoms with an increasing number of protons, boron being the first element to use the first nonspherically symmetric orbital (see Icke 150–162).

This abstract structure of progressively broken spherical symmetries is a beautiful illustration of a Deleuzian multiplicity, and it accounts for the numerical rhythms of the table: the number of available orbitals at each energy level multiplied by two (given that two electrons of opposite spin may inhabit the same orbital) perfectly fits the recurrent sequences of chemical properties. And yet, such an abstract structure is not enough. To this virtual entity one must add an intensive process that embodies it without resembling it and which physically individuates the different chemical species. This intensive process is known as *stellar nucleosynthesis* and, as its name indicates, it occurs today mostly within stars (veritable factories for chemical assembly) although presumably it may also have occurred under the much greater intensities present right after the Big Bang. The latter were sufficient to individuate hydrogen and helium, but the rest of the elements had to wait millions of years until intensive differences in density allowed the individuation of the stars themselves. In this other environment, the next two elements (lithium and beryllium) are individuated via a synthesis of the first two, and these new species in turn become the gateway to reach the conditions under which larger and larger nuclei may be individuated. Just what degree of nucleosynthesis a given star may achieve depends on specific critical thresholds. To quote P. W. Atkins:

> The first stage in the life of a newly formed star begins when its temperature has risen to about 10 million degrees Kelvin. This is the hydrogen-burning stage of the star's life cycle, when hydrogen nuclei fuse to form helium...When about 10 percent of the hydrogen in the star has been consumed a further contraction takes place and the central region of the star rises to over 100 million degrees...Now helium burning can begin in the dense hot core, and helium nuclei fuse into beryllium, carbon, and oxygen...In sufficiently massive stars those with a mass at least four times that of our sun the temperature can rise to a billion degrees, and carbon burning and oxygen burning can begin. These processes result in the formation of elements....including sodium, magnesium, silicon and sulfur. (72–73)

I will not discuss the technical details of just exactly how these different syntheses are carried out. It is enough for my purposes that intensive differences as well as intensive thresholds are crucially involved. But even in this sketchy form the nature of the different chemical species already looks quite different than it does when a naive realist tackles the question of their identity. As mentioned in the introduction to this essay, such a naive realist would remark that, given that if one changes the atomic number of an element one thereby changes its identity, the atomic number must be the essence of the element in question. In a Deleuzian ontology such a statement would be inadmissible. Not only does atomic number by itself not explain why an element has the properties it has (we need also the theory of electron orbitals), but the actual process of creating increasingly heavier atomic nuclei is completely ignored, even though this is what actually creates individual atoms with a given identity. Moreover, the reification of atomic numbers into essences ignores yet other individuation processes, those responsible for the creation of larger-scale entities, such as a sample of gold or iron large enough to be held in one's hand. Naive realists treat the properties of such large samples as being merely the sum of the properties of its atoms and hence reducible to them. But this reducibility is, in fact, an illusion.

Here is where the idea of individual singularities of different spatiotemporal scales nested within one another becomes useful. As I said above, individual organisms are singular, unique, machine-like entities whose component parts are individual cells, while the organisms themselves are the "cogs and wheels" of individual species. However, this statement is incomplete because it ignores the individual entities which bridge any two spatiotemporal scales: between individual cells and individual organisms there are several intermediate structures such as tissues, organs, organ systems (similarly, between organisms and species there are reproductive communities inhabiting concrete ecosystems). Now, to return to the chemical world, between individual atoms of gold or iron and an individual bulk piece of solid material there are several intermediate components: individual atoms form crystals; individual crystals form small grains; individual small grains form larger grains, and so on. Each intermediate component bridging the micro and the macro world is individuated following concrete causal processes, and the properties of an individual bulk sample emerge from the causal interactions between these intermediate structures (see Duncan and Rouvray 113).

There is, in fact, even more to this intensive story, given that what I just said relates only to tendencies and says nothing about capacities: the different capabilities of the chemical elements to enter into heterogeneous assemblages with each other. Although the simplest such assemblages (dyadic molecules) may be put into tables displaying rhythms of their own, the table approach is hopeless when dealing with, say, the seemingly unlimited number of possible carbon compounds. I will let my case rest here, however, since the contrast with the picture that the naive realist holds is already clear enough. Instead let me move on from chemical to biological species, the component parts of ecosystems. In this case too we have inherited complex static classifications exhibiting regularities that cry out for further explanation. I must say in advance that the picture here is much less clear than that of chemistry, given that the classification is much more complex and that history (evolutionary history) plays an even greater role. I will tackle the question in the opposite order than I did for Mendelev's table, starting with the intensive aspects.

As I said above, the individuation of a species consists basically of two separate operations, a sorting operation performed by natural selection and a consolidation operation performed by reproductive isolation. The mechanism of these two operations may, in turn, be characterized in terms of intensive differences and thresholds. The idea that, for example, a given predator species exerts selection pressures on a prey species needs to be explained in terms of the relations between the *densities* of the populations of predators and prey. In many cases these two populations form a dynamical system which exhibits endogenous equilibria such as a stable cycle of boom and bust. The other operation, reproductive isolation, also needs to be defined intensively, in this case in terms of the rates of flow of external genetic materials into a species' gene pool. Philosophically, this translation eliminates the temptation to characterize species in terms of a static classification where their identity is simply assumed and all one does is record observed similarities in the final products.

In a Deleuzian ontology resemblance and identity must be treated not as fundamental but as derivative concepts. If selection pressures happen to be uniform in space and constant in time we should expect to find more resemblance among the members of a population than if those selection forces are weak or changing. Similarly, the degree to which a species possesses a clearcut identity will depend on the degree to which a given reproductive community is effectively isolated. Many plant species, for example, retain their capacity to hybridize throughout their lives (they can exchange genetic materials with other plant species) and hence possess a less clear-cut genetic identity than perfectly reproductively isolated animals. In short, the degree of resemblance and identity exhibited by organisms of a given species depends on contingent historical details of the process of individuation, and is therefore not to be taken for granted.

What would the virtual component of the process of speciation be like? This is a highly speculative question given the current state of evolutionary biology. It involves moving from the lowest level of the classification (the species) to a level much higher up, that of the *phylum*. Phyla represent the level of classification just underneath kingdoms. The animal kingdom, for example, divides into several phyla including chordata, the phylum to which we as vertebrates belong. But phyla may be treated not just as labels for static categories but also as *abstract body plans*. We know little about what these body plans are, but some progress has been made in defining some of its parts, like the "virtual vertebrate limb," or more technically the *tetrapod limb*, a structure which may take many divergent forms, ranging from the bird wing, to the single-digit limb in the horse, to the human hand and its opposed thumb. It is very hard to define this abstract structure in terms of the common properties of all the adult forms, that is, by concentrating on homologies at the level of the final product. But, focusing instead on the embryological intensive processes which produce organisms, one can hypothesize that there is a symmetry-breaking sequence of bifurcations subject to genetic control. The genes of horses, birds and humans (or rather the enzymes that they code for) control parameters which determine which bifurcations will occur and which will be blocked, this in turn determining what distributions of attractors will be there to channel processes in one or another direction (see Hinchliffe).

When embryological development reaches the budding of fingers in a limb's tip, for example, some genes (those of horses) will inhibit the occurrence of the event (and so prevent fingers from forming) while others (those of humans) will not. In the case of snakes or dolphins the very branching of the limb may be inhibited. So what we would need is a topological description of a virtual vertebrate (not just the limbs) and an account of how different genetic materials select certain pathways within this body plan and inhibit certain others, thus resulting in different species' morphologies. While we are far from having such a topological model (not to mention far from being able to express it as a symmetry-breaking cascade) the work that has already been done on the subject suggests that we can be optimistic about its prospects. At any rate, within a Deleuzian ontology one is forced to pursue this empirical research since we cannot be satisfied with a static classification recording resemblances and identities. For the same reason, one cannot view selection pressures as sculpting animal and plant forms in detail (a conception that assumes an inert matter receiving form from the outside) but as teasing a form out of a morphogenetically pregnant material. And similarly for genes: they cannot be seen as defining a blueprint of the final product (and hence its essence) but only as a program to guide self-organizing embryological processes towards a given final state.

Finally, to this account of the tendencies of embryonic matter that yield, in conjunction with genes, the bodies of animals and plants we must add the capacities to affect and be affected that those bodies have relative to one another. Deleuze was particularly fond of the relations of symbiosis into which animals (such as pollinating insects) may enter with plants. But all other ecological relations (predator–prey, parasite–host) involve the same coupling of heterogeneous components into an assemblage. An ecosystem as a whole is, in fact, a complex assemblage of those assemblages, a complex whole that demands a philosophy of difference in order to be properly conceptualized. The ontology sketched here can be only an introduction to such a philosophy of difference, a first step in the direction of breaking the monopoly on explanation that the general and the particular have held for so long, replacing these terms and the static and closed classifications they yield with an open-ended becoming based on individual and universal singularities.

Notes

1. See Deleuze and Guattari:
   What is the significance of these indivisible distances that are ceaselessly transformed and cannot be divided or transformed without their elements changing in nature each time? Is it not the intensive character of this type of multiplicity's elements and the relations between them? Exactly like a speed or a temperature, which is not composed of other speeds or temperatures, but rather is enveloped in or envelops others, each of which marks a change in nature. (31)
2. The mathematical concept of "symmetry" is defined in terms of "groups of transformations." For example, the set consisting of rotations by 90 degrees (that

is, a set containing rotations by 0, 90, 180, 270 degrees) forms a group, since any two consecutive rotations produce a rotation also in the group, provided 360 degrees is taken as zero. (Besides this "closure," sets must have several other formal properties before counting as groups.) The importance of groups of transformations is that they can be used to classify geometric figures by their *invariants*: if we performed one of this group's rotations on a cube, an observer which did not witness the transformation would not be able to notice that any change had actually occurred (that is, the visual appearance of the cube would remain invariant relative to this observer). On the other hand, the cube would not remain invariant under rotations by, say, 45 degrees, but a sphere would. Indeed, a sphere remains visually unchanged under rotations by *any amount* of degrees. Mathematically this is expressed by saying that the sphere has *more symmetry* than the cube relative to the rotation transformation. That is, degree of symmetry is measured by the number of transformations in a group that leave a property invariant, and relations between figures may be established if the group of one is included in (or is a subgroup of) the group of the other.

Works cited

Atkins, P. W. *The Periodic Kingdom*. New York: Basic Books, 1995.

DeLanda, Manuel. *Intensive Science and Virtual Philosophy*. London: Continuum Press, 2002.

Deleuze, Gilles. *Difference and Repetition*. Trans. Paul Patton. New York: Columbia University Press, 1994.

Deleuze, Gilles, and Félix Guattari. *A Thousand Plateaus: Capitalism and Schizophrenia*. Trans. B. Massumi. Minneapolis: University of Minnesota Press, 1987.

Duncan, Michael A., and Dennis H. Rouvray. "Microclusters." *Scientific American* (December, 1989): 110–115.

Hinchliffe, Richard. "Toward a Homology of Process: Evolutionary Implications of Experimental Studies on the Generation of Skeletal Pattern in Avian Limb Development." *Organizational Constraints on the Dynamics of Evolution*. Eds. J. Maynard Smith and G. Vida. Manchester: Manchester University Press, 1990, 119–131.

Icke, Vincent. *The Force of Symmetry*. Cambridge: Cambridge University Press, 1995.

Nicolis, Gregoire, and Ilya Prigogine. *Exploring Complexity*. New York: W.H. Freeman, 1989.

Stewart, Ian, and Martin Golubitsky. *Fearful Symmetry*. Oxford: Blackwell, 1992.

van Wylen, Gordon. *Thermodynamics*. New York: John Wiley & Sons, 1963.

3
A Thousand Ecologies

Ronald Bogue

It is evident from the title alone of Guattari's *The Three Ecologies* (1989) that his thought had taken an ecological turn in his later writings, but what of his works jointly authored with Deleuze, especially the earlier *Anti-Oedipus: Capitalism and Schizophrenia, I* (1972) and *A Thousand Plateaus: Capitalism and Schizophrenia, II* (1980)? Are those texts ecological, or at least eco-friendly? They are, I believe, though in a sense that requires careful specification, in part due to the contested nature of many concepts and models in ecology studies, but in part as well due to Deleuze and Guattari's idiosyncratic terminology and the ends to which it is addressed in their writings—ends that are not incompatible with those of an ecological philosophy, but that subordinate such ends to other concerns. Deleuze and Guattari's primary goal in the two volumes of *Capitalism and Schizophrenia*, as well as *What Is Philosophy?* (1991), is to develop and practice a mode of thought, and in the process engage issues in a wide range of fields, some of which have an obvious bearing on the sciences of ecology (biology, chemistry, genetics, ethology), others of which would seem far removed from that domain (aesthetics, sociology, political theory, history). In a 1988 interview, Deleuze said that he planned to write a book titled *"What Is Philosophy?"*, but he added that he and Guattari also wanted "to get back to our joint work and produce a sort of philosophy of Nature, now that any distinction between nature and artifice is becoming blurred" (*Negotiations* 155). Deleuze and Guattari never completed this projected exposition of a "philosophy of Nature," but the outlines of that philosophy may be discerned in the works they did produce, and those outlines may be regarded as ecological in many respects.

At the beginning of *Ecology, Community and Lifestyle* (1989), Arne Naess helpfully distinguishes ecology from what he calls "ecophilosophy" and "ecosophy." Ecology he defines as "the interdisciplinary scientific study of the living conditions of organisms in interaction with each other and with the surroundings, organic as well as inorganic" (Naess 36). He characterizes the methodology inherent in the study of ecology as one "suggested by the simple maxim 'all things hang together.'" The field of "ecophilosophy" he describes as the study of "problems common to ecology and philosophy." Ecophilosophy is "a descriptive study" that "does not make a choice between fundamental value priorities, but merely seeks to examine a particular kind of problem" at the juncture of ecology and philosophy. Unlike ecophilosophy, "ecosophy" addresses questions concerning humans and nature from the perspective of "one's own personal code of values and a view of the world which guides one's own decisions" (36). Clearly, Deleuze and Guattari are not ecologists engaged in the interdisciplinary scientific study of organisms and their surroundings, but their thought about nature is decidedly ecophilosophical. And, as we shall see, that thought is ecosophical—that is, value-laden and value-directed—though in a way that Naess himself, with his commitment to deep ecology, would no doubt find unacceptable.

Deleuze and Guattari's most focused consideration of organisms and their surroundings is Plateau Eleven of *A Thousand Plateaus*, which is devoted to the concept of the refrain. Using song in territorial birds as a paradigmatic instance of a natural refrain, Deleuze and Guattari expand the sense of the refrain to include any organizing rhythmic pattern that brings together heterogeneous entities that function in concert within the natural world. Their chief inspiration for this musical model of nature is the pioneering ecologist Jacob von Uexküll, who speaks of a grand symphony of nature, in which relations of point and counterpoint structure organisms and their milieus.[1] In von Uexküll's analysis, each organism traces a developmental melody in its gestation, growth and eventual demise. Its form unfolds in counterpoint to diverse elements of its surroundings, the gutter-like contour of the oak leaf in counterpoint to falling rain, the octopus's muscular pocket in counterpoint to the incompressible water it squeezes in order to propel itself. Relations among conspecifics (in sexual organisms, for example, between male and female, between parents and progeny, among territorial competitors, and so on), between predators and prey (spider and fly), among symbionts of various sorts (orchid and wasp, liver fluke and sheep, termite and intestinal parasite), all are so many contrapuntal arrangements of mutually corresponding melodies, the totality of such motifs forming the symphony of nature.

Deleuze and Guattari's concept of the refrain is essentially an elaboration on von Uexküll's notion of the contrapuntal melody. Like von Uexküll, they stress the interconnections among developmental, ethological, and environmental processes, treating the growth trajectories of individual organisms, the regular patterns of their behavior, and the configurations of their interactions with their organic and inorganic surroundings as intertwined and interdependent refrains. Hence, when they consider the song of the brown stagemaker bird, they regard the sonic refrain as but one component of "a veritable *machinic opera* tying together orders, species, and heterogeneous qualities" (*Thousand Plateaus* 330). The male stagemaker sings while perched within its territory on a stick selected as the stage of its performance. The male delineates its territory by sawing leaves from surrounding trees and distributing them bottom side up on the periphery of its staked area. It fluffs its throat feathers while signaling through song its presence to other males and to prospective mates. The sonic refrain of the bird's song, then, is only one of a multitude of refrains—the refrain of its development from egg to adult, the refrains of its behavioral patterns (feather display, leaf gathering, stick perching), the refrains of interaction among other males and with female stagemakers, and so on.

Implicit as well in Deleuze and Guattari's treatment of the refrain is a specific approach to evolutionary theory, something that is less evident in von Uexküll's work. In their discussion of birdsong and territoriality, Deleuze and Guattari insist that the formation of territories is not a secondary product of basic drives (as Lorenz and others posit), but instead is its own explanation. They recognize that various functions (sexual, alimentary, aggressive, predatory) are organized within a territory, but no single function causes the territory to come into existence. Various functions

> are organized or created only because they are *territorialized*, and not the other way around. The T factor, the territorializing factor, must be sought elsewhere: precisely in the becoming-expressive of rhythm or melody, in other words, in the emergence of proper qualities (color, odor, sound, silhouette...). (*A Thousand Plateaus* 316)

In the case of the stagemaker, the plucking of a leaf and its inverted placement on the ground mark the emergence of a quality proper to the bird (the bird's action serving to extract the leaf from the natural surround and make the color of its inverted surface a territorial quality), and that leaf is a component of a rhythm that expresses the territory. One might see this analysis as a circular explanation, in that the bird's territorializing activities explain the formation of the territory. Deleuze and Guattari's point, however, is (1) that the territory is not some inert stretch of ground staked out by a bird, but instead the complex of patterns and rhythms involving the bird, other birds, other animals, trees, plants, earth, water, air in a territorial ensemble—in short, the territory is the ecosystem of bird–environment; and (2) that the emergence of a territorial animal in evolutionary history is the manifestation of a particular configuration of organism–environment that results simply from a nonteleological process of creative experimentation exhibited throughout the world.

In this regard, I see Deleuze and Guattari as promoting a view of nature similar to that put forward by Varela, Thompson and Rausch in *The Embodied Mind* (1991).[2] Varela et al. argue at length against any separation of organism and environment in evolutionary theory, asserting that organisms and their milieus are dynamic entities involved in an interactive process of coevolution.

Organisms and environments are "mutually unfolded and enfolded structures" (Varela et al. 199) that together bring forth a world. This process of reciprocal coevolution Varela et al. refer to as a "structural coupling" (151) of organism and environment. Varela et al. also oppose the standard neo-Darwinian notion of the survival of the fittest, claiming that natural selection does not produce the best organism for a given environment, but instead merely eliminates those organisms that are not viable. Nature affords a wide range of possibilities for viable organism–environment structural couplings, and all that is required is that an organism be good enough for survival in its milieu, not that it be the milieu's optimal inhabitant (whatever that might be). In their terms, the evolutionary process "is *satisficing*, (taking a suboptimal solution that is satisfactory) rather than optimizing," and it proceeds via "*bricolage*, the putting together of parts and items in complicated arrays, not because they fulfill some ideal design but simply because they are possible" (196).

This notion of Nature as *bricoleur* is widespread in *A Thousand Plateaus*, evident not only in the concept of the refrain, but also in that of the "assemblage" ' as collection of heterogeneous entities that somehow function together and that of the interpenetrating "strata" of inorganic, organic and anthropological forms (developed in Plateau Three). A world of *bricolage* is evident as well in *Anti-Oedipus*, where the cosmos is described as a realm of universal desiring-production, in which ubiquitous desiring-machines are connected in circuits through which pass diverse flows and fluxes. Such circuits of desiring-production may include what we normally think of as machines (referred to as "technical machines" by Deleuze and Guattari), but they may also connect human organs, diverse organisms and inorganic entities and processes in makeshift, variable relations, and the flows may consist of matter, energy, information, sensations, thoughts, fantasies, and so on. Unlike "technical machines," desiring-machines do not aspire to a maximum efficiency, but instead "work only when they break down, and by continually breaking down" (*Anti-Oedipus* 8). Desiring-production is "satisficing," in the language of Varela et al., good enough to function but only in a way that makes room for glitches, pauses, divagations, and impromptu variations. Perhaps the best images of desiring-production are those of the Rube Goldberg cartoons included in the appendix of the second edition of *Anti-Oedipus*, each of which presents a comically improbable sequence of human, animal, and inorganic entities linked in diverse causal relations (perceptual, behavioral, mechanical, chemical, and so on), the assemblage of entities forming a complex machine of decided inefficiency dedicated to the performance of the simplest of tasks (eating less, remembering to mail a letter). The satisficing Nature of desiring-production is likewise filled with shifting circuits of heterogeneous elements connected in multiple, unexpected combinations that function with varying degrees of efficiency.

Throughout *Anti-Oedipus* Deleuze and Guattari use the figure of the machine to describe the world, and, though machine imagery is somewhat less prevalent in *A Thousand Plateaus*, it is still present. In *Anti-Oedipus*, Deleuze and Guattari wish to counter psychological models that stress interpretation, replacing a psychoanalysis that uncovers the meaning of the unconscious with a schizoanalysis that simply charts its functioning. A machine does not express any meaning as it operates but simply does what it does, and in this regard the unconscious may be regarded as a machine. However, Deleuze and Guattari adopt the machine model not simply to undo psychoanalytic presuppositions but also to subvert all distinctions between the natural and the artificial, between the nonhuman world and the social, cultural and technological world constructed by humans. As they assert early in *Anti-Oedipus*,

> we make no distinction between man and nature: the human essence of nature and the natural essence of man become one within nature in the form of production or industry, just as they do within the life of man as a species.... man and nature are not like two opposite terms confronting each other...; rather, they are one and the same essential reality, the producer-product. (4–5)

All the elements of culture and nature, of the human and the nonhuman world, are desiring-machines engaged in desiring production. "Everything is a machine.... Producing-machines, desiring-machines, everywhere, schizophrenic machines, all of species life: the self and the non-self, outside and inside, no longer have any meaning whatsoever" (*Anti-Oedipus* 2). The nature-machine model, of course, goes back at least to La Mettrie, but Deleuze and Guattari insist that their desiring-machines are not mere mechanical devices. They propose their "machinism" as an alternative to both mechanism and vitalism, rejecting paradigms that reduce everything to a rudimentary physics as well as those that spiritualize matter with a separate life force. Their machinism instead posits the existence of an "anorganic life" (*Thousand Plateaus* 503) that spans the human, nonhuman, organic and inorganic in a single process of interfused cofunctioning.

Such a machinism should not be construed as an antinature or antiecology stance, though it does suggest opposition to any conception of environmentalism as a means of restoring a corrupted nature to its prelapsarian, nonhuman purity. One might see such a stance as merely a realistic response to present circumstances. As the conservation biologist Michael Soulé argues,

> [s]oon, the distinctions between preservation, reintroduction, and restoration will vanish. In 2100, entire biotas will have been assembled from (1) remnant and reintroduced natives, (2) partly or completely

engineered species, and (3) introduced (exotic) species. The term *natural* will disappear from our working vocabulary. This term is already meaningless in most parts of the world because anthropogenic fire, chemicals, and weather, not to mention deforestation, grazing, and farming, have been changing the physical and biological environment for centuries, if not millennia. (301)

However, Deleuze and Guattari's machinism is less a concession to an unfortunate reality than a commitment to exploring the possibilities of human invention. Their language of machines and desiring-production seeks to avoid the extremes of either technophobia or technophilia, of either a Rousseauistic return to nature or a science-driven construction of an artificial paradise. Their position in this regard seems close to that of Donna Haraway, whose Cyborg Manifesto echoes Deleuze and Guattari's machinic thematics at several points. For her, as for Deleuze and Guattari, the natural–artificial distinction is untenable, and the figure of the cyborg, like that of the desiring-machine, serves to counter any conception of humans and their relationship to the world in terms of a stable, unproblematic nature (whether human or nonhuman). The cyborg appears "precisely where the boundary between human and animal is transgressed," and hence "far from signaling a walling off of people from other living beings, cyborgs signal disturbingly and pleasurably tight coupling" (Haraway 152). Haraway does not minimize the dangers emergent in our increasingly technological world, but, like Deleuze and Guattari, she finds in that world potential for new ways of configuring social and environmental relations in terms that blur clear distinctions among humans, machines and nature.

Deleuze and Guattari's focus on interconnecting flows of desiring-production, on interfolded refrains that constitute organism–environment complexes, and on the cosmos as the domain of a single anorganic life might suggest that they advocate a kind of ecological holism, and in a limited sense this is true. They certainly oppose any atomistic analysis of nature that ignores the constitutive relations of living systems, and often their treatment of circuits of flows and refrains reinforces the basic insight that "all things hang together" (Naess 36). But they repeatedly assert as well that all circuits of desiring-production are irreducible multiplicities that cannot be subsumed within the traditional logic of the one and the many, or the whole and its parts. In *Anti-Oedipus*, they observe that "the problem of the relationships between parts and the whole continues to be rather awkwardly formulated by classic mechanism and vitalism, so long as the whole is considered as a totality derived from the parts, or as an original totality from which the parts emanate, or as a dialectical totalization" (44). If there is a whole, they claim, it is "a totality alongside various separate parts, it is a whole *of* these particular parts but does not totalize them, it is added to them as a new part fabricated separately" (42).

Deleuze himself, however, does not categorically reject all reference to the "whole" (*le tout*), for in *Cinema 1* he assigns the concept of the whole a prominent place in his film theory, positing the existence of a whole immanent within every film shot that "is like thread which traverses [the shot] and gives each one the possibility, which is necessarily realized, of communicating with another, to infinity" (*Cinema I* 16–17). But this whole is not a closed totality. Rather, "the whole is the Open," by which "every closed system opens to a duration which is immanent to the whole universe" (17). Deleuze and Guattari make similar use of the notion of an open whole in *What Is Philosophy?* when they describe philosophy's plane of immanence as "a powerful Whole that, while remaining open, is not fragmented: an unlimited One-All, an 'Omnitudo' " (*What Is Philosophy?* 34). And though they nowhere refer explicitly to the cosmos as an open whole, their speculations about the existence of an ultimate cosmic "totality of all BwO's [Bodies without Organs], a pure multiplicity of immanence...the plane of consistency (*Omnitudo*, sometimes called the BwO)" (*Thousand Plateaus* 155–156) suggests that they would not object to such an appellation.

In one sense, this stress on the openness of the whole may be seen as a reminder of the basic fact of any real-world systems analysis, including an ecological analysis, that no system is finally self-contained. In Anthony Wilden's terms, context is always a function of "punctuation," of a provisional closure of a system and its components (see Wilden 111–113). If one studies a specific ecosystem, say an Australian coral reef, one delimits the area, the life forms and processes one wishes to study, ignoring the larger context of the Pacific Ocean as a whole, of which the coral reef is a part, and the even larger context of the earth's ocean system, that of the solar system as movement–energy source, that of the galaxy, and so on. At every scale of analysis, what counts as a system is a function of its punctuation, the determination of a provisional closure in a theoretically unlimited and always open totality (especially if the universe is indeed expanding).

Yet Deleuze and Guattari also advocate not simply a spatial, methodological openness but also a temporal openness, and in this regard their thought runs counter to some tendencies in discussions of ecology, especially those regarding questions of conservation and habitat restoration. Throughout his career, Deleuze remained a proponent of the Bergsonian intuition that time matters and that the future is genuinely new. Bergson observes that the dominant conception of time in Western philosophy is essentially spatial, one in which time functions merely as a measure of successions of static slices of space. As a result, the dynamic thrust of the world's becoming is ignored and time adds nothing significant to our accounts of reality. Bergson objects as well to deterministic models of the universe, arguing that in them "the whole" is already "given," that is, already knowable and hence closed. Such is the view of classical science, summed up in the Laplacean conception of the universe as one in which linear causal relations govern all phenomena and any future state of the world may be predicted from a complete knowledge of past and present configurations of matter and energy. Bergson argues that the whole is never given, for the future is unpredictable and unknowable. The movement of becoming is fundamental to any conception of the universe, as is its thrust into an undetermined future.[3] This Bergsonian view of becoming and the openness of the future is evident throughout Deleuze and Guattari's writings, especially in the section on "becomings" in *A Thousand Plateaus* (232–309).

Such a stress on becoming and indeterminate wholes may not be incompatible with ecological thought, but it does call into question tacit assumptions common in public discussions of environmental issues. Chief among these is the postulate of a "balance of nature" that humans have disrupted and that conservation efforts must restore. Though ecosystems analyses include a temporal dimension, the postulate of a balance of nature suggests that in natural (that is, nonhuman) ecosystems the interaction of elements is essentially homeostatic—regular, stable, and unchanging overall. In such a conception, time ultimately makes no difference, the future is but a repetition of the present system of kinetic relations, and the whole is closed. Obviously, Deleuze and Guattari would find unacceptable any postulated balance of nature, but in this they are not alone, for many ecologists have also questioned this hypothesis. Stuart Pimm, for example, in *The Balance of Nature?*, argues at length that natural systems seldom operate in equilibrium, and indeed often function in states far from equilibrium. Stability in the strict sense is rare in ecosystems, Pimm shows, and if there is a degree of regularity in natural systems it must be analyzed in terms of the system's response to variables—in terms of resilience (the speed with which a variable within the system is returned to equilibrium), persistence (the rate at which a variable is changed to a new value within the system), resistance (the ability of the system to absorb a permanently changed variable), and variability (the degree to which its variables vary over time).[4] Yet despite such regularity within variability, says Pimm, ultimately all natural systems change over time, and if ecological models are to be adequate they must include a historical, evolutionary dimension in their analyses:

> Species are added to and lost from communities, at rates that depend on features of the species themselves and also on the features of the community's food web structure and on how that structure developed....Community structure changes, and as it changes it alters the framework against which population changes take place. (Pimm 4)

What Pimm's work makes clear is that environmental issues are less about natural stability and human-induced instability than about rates of change and kinds of change. The question is not whether humans have induced change in ecosystems, but whether they have inordinately accelerated or inhibited change and in ways that are deleterious, whether to humans specifically or to terrestrial life forms in general. I would argue that Deleuze and Guattari's concept of the body as a configuration of speeds and affects is germane to the issues of both rates of change and kinds of change. In *A Thousand Plateaus* Deleuze and Guattari state that:

> on the plane of consistency, *a body is defined only by a longitude and a latitude*: in other words the sum total of the material elements belonging to it under given relations of movement and rest, speed and slowness (longitude); the sum total of the intensive affects it is capable of at a given power or degree of potential (latitude). (*Thousand Plateaus* 260)

What constitutes a body may vary from subindividual configurations of speed and affect (individual organs, the circulatory system, digestive system, and so on), through the entities that we commonly refer to as bodies (organisms), to supraindividual configurations of speed and affect (male–female, predatory–prey, species communities, ecosystems as a whole). In Plateau Six, "How Do You Make Yourself a Body without Organs?," Deleuze and Guattari formulate a diagnostics and an ethics of speeds, asserting that in producing a Body without Organs (that is, a body on the plane of consistency), one always runs the danger of entering into a suicidal black hole or of fostering a fascistic, cancerous Body without Organs. In the case of the black hole, a precipitous speed induces self-destruction; in that of the cancerous Body without Organs, accelerated components engulf and eventually destroy other components, supplanting them and rendering the whole diseased and pathogenic. Although Deleuze and Guattari frame their discussion largely in sociopolitical terms, it is not difficult to extend the concepts of the black hole and the cancerous Body without Organs to a treatment of ecological issues, especially given their opposition to any definitive separation of the social, cultural and technological world of humans from the nonhuman world. If ecosystems are construed as bodies, clearly such phenomena as global warming may be regarded as black holes of accelerated self-destructive speeds; and systems in exacerbated disequilibrium, such as those induced by the introduction of rabbits in Australia or kudzu in the American south, may be seen as cancerous bodies, each possessed of pathogenic differential speeds among its components.

Deleuze and Guattari's concept of corporeal affect, in addition to that of speed, may also be brought to bear on an ecological diagnostics and ethics, but before addressing this point I must return briefly to Deleuze and Guattari's qualified holism. Besides the concepts of a spatially open whole of ever-expanding contexts and that of a temporally open whole of becoming and an undetermined future, one may also discern in Deleuze and Guattari what might be called an epistemologically and methodologically open whole. In Plateaus Twelve and Fourteen Deleuze and Guattari distinguish striated from smooth space, describing the one as a "relative global" and the other as a "local absolute." The striated space of the relative global "is limited in its parts, which are assigned constant directions, are oriented in relation to one another, divisible by boundaries, and can interlink," whereas the smooth space of the local absolute is "an absolute that is manifested locally, and engendered in a series of local operations of varying orientations" (*Thousand Plateaus* 382). A striated space is a discretely bounded space, and hence a totalizing, "global" space, and within its borders all points may be charted in a fixed coordinate system—hence each interior component has a "relative" existence in relation to the global totality. A smooth space, by contrast, has no demarcated boundaries. It is an area of ever-expanding horizons, an open whole, and, as such, a whole that can never be grasped and mastered in its totality. It can only be experienced and conceived from within, from a "local" perspective. Yet when it becomes manifest, it does so as an "absolute," as an unqualified, perpetually expanding whole-in-becoming.

Striated and smooth spaces are inseparable from the processes that take place within them. They are produced by their inhabitants, and in this sense they are functions of modes of being and relations of power. One produces a striated space when one striates it, when one delimits an area and charts it in a coordinate system. Conversely, one produces smooth space when one follows flows of becoming, when one smoothes over boundaries and engenders a metamorphic movement toward an ever-unfolding horizon. Although Deleuze and Guattari discuss striated and smooth space primarily in human terms, their treatment of territoriality and the refrain suggests that other organisms produce and inhabit striated and smooth spaces, and it is evident as well that priority must be granted to the smooth over the striated space. Humans, like all other life forms, seek to master their ambient space-time, to organize it, control it and render it a habitable milieu, territory, social sphere, and so on. Yet they also participate in flows that unsettle organizational patterns and thereby open up mutative lines of potential development. When they do so, organisms produce the smooth space of an open whole, one in which the organisms' actions coalesce with the becoming of the world in its thrust toward an undetermined future.

If the cosmos is an open whole, and if that whole presents itself only as a smooth "local absolute," the methodological implication is that the whole cannot be conceived from the outside, but must always be understood from a perspective within the whole. Deleuze and Guattari's valorization of smooth space, then, implies a commitment to a kind of Nietzschean perspectivism (hardly a surprise, given Deleuze's longstanding Nietzschean proclivities) and a rejection of the notion that there is any "view from nowhere." In this regard, Deleuze and Guattari's position is consonant with that of Varela and his frequent collaborator Humberto Maturana, whose collective work on natural systems Cary Wolfe aptly situates within what is sometimes called "second-order cybernetics," which may be characterized as a disciplinary matrix within which the study of the organization of complex systems (cybernetics) includes a meta-level recognizing the existence of the observer–scientist as a complex system within the observed system itself. As Wolfe notes, Maturana and Varela's second-order cybernetics avoids any simplistic relativism, for it "does not dispense with systematic description altogether," but instead "recasts the relationship between a system and its elements...as open-ended and yet not random, fundamental and yet not foundational in the usual ontological sense" (Wolfe 55). This second-order cybernetics does, however, require that all knowledge claims be situated within an emergent context.

Wolfe sees Maturana and Varela's second-order cybernetics as providing a means of moving beyond the traditional humanistic dualism of mind and body and its related representational model of cognition, in that Maturana and Varela's approach commits them to a notion of cognition as embodied action. Later in his analysis, however, Wolfe reproaches Maturana and Varela for engaging in "speciesism" (66), or a valorization of *Homo sapiens* over other species, an activity that he regards as inimical to second-order cybernetics. I find this critique puzzling, if Wolfe is using "speciesism" as many deep ecologists do—that is, as a label for a bias that denies the intrinsic value and equal rights of all species. The hard-line position on equal rights and equal value for all species, I would argue, entails precisely the "view from nowhere" that second-order cybernetics makes untenable. If the observer is always part of what is observed, and if the observer is a human being, it is difficult to see how the embodied action of that observer can be anything but observer-centered, and hence "species-centric."

In this characterization of second-order cybernetics and the ethics of deep ecology I have perhaps moved too swiftly from observation to valorization, from "is" to "ought," but, though I would not conflate epistemology and ethics, I would argue that they are not unrelated. Both knowledge and values are perspectival, even if the relationship between the two is not fixed. Maturana and Varela's second-order cybernetics, like Deleuze and Guattari's perspectivism, does indeed problematize the assumptions of traditional representational realism, but so also does it call into question the postulate of nonsituated, transcendent values. As Sahotra Sarkar demonstrates at some length, deep ecology's attribution of intrinsic value to every species implies such a postulate, and I find convincing his argument that this postulate is incoherent.[5] Whatever one's stance on biodiversity, conservation, habitat restoration, and so on, he argues, the values one promotes must be "anthropocentric," at least in the sense that they cannot be entirely divorced from our existence as members of the species *Homo sapiens*. In this regard, then, Deleuze and Guattari's perspectivism, like Maturana and Varela's second-order cybernetics, must be seen as anthropocentric.

Yet it is precisely at this juncture that Deleuze and Guattari's thought might help us reconfigure this issue. In *What Is Philosophy?* Deleuze and

Guattari assert that the goal of thought is to create "possibilities of life or modes of existence" (*What Is Philosophy?* 73) and thereby help constitute a "people to come" and a "new earth" (*What Is Philosophy?* 109). No genuine, viably functioning human collectivity exists at present, they argue, and hence the formation of such a collectivity must involve a future people, a "people to come." But the formation of such a people is inseparable from the creation of a new earth, for there is no ultimate distinction between humans and nature, and hence no means of transforming humans without simultaneously transforming the world they inhabit. The only means of fashioning a new people and a new earth is to engage in "becomings," or processes of "becoming-other"—becoming-woman, becoming-child, becoming-animal, becoming-molecular. Such processes undo hierarchical binaries that privilege male over female, adult over child, human over nonhuman, and the macro over the micro, not by simply inverting them, but by inducing a passage *between* boundaries such that both poles of a binary opposition dissolve in zones of indiscernibility as something new and uncharted emerges.

> To think is to experiment, but experimentation is always that which is in the process of coming about—the new, remarkable and interesting…. The new, the interesting, are the actual. The actual is not what we are but, rather, what we become, what we are in the process of becoming—that is to say, the Other, our becoming-other. (*What Is Philosophy?* 111–112, translation modified)

Most significant for our purposes are processes of becoming-animal (and perhaps also those of becoming-plant or becoming-mineral), for such processes call into question our species identity. Hence, to say that Deleuze and Guattari's perspectivism is necessarily anthropocentric is not to presume that we know what "anthropos" is or can become. The future of the human in becomings is other-than-human, whatever that may be, and one of its trajectories passes through a becoming-animal. "We become animal so that the animal also becomes something else. The agony of a rat or the slaughter of a calf remains present in thought not through pity but as the zone of exchange between man and animal in which something of one passes into the other" (*What Is Philosophy?* 109). Thus, Deleuze and Guattari's perspectivism, rather than reinforcing an opposition of the human and the nonhuman, instead calls for a dissolution of that divide.

And it is here that we may return to the question of corporeal affects and their valorization. In processes of becoming, bodies are defined by their speeds and their affects, the latter characterized by the given body's powers of affecting and being affected. Important to note is that this affectivity is not unidirectional—a capability not simply of affecting but also of being affected. Some organisms have limited ranges of affectivity, capable of perceiving, acting on and reacting to their surroundings in only a few ways, whereas others have more expanded channels of interaction with their world. Humans seem to have multiple modes of affectivity, yet, as Deleuze is wont to remark in a phrase he adopts from Spinoza, we do not yet know what a body is capable of.[6] Only through an experimentation on the body do we discover what a body is—that is, what its powers of affecting and being affected might become. Nor is it even obvious what constitutes a body, for, in processes of becoming, a configuration of speeds and affects might form a body within an organism (an organ, a metabolic system) or a collective body (multiple organisms, communities, ecosystems). The creative movement of thought involves a process of becoming-other, and such a becoming-other is inseparable from an experimentation of bodies, a process of formation and transformation of configurations of speeds and affects that opens up new possibilities of interaction.

If the object of thought is to invent "possibilities of life or modes of existence" and thereby help constitute a "people to come" and a "new earth," and if such invention must proceed through a becoming-other, including a becoming-animal (or plant, bacterium, and so on), then a general reduction in the number of organisms available for such becomings would constitute a reduction in the possibilities of life. For this reason, I would argue that Deleuze and Guattari's ethics of inventing a people to come and a new earth is compatible with the promotion of biodiversity. Modes of existence that destroy habitats, induce pandemics, or foster pathogenic disequilibrium inhibit a creative exploration of the possibilities of bodies and decrease the options for a reconfiguration of humans and the earth. The fewer the life forms available for becoming-other, the fewer the trajectories available for creative transformation.[7]

If, then, we return to Naess's categories of ecology, ecophilosophy and ecosophy, we may say first that, though Deleuze and Guattari are obviously not ecologists, that is, scientists engaged in empirical ecological research, their thought about nature has a decidedly ecological orientation, in that their machinism promotes the view of nature as a complex of interactive organism–environment systems. Second, though they do not directly examine the relationship between the disciplines of philosophy and ecology, and hence they are perhaps not ecophilosophers in the strict sense Naess gives the term, their treatment of nature invites a reconceptualization of fundamental philosophical issues, especially those of the relationship between the human and the nonhuman, the natural and the artificial, organism and environment, observer and context, parts and the whole. And third, though they do not engage directly in the construction of an ecological ethics, they are ecosophers in that the ethics of their thought informs their views of the relationship of humans to the world. Their machinism calls into question any blanket condemnation of technology, any unqualified valorization of a pristine, nonhuman wilderness, and any promotion of the intrinsic value of all organisms. Their qualified holism posits a whole that is open, both spatially and temporally, and hence one that undermines any conception of nature as fundamentally balanced and stable in its functioning. And that whole is only manifest as a "local absolute," and thus it can only be conceived and experienced from a specific perspective within the whole. As a result, their perspectivism commits them to an anthropocentric ethics, yet one that subverts received notions of the human and calls for the creation of a new collectivity and a new earth. Finally, that perspectivism implies that their holism is a pluralism, in that the open whole revealed in each perspectival view is ultimately irreducible to any "perspective of all perspectives" that might unify the plurality of perspectives that may emerge from that whole. In this sense, if thought may unfold across a thousand plateaus, there are a thousand ecologies that would unfold within those plateaus, a thousand ways of attempting to create a new collectivity and a new earth.

Notes

1. Deleuze and Guattari rely primarily on von Uexküll (1956) for their account of his symphony of nature. See Bogue 58–62 for a more extended discussion of von Uexküll and Deleuze|Guattari.
2. For a more detailed examination of the relevance of Varela et al. (1991) to Deleuze and Guattari's thought, see Bogue 66–69.
3. See Deleuze's *Bergsonism* (96–98) for a presentation of Bergson's critique of the whole as "given" in deterministic models.
4. See Pimm 13–14 for a exposition of these terms; see also Sarkar 118–119 for an expansion of Pimm's taxonomy.
5. See Sarkar, especially 45–74.
6. For an exposition of this Spinozistic conception of the body, see Deleuze's *Spinoza: Practical Philosophy* 122–130.
7. It should be noted, however, that such a commitment to biodiversity need not imply a commitment to wilderness restoration, as is sometimes assumed. Not only is the concept of wilderness as "territory devoid of human presence" suspect in its implicit opposition of humans and nature, but the goal of wilderness restoration may at times be in conflict with that of the maintenance of biodiversity. Sarkar offers an instance of such a conflict in India, where the creation of a wilderness area through the removal of human herders and their cattle from a marshland led to a catastrophic growth of grasses that made the area uninhabitable by migrating birds and many other species (see Sarkar 42–43). For this and many other reasons Sarkar argues that conservation biology, with its focus on biodiversity, should be recognized as a discipline distinct from that of general ecology.

Works cited

Bogue, Ronald. *Deleuze on Music, Painting, and the Arts.* New York: Routledge, 2003.
Deleuze, Gilles. *Cinema 1: The Movement-Image.* Trans. H. Tomlinson and B. Habberjam. Minneapolis: University of Minnesota Press, 1986.
——. *Spinoza: Practical Philosophy.* Trans. R. Hurley. San Francisco: City Lights, 1988.

Deleuze, Gilles. *Bergsonism*. Trans. H. Tomlinson and B. Habberjam. New York: Zone Books, 1991.

———. *Negotiations*. Trans. Martin Joughin. New York: Columbia University Press, 1995.

Deleuze, Gilles, and Félix Guattari. *Anti-Oedipus: Capitalism and Schizophrenia, I.* Trans. R. Hurley, M. Seem, and H. R. Lane. Minneapolis: University of Minnesota Press, 1983.

———. *A Thousand Plateaus: Capitalism and Schizophrenia, II.* Trans. B. Massumi. Minneapolis: University of Minnesota Press, 1987.

———. *What Is Philosophy?* Trans. H. Tomlinson and G. Burchell. New York: Columbia University Press, 1994.

Guattari, Félix. *The Three Ecologies*. Trans. Gary Genosko. London: Athlone Press, 2000.

Haraway, Donna. "A Cyborg Manifesto: Science, Technology, and Socialist-Feminism in the Late Twentieth Century" in *Simians, Cyborgs, and Women: The Reinvention of Nature*. London: Free Association Books, 1991, 149–181.

Naess, Arne. *Ecology, Community and Lifestyle*. Trans. and ed. D. Rothenberg. Cambridge: Cambridge University Press, 1989.

Pimm, Stuart. *The Balance of Nature? Ecological Issues in the Conservation of Species and Communities*. Chicago: University of Chicago Press, 1991.

Sarkar, Sahotra. *Biodiversity and Environmental Philosophy: An Introduction*. Cambridge: Cambridge University Press, 2005.

Soulé, Michael. "Conservation Biology in the Twenty-first Century: Summary and Outlook" in *Conservation for the Twenty-first Century*. Eds D. Western and M. Pearl. New York: Oxford University Press, 1989, 297–303.

Uexküll, Jakob von. *Mondes animaux et monde humain, suivi de Théorie de la signification*. Trans. P. Muller. Paris: Gonthier, 1956.

Varela, Francisco J., Evan Thompson, and Eleanor Rosch. *The Embodied Mind: Cognitive Science and Human Experience*. Cambridge, MA: MIT Press, 1991.

Wilden, Anthony. *System and Structure: Essays in Communication and Exchange*. 2nd edn. London: Tavistock, 1980.

Wolfe, Carey. "In Search of Post-Humanist Theory: The Second-Order Cybernetics of Maturana and Varela." *Cultural Critique* 30 (1995): 33–70.

4

Structural Couplings: Radical Constructivism and a Deleuzian Eco*logics*

Hanjo Berressem

> Our nervous system computes invariants from constantly changing stimuli, because we act as if the future were like the past, and we are embedded in a culture that loves permanence and duration above all else. Maybe this is why there are so few voices that talk about becoming, about beginning and about change.*
>
> (von Foerster, Wissen 370)
>
> horizontal weather.
>
> (Deleuze, Logic of Sense 11)
>
> In the pronoun we, I of course included the starfish and the redwood forest, the segmenting egg, and the Senate of the United States.
>
> (Bateson, Mind and Nature 4)

deleuze   **guattari**

Why Gilles Deleuze and ecology? Would it not have been the more obvious choice to write about Félix Guattari, whose post-1980s work is so directly ecological that one might well talk of a general shift from "molecular revolutions" to "molecular evolutions?" The answer, of course, is both "yes" and "no." On the one hand, it would be strange to develop a Deleuzian ecology—or, as I will call it, a Deleuzian eco*logics*—without Guattari. Recent attempts to isolate a "pure" Deleuze notwithstanding, I think that it is impossible, and even if it were possible, that it would be counterproductive to try to untangle their works.[1] On the other hand, despite the elective affinity between Guattari and "the ecological"—as a set that includes the whole ecological spectrum from hardcore science to pop ideology—Deleuzian philosophy offers something to the ecological project that Guattari's *"ecosophy"* (*Three Ecologies* 28) carries with it only implicitly: *a radical philosophy for a similarly radical ecology.*

---

* All quotes from German sources translated by Hanjo Berressem.

This radical philosophy, which in turn contains Guattari's "ethico-political" (*Three Ecologies* 28) project only implicitly, conceptualizes humans as radically immanent to a productive, machinic field made up of what is commonly differentiated into "natural" and "artificial" machines; a differentiation that Deleuze's philosophy in actual fact undoes because it considers nature as itself artificial: nature is not "natural," one might say, it is "machinic." Programmatically, Deleuze|Guattari write in the beginning of *Anti-Oedipus*:

we make no distinction between man and nature: the human essence of nature and the natural essence of man become one within nature in the form of production or industry [...] man and nature are not like two opposite terms confronting each other—not even in the sense of bipolar opposites within a relationship of causation, ideation, or expression (cause and effect, subject and object, etc.); rather they are one and the same essential reality, the producer-product. (4–5)

From within the notion of this productive field, Deleuze's philosophy negotiates the radically paradoxical interplay of energetic, intensive, perceptual, cognitive and conceptual parameters in the human subject. In Deleuzian terms, it addresses humans and "their" world as part of the world's "plane of immanence|consistency."[2]

To avoid misunderstandings, let me note that when I use the terms "radical ecology" or "radical philosophy," these do *not* immediately concern what is generally considered as a "radical ecology," a "radical philosophy" or a "radical politics."[3] Rather, I mean the term "radical" to refer quite narrowly to the theoretical attractors into which I will suspend a Deleuzian eco*logics*. The first of these is the theory of radical constructivism, in particular biologists Humberto R. Maturana and Francisco J. Varela's theory of autopoiesis and Heinz von Foerster's work in second-order cybernetics. The second attractor concerns the adaptation of nonlinear dynamics by historian of science Michel Serres. While the reference to nonlinear dynamics has been made useful for Deleuze studies, for example, in the work of Manuel DeLanda and John Protevi, references to the theory of autopoiesis have remained surprisingly unexplored, although the theory of autopoiesis is closely related to that of nonlinear dynamics. The third theoretical attractor concerns what one might call the "radical medio*logics*" of Czech philosopher of communication Vilém Flusser and German sociologist Niklas Luhmann. While Flusser's work is still widely unknown, Luhmann's work, which develops a radically constructivist sociology from within both autopoietic and cybernetic registers, has become an important reference in ecocriticism.[4] As all of these attractors can be subsumed under the umbrella of systems theory, systems theory makes up the most general theoretical reference of "the meeting of Guattari and Deleuze on the operating table of ecology."[5]

One reason why it is helpful to have Guattari on the operating table is that, while all of these attractors are at work in Deleuze's singular, as well as in Deleuze|Guattari's communal work, they are easier to identify and isolate in Guattari's ecosophy. Bringing ecosophy to bear on Deleuzian philosophy serves three purposes: (1) a reading of Guattari's ecosophy provides the theoretical parameters for a reading of Deleuzian philosophy as "radically ecological;" (2) beyond the register of the ecological and beyond the specific uses Guattari puts them to, the theoretical attractors Guattari brings into play provide a conceptual frame for a reading of Deleuzian philosophy as a "radical philosophy;" (3) a reading of Deleuze "with" Guattari allows for a conceptual consolidation of a "radical Guattari," a "radical Deleuze" and a "radical ecology."

Guattari develops his ecosophy in *The Three Ecologies* and *Chaosmosis*, the former of which addresses the ecological with only a small number of scientific references, while the latter addresses mainly science, with only a small number of references to ecology.[6] Symptomatically, however, both texts contain explicit references to Maturana|Varela's concept of autopoietic systems: living, "self-organizing" systems that second-order cybernetics calls "eigenorganizations [*Eigenorganisationen*]" (von Foerster, *Wissen* 296).[7] Eigenorganizations are operators|agents in an "eigenspace" in which they are defined as the results of recursive operations that have settled into stable attractors, or, in mathematical terms, as the results of functional processes that feed, with every reentry, the same, or a very similar value into the system. "[U]nder certain conditions there exist indeed solutions which, when reentered into the formalism, produce again the same solution. These are called 'eigenvalues'" (von Foerster, quoted in Segal, *Dream of Reality* 145), von Foerster notes. As systemic stabilizers, eigenvalues denote "functional [...], operational [...], structural *equilibria*" (von Foerster, *Understanding* 265).

Autopoietic or, as one might also say, "eigen"organizing systems are operationally and informationally closed because "their identity is specified by a network of dynamic processes whose effects do not leave that network" (Maturana|Varela, *Tree* 89). At the same time, however, they are open to their surroundings in terms of energy and energy transfers. As von Foerster notes, "[t]he concept of self-organization is probably the most general concept for the description of these fascinating processes, which take place in systems that are organizationally closed but energetically (thermodynamically) open" (*Wissen* 296; see also *Understanding* 281). Translated into Guattari's terminology, they are "auto-referential existential assemblages engaging in irreversible duration" (*Three Ecologies* 44). These "eigen"organizing systems ("auto-referential assemblages"), which slide down an evolutionary slope ("engaging in irreversible duration") even while they are, as "living organisms" and as "regulators," what von Foerster calls "entropy retarders" (*Understanding* 193), make up, quite literally, the subjects of Guattari's ecosophy.

Although Guattari relies in many ways on the theory of autopoiesis, he also notes that he uses it "in a somewhat different sense from the one Francisco Varela gives this term" (*Chaosmosis* 7). In particular, he proposes to widen its operational frame beyond that of the biosphere, to which Maturana|Varela—at least according to Guattari—tend to restrict it, toward the more inclusive "mechanosphere"; the field of Deleuze|Guattari's fully machinic|machined world.[8] Answering Maturana|Varela's question "whether human societies are or are not themselves biological systems" (*Autopoiesis* 177), which they agree to disagree on and "leave pending," with an emphatic yes, Guattari proposes that "their notion of autopoiesis—as the auto-reproductive capacity of a structure or ecosystem—could be usefully enlarged to include social machines, economic machines and even the incorporeal machines of language" (*Chaosmosis* 93). In fact, according to Guattari, one can fold the oftentimes conceptually separated realms of nature and culture onto each other by "view[ing] autopoiesis from the perspective of the ontogenesis and phylogenesis proper to any mechanosphere superposed on the biosphere" (*Chaosmosis* 40).[9]

In a similar way to that in which it "differs" from the theory of autopoiesis, Guattari's ecosophy differs from the ecological: It aims to be "comprehensive." In fact, within Guattari's overall project of the implementation of a radical reconfiguration of the registers of thought and life, the creation of an ecosophy might be considered as ultimately nothing but a "collateral effect." The scope of his general project, at least, goes well beyond the more limited ecological sense of saving the environment, cutting back on the use of natural resources, maintaining biodiversity by saving "trash species," or fighting the greenhouse effect, and also beyond the common "subject of ecology." As a fallout of his general project, ecosophy is more comprehensive than the more visible and mainstream versions of ecology in that it is based on the radical inclusion of the invisible levels of the biochemical and the physical mechanosphere. While these are addressed by the more invisible, science-oriented versions of ecology, it is also more comprehensive than these, because they in turn tend to eclipse the cultural conclusions Guattari draws from their investigations. The power of Guattari's ecosophy, then, lies in its radical alignment of a cultural and a "natural" machinics.

In the most general sense, in fact, Guattari's critique aims at any form of the structural uncoupling of these two machines; an uncoupling that concerns, in Guattari's terminology, that of "incorporeal Universes of reference" (*Three Ecologies* 38) and "value" from "existential Territories" and their "idiosyncratic territorialized couplings" (*Chaosmosis* 4) (symptomatically, the term "coupling" evokes Maturana|Varela's concept of "structural couplings," of which more later), while in Deleuze's terminology it concerns the uncoupling of an "originary world" (*Cinema 1* 123) from a "derived milieu" (125). In an image that is once more evocative of the creation of eigenorganizations, Guattari describes these territories as "never given as object but always as intensive repetition, as piercing existential affirmation" (*Chaosmosis* 28). As he notes,

[p]rocess, which I oppose here to system or to structure, strives to capture existence in the very act of its constitution, definition and

deterritorialization. This process of fixing-into-being relates only to expressive subsets that have broken out of their totalising frame and have begun to work on their own account, overcoming their referential sets and manifesting themselves as their own existential indices, processual lines of flight. (*Three Ecologies* 44)

In this radical processualization, in which a "stable structure" is in actual fact a dynamics that has been decelerated to zero, "the [ecological] event [...] becomes a nucleus of processual relay" (Guattari, *Chaosmosis* 105–106), taking on the function and the position of those things formerly known as substances. As Deleuze notes about his philosophical project, "to reverse Platonism is first and foremost to remove essences and to substitute processes in their place, as jets of singularities" (*Logic of Sense* 53).

Both Guattari's ecosophy and Deleuze's radical philosophy, then, are directed against the "fateful" uncoupling of culture (the plane of transcendence|organization) from nature (the plane of consistency|composition, which is a plane of immanence|univocality)[10]; an uncoupling that they had already critiqued in *Anti-Oedipus*, in which they substituted the production of production and the production of representation for the common differentiation between production and representation.[11] As in *Anti-Oedipus*, in *The Three Ecologies* the aim is to "conjoin pre-personal traits with social systems or their machinic components" (61); in other words, to conjoin the natural and the cultural into a generalized machinics that allows for the implementation of a "mental, a natural and a cultural ecology" that combines "the environment, the socius and the psyche" (Guattari, *Chaosmosis* 20). In this implementation, "[t]ranscendence is always a product of immanence" (Deleuze, *Immanence* 31); a genesis that can only be thought from within a radical reversal of the common distribution of the registers of thought and the registers of life. As Deleuze notes, "[l]ife will no longer be made to appear before the categories of thought; thought will be thrown into the categories of life" (*Cinema 2* 189). Although this statement refers to "cinematic naturalism," it also applies to a radical eco*logics*: *If Guattari provides the politics|sciences for such a radical* eco*logics, Deleuze provides for it a fully worked-out philosophy.*

If one wanted to immerse Guattari's as well as Deleuze's work into the ecological, therefore, it would have to be in the very general sense of Ernst Haeckel, who defined ecology in 1866 as the "economies of living forms" (composed from *oikos* (household) and *logos*, the term literally denotes the study of "the household of nature"); a definition that translates quite easily into a "systems theory of the living." Such a theory relies on the radical inclusion of the realm of "the virtual;" a realm that is so important to Guattari that he calls his project "an ecology of the virtual" (*Chaosmosis* 91).

What is an ecology of the virtual? As a field of pure potentiality, the virtual ties the ecological to the genesis of both human and nonhuman autopoietic systems from a field of "ontological heterogeneity" (*Chaosmosis* 61) and "ontological intensity" (*Chaosmosis* 29). In a reference to nonlinear dynamics, Guattari calls this field of "non-discursive intensities" (*Chaosmosis* 22), which denotes in a more Deleuzian terminology the multiplicity of the plane of immanence, the field of virtual chaos: "a deterministic chaos animated by infinite velocities" (*Chaosmosis* 59). As Serres does in *Genesis*, Guattari argues that every autopoietic system—from single cell to city—is actualized from this multiplicity of "virtualities" (27): "It is out of this chaos that complex compositions, which are capable of being slowed down in energetico-spatio-temporal coordinates or category systems, constitute themselves" (59).

Deleuze|Guattari define the machinic processes of the constitution and the consolidation of life into stabilized|sustained, or, in more philosophical terms, into "habitual" systems, by way of the relation between processes of territorialization|striation, which involve reductions, contractions and framings, and processes of deterritorialization|smoothing. According to these twofold dynamics, any living system is nothing but "a *habitus*, a habit, nothing but the habit in a field of immanence," and processes of systemic|organic eigenorganization are those of "contracting a habit" (Deleuze, *Difference* 74).[12] In fact, "we must regard habit as the foundation from which all other psychic phenomena derive" (78).[13] In the case of the human being, this includes "the habit of saying I" (Deleuze|Parnet, *Dialogues II* 48). The reference to habitual processes ties Deleuze's philosophy not only to the empiricism of David Hume, but also to the pragmatism of Charles Sanders Peirce and William James. Peirce, in fact, develops a general theory of habits in which even natural laws are nothing but ecological habits. As Peirce notes, "habit is by no means exclusively a mental fact. Empirically, we find that some plants take habits. The stream of water that wears a bed for itself is forming a habit" ("Survey" 342); a sentiment that is echoed by James, who uses processes of habit formation to describe an ecology of neurological activities.[14] Deleuze echoes such pragmatic ideas when he ties his philosophy to a theory of the emergence of the subject in which both physical and psychic habit formation is the operation according to which "the subject is constituted within the given" (*Empiricism* 104).[15] Within this constitution, Deleuze aligns biophysical registers (the registers of life and production) and psychic registers (the registers of thought and representation): Biologically, subjects are formed through "the primary habits that we are; the thousands of passive syntheses of which we are originally composed" (*Difference* 74). These passive, organic syntheses constitute our "habit of living" (*Difference* 74) and "the *living present*" (81, emphasis added).

Even during these passive processes of contraction, however, an "emergent" (Guattari, *Chaosmosis* 65) self, which can be understood as a set of "harmonic resonances" (26) ("eigenfrequencies")[16] of specific habits, begins to feedback with these processes through subindividual processes of "contemplation" (these can be seen as analogous to Peirce's "inattentive habit[s]" ("Survey" 328), which also operate on the same unconscious level as what Peirce calls "perceptual judgements"[17]). These feedback loops lead to the constitution of a "passive self" that "contemplates and contracts the individuating factors of such fields [the "pre-existing fields of individuation"] and constitutes itself at the points of resonance of their series" (Deleuze, *Difference* 276). It is therefore "simultaneously through contraction that we are habits, but through contemplation that we contract" (74).[18] While unconscious reasonings and habit formations are related to passive syntheses and to the registers of life, conscious reasonings and habit formations are related to active syntheses and to the registers of thought, which is why "[t]he term 'reasoning' ought to be confined to [...] fixation of one belief by another as is reasonable, deliberate, self-controlled" (*Difference* 293).

> [P]erceptual syntheses refer back to organic syntheses [...] We are made of contracted water, earth, light and air—not merely prior to the recognition or representation of these, but prior to their being sensed. Every organism, in its receptive and perceptual elements, but also in its viscera, is a sum of contractions, of retentions and expectations. [...] [B]y combining with the perceptual syntheses built upon them, these organic syntheses are redeployed in the active syntheses of a psycho-organic memory and intelligence. (*Difference* 73)

The notion that habitual processes bring about a processual self allows for a first conceptual reaction between Guattari's ecosophy, Deleuze's radical philosophy and the ecological; a reaction in which systems theory functions as a kind of catalyst: Guattari's emergent self and Deleuze's "habitual" self are congruent to "eigenorganizations," which are, in turn, congruent to what nonlinear dynamics calls "strange attractors" (von Foerster, *Wissen* 296). In other words, when Guattari talks of auto-referential assemblages, or Deleuze of processes of singularization, they are, quite literally, talking about eigenorganizations.[19] Such a terminological consolidation facilitates shifting the different fields into each other without depriving them of their respective singularity.[20] If recursive biophysical and "computational" processes are indicators of the creation of informationally|operationally closed systems that are defined as the product of these operations, for instance, the modeling of organic eigenorganization as the contraction of habits—"What organism is not made of elements and cases of repetition, of contemplated and contracted water, nitrogen, carbon, chlorides and sulphates, thereby intertwining all the habits of which it is composed?" (Deleuze, *Difference* 75)—turns habit formation into the philosophical version of what von Foerster had described as the systemic eigenorganization through "recursive mathematical formalisms," which might well be called habitual computations or computational habits. While this turns physics into a branch of philosophy, it also turns philosophy into a branch of physics: As James noted, "the law of habit [...] is a material law" (*Psychology* 126), which means that "[t]he philosophy of habit is thus, in the first instance, a chapter in physics rather than in physiology or psychology" (105).

In *The Garden in the Machine*, Claus Emmeche relates the consolidation of such habitual systems directly to an *"ecology* of computation" (128, emphasis added) in which "the emergent properties, instead of being represented in any central master code, are constructed anew each time an organism is created" (80). Such consolidations are extremely precarious machinic processes in which "random developmental noise or more violent environmental disturbances [...] push equilibrium over into another path, resulting in a very different final product" (84) and that oscillate between fractal self-similarity and negative feedback loops on the one hand and a dynamics of positive feedback loops on the other, and in which "random developmental noise or more violent environmental disturbances" can "push equilibrium over into another path, resulting in a very different final product" (84). To quote the title of a book by René Thom, they oscillate between "structural stability and morphogenesis."[21]

Through these habitual|recursive operations, autopoietic systems separate themselves operationally|informationally from the infinitely complex multiplicity of the world's plane of immanence, a process by which they take on, for other autopoietic systems, the character|contour of stable "objects|subjects" with specific properties. In fact, what one calls "objects" are merely the result of the repeated eigenbehavior of eigenorganizations. Within the operational eigenspace that they have developed they become habitual clusters contoured against the background of an infinite multiplicity: In ecological terms, autopoietic systems are eigenorganizations that, after a *process* of systemic consolidation, have settled into a temporary energetic *steady state* with|in their milieu, or, in other words, systems that show eigenbehavior, which means that they have created both a physical and a psychic border between themselves and the larger field to which they are energetically immanent. They are "the things formerly known as subjects|objects" or also "subjects|objects under radical conditions." A first "radical|theoretical" consolidation lies in that *for the development of a "radical philosophy" for a "radical ecology," Guattari's sociopolitical subject (the emergent subject), Maturana|Varela's biological subject (the autopoietic subject), the subject of systems theory (the eigenorganized subject) and Deleuze's philosophical subject (the habitual subject), form a conceptual ecology.*

The creation of each of these "subjects" involves a local slowing down of the infinitely fast and chaotic movements on the plane of immanence; or, in Guattari's terms, of the deterministic chaos of the plane of "virtual chaos." Stressing the implicit hardening of intensities|waves into objects| particles that accompanies these decelerations|territorializations, Guattari describes the formation|consolidation of stable systems as a "crystallization of intensity" (*Chaosmosis* 30). The results of these processes of singularization are eigenorganizations with an *"autopoietic* consistency"

(78, emphasis added): operationally|informationally self-referential systems that are, simultaneously, immanent to the energetics of "intensive and processual becomings" (117). As the result of a reduction of a virtual, open chaos to a closed-off, actualized system, "every species of machine is always at the junction of the finite and the infinite, at this point of negotiation between complexity and chaos" (111). In fact, becoming eigenorganized involves the reduction of an infinite causal nexus to a finite one: "The essential contribution of cybernetics to epistemology is the ability to change an open system into a closed system, especially as regards the closing of a linear, open, infinite causal nexus into a closed, finite, circular causality" (von Foerster, *Understanding* 230).

The radical immanence of every eigenorganization to an anonymous field of intensities|energetics is why the complex genesis of autopoietic systems from an infinitely complex, imperceptible "virtual ecology" is so crucial for Guattari. In fact, the "ecology of the virtual is thus just as pressing as ecologies of the visible world" (*Chaosmosis* 91) and although a "generalized ecology—or ecosophy—will work as a science of ecosystems" (91) and address the overall politics of the ecological project, it cannot be limited to the operational|informational field and to specific political|cultural agendas. Ultimately, the "primary purpose of ecosophic cartography is not to signify and communicate but to produce assemblages of enunciation capable of capturing the points of singularity of a situation" (128).

Both Guattari's ecosophy and Deleuze's philosophy, then, start long "before" the finished consolidation of subjects, objects and "cultures," and they do so without taking recourse to an essentialist agenda or the notion of an uncontaminated origin. Rather, they address directly the multiplicity of anonymous, communal ecologies operative on nonhuman levels, as well as the forms of machinic assemblages and cultures—in the sense that one talks of bacterial cultures—from which subjects and objects are produced in the first place: "While the logic of discursive sets endeavours to completely delimit its objects, the logic of intensities, or eco-logic, is concerned only with the movement and intensity of evolutive processes" (Guattari, *Three Ecologies* 44). In its most basic definition and its most general logic, therefore, Guattari's ecology is radically intensive rather than *objec*tive. And, although Guattari is indeed "a deep ecologist of sorts" (Conley 22), in particular as concerns the references, in both projects, to systems theory, nonlinear dynamics and to the idea of a *"relational, total-field"* (Naess, "The Shallow," emphasis added)—symptomatically, Guattari hijacks the term ecosophy from the field of deep ecology—the radical inclusion of the intensive register into the logic of ecology is why one should also stress the "of sorts."[22]

Let me illustrate this "of sorts" by means of the radical ethics|politics that Guattari's ecosophy implies and the way in which they differ from both a "shallow" and a "deep" ethics|politics. Most importantly, neither a shallow nor a deep ecology goes beyond the basic differentiation between subject and object. While a shallow ecology might wish to save a landscape because doing so serves specific ideological agendas, deep ecology would save it not only because it sees it as a living entity, but because it considers it as a subject with specific unalienable rights: While shallow ecologies treat nonhuman eigenorganizations quite openly as objects, deep ecology treats them as subjects, granting them a similar ontological and legal status according to what von Foerster sees as a "subjective fallacy:" "[p]rojecting the image of ourselves into things or functions of things in the outside world" (*Understanding* 169). To succumb to this subjective fallacy is especially unfortunate in an ecological context, because it obscures the fact not only that landscapes are not human, but that humans themselves, as nested aggregates of autopoietic systems, are not completely human. They are assemblages of an infinite number of heterogeneous, both human and non-human, both material (bodies) and immaterial (habits|routines) parts|series that are organized in a specifically "human" manner, similar to the way that animals are made up of an infinite number of heterogeneous series, both material (bodies) and immaterial (habits|routines), organized in what we call a specifically "animalistic" manner.

A radical ecological ethics|politics—what one might also call an "eigen"ethics or "eigen"politics—is thus neither to do with providing a human|*subject*ive status to other species or to nonhuman aggregates, nor with subsuming these aggregates fully under specific cultural agendas and thus to fully *object*ify them. Rather, it is to do with *intensifying* them and with developing an ethical|political position from within an overall machinics and its locally and temporally bounded architectures. As William E. Connolly notes, "[a]n ethical sensibility [...] is composed through the cultural layering of affect into the materiality of thought. It is a constellation of thought-imbued intensities and feelings" (107). Deleuze relates such an intensive ethics to Hume as well as to Baruch de Spinoza's question of "what a body can do" and to the "*conatus*, the striving for self-preservation" (Zimmerman 39). While the reference to Spinoza seems at first sight to *relate* Deleuze and deep ecology, the Spinozism of deep ecology results in an ethics of self-realization, while Deleuze's Spinozism, which revolves around the concepts of pure immanence and the difference between affects|affections, results in a "radical ethics for an intensive world." Although this ethics is not directly informed by ecology, it is, quite automatically, "radically ecological."

The Deleuzian world, which is made up of an infinite number of machines and machinic arrangements—in the same manner in which the Deleuzian body is not something the subject "is" but something it "has acquired"—is a world of recursively assembled parasitations and symbioses, like the one described by Annie Dillard in *Pilgrim at Tinker Creek* when she notes that "[t]here are so many parasitic wasps that some parasitic wasps have parasitic wasps. One startled entomologist, examining the gall made by a vegetarian oak gall wasp, found parasitism of the fifth order. This means that he found the remains of an oak gall wasp which had a parasite wasp which had another which had another which had another which had another, if I count it aright" (232). *A radical ecology conceptualizes the world according to an "assemblage theory" that understands it as a multiplicity of an infinite number of recursively nested machinic aggregates, all of which are radically immanent to a plane of intensities|energetics, although they are simultaneously radically separated from that plane in terms of their operational| informational registers.*

radical deleuze

In terms of "the subject of ecology," then, Deleuze's radical philosophy offers the ecological project a philosophical figure of thought in which the relation between subject and object or between subject and subject is the "result" of the more radical relation between eigenorganization and intensity|energetics. As I noted above, *if deep ecology subjectifies nature while shallow ecology objectifies it, a radical ecology intensifies it, treating it as a complex arrangement of recursively operating, both human and nonhuman eigenorganizations that operate within an energetic environment.* In fact, the elective affinity between Guattari's ecosophy, Deleuze's radical philosophy and the theories of autopoiesis and second-order cybernetics lies in that all of them negotiate, in a radically constructivist manner, the relation between energetic and operational|informational registers, which, in autopoietic terms, concern the qualitative categories of psychic computation|integration (the field of perception, cognition and observation) and the quantitative categories of physical affection|propagation-of-forces (the field of energy and intensity).

Depending on the specific context, both Deleuze and Deleuze|Guattari set up a number of terminologies for these negotiations. In machinic terms, which address, amongst others, the fields of biology, physics and chemistry, they are the coupling of "representations" (the theater) and "productions" (the factory). In linguistic terms, they are the relation between words ("propositions") and things ("matters-of-fact"|"states-of-affairs"). In psycho- or schizoanalytic terms, which address the threshold between "conscious" and "unconscious" registers, they are the relation between "psychic reality" and "lived reality." In ontological registers, finally, they concern the difference between "coded milieus" and "originary worlds." In all of these contexts, the project is to think the complex, often nonlinear and far-from-equilibrium processes|dynamics according to which the distinct formalizations|logics of these respective registers are operating in the human being within the frame of a radical philosophy. As Deleuze|Guattari note in *A Thousand Plateaus*, "we should never oppose words to things that supposedly correspond to them [...] what should be opposed are distinct formalizations, in a state of unstable equilibrium" (67). In *What is*

Philosophy?, in fact, Deleuze|Guattari's description of the evolution of machinic sets differentiates between matters-of-fact, things and bodies precisely along the lines that define the world of autopoiesis. While matters-of-fact refer to a virtual realm of pure directionality and geometrical vectors, things refer to systemic energetics, while bodies refer to systemic operationality|informatics: matters-of-fact "refer to geometrical coordinates of supposedly closed systems, things refer to energetic coordinates of coupled systems, and bodies refer to the informational coordinates of separated, unconnected systems" (123).[23]

Although one tends to identify, almost by default, autopoietic systems as "humans," humans are by no means the only form of autopoietic system. They are, however, the only species of autopoietic system that includes the agency of what Maturana|Varela call the "observer," which is the autopoietic version of self-consciousness. All forms of autopoietic systems differ in their structural assembly, in the degree|form of their cognitive abilities|operations, in the way they pick up, monitor and read intensities and thus in how they informationally|operationally negotiate intensive dis|equilibria [A few reminders of how different the perception of the world is for different species and levels of organization are: Herman Melville, who notes in *Moby Dick* that the whale sees two different|complementary worlds at the same time; Dillard, who notes that "butterflies taste with their feet" (255); or Richard Lewontin, who notes that "[a]s the temperature in a room rises, my liver detects that change, not as a rise in temperature, but as a change in the concentration of sugar in my blood and the concentration of certain hormones" (quoted in McMurry 77)]. It is only with humans, however, that the system's cognitive apparatus includes the plane of subjectivity and language, on which the informational|operational agency of the observer is created; an agency that is positioned on the surface of a larger, or, as Dillard calls it, "layered" (84) consciousness that Serres describes as a set of recursive integrations:

> At this point the unconscious gives way from below; there are as many unconsciousnesses in the system as there are integration levels. It is merely a question, in general, of that for which we in generally possess no information. [...] Each level of information functions as an unconscious for the global level bordering it [...]. What remains unknown and unconscious is, at the chain's furthermost limit, the din of energy transformations: this must be so, for the din is by definition stripped of all meaning, like a set of pure signals or aleatory movements. These packages of chance are filtered, level after level, by the subtle transformer constituted by the organism [...]. In this sense, the traditional view of the unconscious [the Lacanian unconscious that is structured like a language!] would seem to be the final black box, the clearest box for us since it has its own language in the full sense. ("Origin" 80)

Or, in the words of James, "[t]he highest and most elaborated mental products are filtered from the data chosen by the faculty next beneath, out of the mass offered by the faculty below that, which mass in turn was sifted from a still larger amount of yet simpler material, and so on" (*Psychology* 288).

While all autopoietic systems make perceptual|cognitive distinctions—in fact, making distinctions is the prerequisites for being an autopoietic system, which is why both human and nonhuman machines are quite literally "system[s] of interruptions or breaks [...]. Every machine, in the first place, is related to a continual material flow (*hylè*) that it cuts into" (Deleuze|Guattari, *Anti-Oedipus* 36)—only humans make "distinctions of distinctions." As Maturana notes,

> [w]hen a metadomain of descriptions (or distinctions) is generated in a linguistic domain, the observer is generated. Or, in other words, to operate in a metalinguistic domain making distinctions of distinctions is to be an observer. An observer, therefore, operates in a consensual domain and cannot exist outside it, and every statement that he makes is necessarily consensual. [...] *An observer operates in two nonintersecting phenomenic domains. As a living system he operates in the domain of autopoiesis. As an observer proper, he operates in a consensual domain that only exists as a collective domain defined through the interactions of several (two or more) organisms.* ("Cognition," emphasis added)

Deleuze is at his most eco*logical* in extending the processes of observation, contemplation, perception and cognition deeply into subindividual levels; into biochemical processes of contemplation that are imperceptible to the agency of the observer (on these levels, they "shade into" what Peirce calls the realm of unconscious reasoning):

> These thousands of habits of which we are composed—these contractions, contemplations, pretensions, presumptions, satisfactions, fatigues; these variable presents—thus form the basic domain of passive syntheses. The passive self is not defined simply by receptivity—that is, by means of the capacity to experience sensations—but by virtue of the contractile contemplation which constitutes the organism itself before it constitutes the sensations. This self, therefore, is by no means simple: it is not enough to relativise or pluralise the self, all the while retaining for it a simple attenuated form. Selves are larval subjects; the world of passive syntheses constitutes the system of the self, under conditions yet to be determined, but it is the system of a dissolved self. There is a self wherever a furtive contemplation has been established, whenever a contracting machine capable of drawing a difference from repetition functions somewhere. The self does not undergo modifications, *it is itself a modification*. (*Difference* 78–79, emphasis added)

As I noted above, active, individual(izing) syntheses emerge from the field of the passive syntheses and from "our thousands of component habits" (75): Thought emerges from life and remains its "attribute."

Unlike passive syntheses, which consist of preindividual, unconscious| inattentive and productive micro-integrations and which are related to the field of "expression," active syntheses are related to "the principle of representation" (81) and to conscious processes of integration, which means that they operate on a conscious informational|operational level that has its own agenda.[24] Invariably, however, "below the level of active syntheses, [there is] the domain of passive syntheses which constitutes us, the domain of modifications, tropisms and little peculiarities" (79); "[b]eneath the general operation of laws, [...] there always remains the play of singularities" (25) and beneath the psychic reality lies "the lived reality of a sub-representative domain" (69). Below reality, there is "the real." As Deleuze notes, "[a] soul must be attributed to the heart, to the muscles, nerves and cells, but a contemplative soul whose entire function is to contract a habit" (74).

Such an "animation" of nature also defines Henry David Thoreau's conceptualization of the world as an aggregate that consists of an infinite number of levels of what might be called "intelligent matter:" matter suffused with perceptual|cognitive powers. "[T]he subtle powers of Heaven and of Earth," Thoreau notes in *Walden*, make up "an ocean of subtle intelligences" that "are every where, above us, on our left, on our right; they environ us on all sides" (87). In fact, Thoreau is not only one of the patron saints of ecology, he is also one of the first radical ecologists in that he describes the relation between humans and nature precisely according to parameters that (p)recapitulate those of the different levels of autopoietic cognition and of the effects of the function of the observer in|on the human:

> We are not wholly involved in Nature [...]. I only know myself as a human entity; the scene, so to speak, of thoughts and affections; and am sensible of a certain doubleness by which I can stand as remote from myself as from another. However intense my experience, I am conscious of the presence and criticism of a part of me, which, as it were, is not a part of me, but spectator, sharing no experience, but taking note of it; and that is no more I than it is you. When the play, it may be the tragedy, of life is over, the spectator goes his way. It [life] was a kind of fiction, a work of the imagination only, as far as he was concerned. (88)

Once again: according to radical constructivism, humans are part of this "intelligent ecosystem" in terms of energetics, and, as consensual humans| observers, fundamentally separated from it. One of the most important elements of Deleuzian philosophy for a radical ecology is that it assembles a radical topo*logics* in which to capture this radically paradoxical logic.

haecology   affect

If the "subject" of a radical ecology can be extrapolated from the concept of humans as nested aggregates of human and nonhuman eigenorganizations, such a radicalization fundamentally reframes many ecological debates, for instance the one about "wilderness." Defined ecologically as a field that is characterized by the absence of human stressors, the shift from the dichotomy of subject|object to that of eigenorganization|energetics allows the reformulation of (moments of) wilderness as what Deleuze calls, in reference to medieval philosophy, "*haecceities.*"[25] In autopoietic terms, *haecceities* are "unobserved" ecological processes within a field of living, processual matter that are defined by an imperceptibly low level of perception|cognition. As Deleuze|Guattari note, "unformed matter, the phylum, is not dead, brute, homogeneous matter, but a matter-movement bearing singularities or haecceities, qualities and even operations" (*Thousand Plateaus* 512). In other words, a *haecceity* is any form of nonhuman dynamic that takes place on the plane of immanence, which makes of the plane of immanence the site a generalized "haecology:"

> there is a mode of individuation very different from that of a person, subject, thing, or substance. We reserve the name haecceity [it-ness] for it. A season, a winter, a summer, an hour, a date have a perfect individuality lacking nothing, even though this individuality is different from that of a thing or a subject. They are haecceities in the sense that they consist entirely of relations of movement and rest between molecules or particles, capacities to affect and to be affected. (261)

Because humans are aggregates of recursively nested eigenorganizations, these biophysical dynamics are, in making up the moving milieu|ecosystem in which the subject is suspended, an integral part of the human realm:

> You are longitude and latitude, a set of speeds and slownesses between unformed particles, a set of nonsubjectified affects. You have the individuality of a day, a season, a year, *a life* (regardless of its duration)—a climate, a wind, a fog, a swarm, a pack (regardless of its regularity) [...] It should not be thought that a haecceity consists simply of a decor or backdrop that situates subjects, or of appendages that hold things and people to the ground. It is the entire assemblage in its individuated aggregate that is a haecceity [...]. Climate, wind, season, hour are not of another nature than the things, animals, or people that populate them, follow them, sleep and awaken within them. (262–263)[26]

According to Deleuze, *haecceities* are "vague essences" (369); systemic vapors that form a biophysical network in which the most microscopic is related to the most macroscopic: "each multiplicity is symbiotic; its becoming ties together animals, plants, microorganisms, mad particles, a whole galaxy" (250). As autopoietic "heterogeneities" (250), they follow complex molecular choreographies, moving in material dances like that of the processing of pebbles by streaming water or of sand by the wind [*Figure 1*: "Close up of Dune Sand in Wind Tunnel," filmstill].

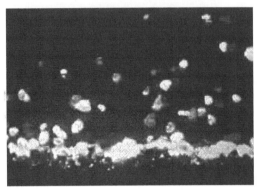

Figure 1

Unlike deep ecology, which subjectifies and humanizes such *haecceities*, Deleuze invariably describes them as radically nonhuman; as tableaus of intensive forces and as the ways that anonymous|communal eigenorganizations affect each other:

> *haecceities*, events, incorporeal transformations that are apprehended in themselves; *nomadic essences*, vague yet riotous; *continuums of intensities* or continuous variations [...] *becomings*, which have neither culmination nor subject, but draw one another into zones of undecidability; *smooth spaces*; composed from within striated space. (507)

When Deleuze asks, "why is expression not available to things? There are affects of things" (*Cinema 1* 97), this does *not*, therefore, lead to an anthropomorphization of *haecceities*. Rather, it involves a dehumanization of the human, because such material affects are singular rather than individual and anonymous rather than subjective: "the affect is impersonal and is distinct from every individual state of things: it is none the less *singular*, and can enter into singular combinations or conjunctions with other affects" (98). As Deleuze notes, the world is the multiplicity of such affects because "[e]ven when they are nonliving, or rather inorganic, things have a lived experience because they are perceptions and affections" (*Philosophy* 154).

Deleuze's philosophy, therefore, differentiates radically between conscious, subjective and perceptible perceptions|affectations and unconscious, nonsubjective imperceptible percepts|affects:

> Percepts are no longer perceptions; they are independent of a state of those who experience them. Affects are no longer feelings or affections;

they go beyond the strength of those who undergo them. Sensations, percepts, and affects are *beings* whose validity lies in themselves and exceeds any lived. They [...] exist in the absence of man because man, as he is caught in stone, on the canvas, or by words, is himself a compound of percepts and affects. (164)[27]

Percepts and affects do *not* coincide with "perceptions or feelings" (24). A molecular sexuality, for instance, plays itself out in the arena of subindividual percepts and "nonsubjective 'affects'" (Deleuze|Guattari, *Thousand Plateaus* 341) rather than in that of individual perceptions and affectations: "[t]he affect is impersonal and is distinct from every individuated state of things: it is none the less *singular*, and can enter into singular combinations or conjunctions with other affects" (Deleuze, *Cinema 1* 98).

In terms of the discussion about wilderness, if one folds the logic of *haecceities* onto the project for a "radically ecological" or a "virtual philosophy," the percept is in the position of wilderness, while perception is in the position of landscape. As Deleuze|Guattari note, "[t]he percept is the landscape before man, in the absence of man" (*Philosophy* 169). The idea of thought as embedded|embodied in an intensive wilderness has important ramifications for the scientific status of a radical ecology. *The ultimate challenge for a radical ecology in terms of its own constitution consists in the development of an ecologics that itself emerges from the immanent level of nonhuman, anonymous natural machines; in becoming an "eigentheory."*[28]

One should note, however, that Deleuze's radical philosophy never advocates the complete loss of "the habit of oneself." In fact, for a human, a move into total *haecceity* would be extremely unecological. What it *does* advocate is to find moments and places that allow for a "healthy" destabilization of habits that leads to the evolution|production of "the new;" what Deleuze|Guattari understand as a desire for de*habitat*ions, both in the sense of changing one's habit and of changing one's habitat—to become nomadic. In advocating nomadisms that remain within the system's range of eigenvalues and its range of "structural stability and morphogenesis," Deleuze's philosophy, Guattari's work at la Borde and Michel Foucault's idea of the care for oneself meet von Foerster's imperative for a radical ethics: "Act in such a way that the number of possibilities of choice increases" (*Wissen* 234).[29] Translated into ecological terms: always maintain the highest coefficient of diversity; be on the move, create "smooth spaces," develop "lines of flight."

When humans become immanent to "smooth space" (Deleuze|Guattari, *Thousand Plateaus* 474) through processes of "becoming imperceptible," without ceasing to exist as eigenorganizations, such deindividualizations and resingularizations bring about moments of "becoming" that are often tied to the concepts of charm or grace, as in Heinrich von Kleist's puppet theater, in the "wholly gratuitous" (7) grace Dillard observes when a bird, for no reason whatsoever, tests and "shows off" its skills: "The mockingbird took a single step into the air and dropped. [...] Just a breath before he would have been dashed to the ground, he unfurled his wings with exact, deliberate care [...] and so floated onto the grass" (8); in William Gibson's depiction of a bicycle messenger who is entirely immanent and attuned to the constantly changing system of a city that is defined less by a flow of disembodied informational traffic than by the material flow of energy|resonances; in the sense that Deleuze speaks of "Rotterdam itself as affect" (*Cinema 1* 111): "She was entirely part of the city, then, one wild-ass little dot of energy and matter, and she made her thousand choices, instant to instant, according to how the traffic flowed, how rain glinted on the streetcar tracks, how a secretary's mahogany hair fell like grace itself, exhausted, to the shoulders of her loden coat" (120); or in the final sentence of Thomas Pynchon's *Against the Day*, which describes balloonists "flying towards grace" (1085). All of these images are defined by an attunement of the autopoietic system to the larger movements of the smooth medium within which it operates. In such moments of surfing the medium, the autopoietic system becomes part of a larger *haecceity*. How is such an immersion possible, however, if the system is informationally|operationally separated from its medium? Which brings me to the difference between "the environment" and "the medium."

medium   **environment**

A radical logic differentiates between the perceived world as constructed| projected|extrapolated|extracted by the autopoietic system's cognitive apparatus, and the energetic world within which the autopoietic system is constructed and to which it is fully immanent; the "informed" rather than "informationed" field of biochemical and physical processes, pressures, tensions and stresses that define life, such as the processes of procreation, metabolism and death. Let me differentiate between these two as the environment—a term that is, in many ecological debates, used rather "innocently"—and what Maturana|Varela call the medium.

When Deleuze differentiates philosophically between a cognitive and an energetic|intensive series within the human subject, this is in tune with the fundamental split that a radical logic introduces into the "system of life." While "psychic mechanism[s]" (Deleuze, *The Fold* 96) work according to the digital operations of differential calculus, physical mechanisms "do not work by differentials, which are always differentials of consciousness, but by communication and propagation of movement" (97). The two operative fields are thus "the psycho-metaphysical mechanism of perception, and the [...] physico-organic mechanism of excitation or impulsion" (97). "Difference," as Bateson notes, "being quite a different thing from force or impact" (*Sacred Unity* 162). Through psychic differenti(aliz)ations, parts of the plane of immanence|composition|consistency are folded into|onto a "plane of transcendence|organization."

The differentiation between eigenorganization and intensity|energetics complexifies the notion of environment in that it separates the world of perception, cognition and observation from a much larger world from which the system emerges through complex processes of eigenorganizations, through the "cut" that underlies radical constructivism. According to George Spencer-Brown, this cut differentiates an organization from the undifferentiated, purely processual multiplicity of an "unformed" phylum:

> *the original act of severance.* The act is itself already remembered, even if unconsciously, as our first attempt to distinguish different things in a world where, in the first place, the boundaries can be drawn anywhere we please. At this stage the universe cannot be distinguished from how we act upon it, and the world may seem like *shifting sand beneath our feet.* (v, emphases added)[30]

As Maturana|Varela note, "[a] universe comes into being when a space is severed into two" (*Autopoiesis* 180). This cut differentiates categorically| radically between the two spaces of the "plane of transcendence" and the "plane of immanence" respectively. While the system's environment concerns the space around the system (*Umwelt*) as the "theatrical" result of informational|operational computations that take place within the autopoietic system, the energetic, productive "factory" made up of the bioecological field to which the autopoietic system is immanent and in which it lives concerns its medium.

> The medium of a unity is always defined by the unity as the domain in which it operates as a unity, not by the observer. The observer specifies a unity by an operation which implies an organization in the distinguished unity if it is a composite one, but the operation of distinction does not characterize the implied organization. Under these circumstances it is the implied organization of the unity that defines its medium, not the operation of distinction performed by the observer. Therefore, when an observer distinguishes a unity he does not necessarily have access to the medium in which it operates as a unity, but he himself defines a domain in which he sees the unity as a separable entity. The domain in which an observer sees a unity as a separable entity I shall call the *environment* of the unity. (Maturana, "Cognition")[31]

As the energetic domain from within which the autopoietic system assembles itself and to which it remains radically immanent, the medium denotes the *inclusive* field of which the system is "part and parcel," while the environment denotes the construct|projection of the medium that the system creates from within itself as the observer of what it experiences, operationally|informationally, as its outside. *The environment follows perceptual|cognitive registers, which differentiates it radically from the medium, which follows intensive|energetic registers.*

In a folding of mediological and philosophical registers onto each other, the plane of immanence (which is an analog to Guattari's site of virtual chaos) is the most loosely coupled, un*formed* and smooth medium: "there is no longer a form," within the plane of immanence "but only relations of velocity between infinitesimal particles of an unformed material. There is no longer a subject, but only individuating affective states of an anonymous force" (Deleuze, "Ethology" 630). Even though complete formlessness is ultimately an ideal and the embodied plane of immanence invariably contains residual, infinitely imperceptible microforms and striations, it is, as a medium, "filled by events or haecceities, far more than by formed and perceived things. It is a space of affects, more than one of properties [...] it is an intensive rather than extensive space" (Deleuze|Guattari, *Thousand Plateaus* 479).

For a radical mediology, which understands the medium as any set of loosely connected|coupled elements that are open to a potentially infinite number of processes of formation and|or combination, the difference between the environment and the medium is that between *information* and *formation*, or between "coded milieus and formed substances" (501).[32] On the basis of this definition, a radical ecology can address not only cultural|technical media, but also nature itself as a scaled ecology of media|form platforms that include such natural media as air, electricity or water.[33] From within the "undivided terrain" (328) of the plane of immanence, which consists of "destratified, deterritorialized matter" (407) which is "laden with singularities" (369), coded milieus emerge, "a code" being defined, as in von Foerster, "by periodic repetition" (313). From within these coded milieus living systems emerge that develop territories, which come into being "when milieu components cease to be directional, becoming dimensional instead, when they cease to be functional to become expressive" (315). Eigenorganizations thus extract "a *territory* from the milieu" (503); in other words, they develop "eigenspaces."

*For a radical eco*logics*, the world is an infinitely complex aggregate of media|form platforms*. Structurally, the medium, as an inherently formless set that has the infinite potential of being informed, functions as a reservoir of an infinite number of virtual forms from which specific forms are actualized. Depending on which plateau is taken as the point of entry, each form thus actualized can become the medium for another form while, simultaneously, each medium is made up of smaller and finer forms, so that *each respective level of mediality is invisible|imperceptible for the level of formation above it, which means that each medium is unconscious.*

If medium|form sets are treated as analog|digital sets, "[a]nalog differences are the 'smallest possible difference,' similar to an infinitesimal or a quantum" while "[d]igital distinctions involve gaps between the discrete elements" (Wilden 189).[34] Cybernetically, the medium is "indistinct" and the infinite density of its differences makes it susceptible to small errors and perturbations (it is "dependent on initial conditions"). The environment, in opposition, is cybernetically distinct and the finite density of its grid of differentiations makes it much more resistant to small errors. Logically, the ultimate medium, as set of pure, infinitely fine differences, would be completely analog and continuous. Because every form of making a distinction "is a rudimentary digitalization of the analog [and] introduces some form of discreteness into a continuum" (159), however, in the embodied world of eigenorganizations, which are themselves the result of making distinctions, the logical purity of a full continuum remains an "ideal" that is comparable to the ideal of a completely smooth space. That notwithstanding, in an embodied world, a space in which the formations that define a medium on a scale that is imperceptible to the specific perceptual apparatus in question *appears* as a pure, continuous medium. Even though a full continuity is ideal, therefore, "it is possible to make systems out of digital [...] elements that will have the *appearance* of being analogic systems" (Bateson, *Mind and Nature* 119).[35]

Every digital|environmental level does not "contain" those differences that are imperceptible to it. In other words, the analog medium contains the digital environmental set:

> The digital system is of a higher level of organization and therefore of a lower logical type than an analog system. The digital system has greater "semiotic freedom," but it is ultimately governed by the rules of the analog relationships between systems, subsystems, and supersystems in nature. *The analog (continuum) is a set which includes the digital (discontinuum) as a subset.* (Wilden 189)[36]

As the medium|form multiplicity from which the world is constructed merely provides analog intensities [disturbances|irritations] that are integrated into conceptual and cognitive assemblages within the autopoietic system, there is no information in the medium that the system has to or can hope to decode and thus there is no flow of information between the two. As von Foerster notes, "[t]he environment contains no information; the environment is as it is" (*Understanding* 252).[37] If sampling denotes the scanning of an analog signal and its subsequent transfer into digital data, autopoietic systems are complex aggregates of sampling machines that translate|integrate analog intensities into digital data. The radical shift from the relation between *subject|object* to *eigenorganization|energetics* is thus immediately related to the fact that *information is generated inside the autopoietic system through the translation|integration of analog intensities into digital data.*

Accordingly, Flusser's radical medio*logics* is not so much concerned with the "contemplation of objects" as with the "computation of concepts" (*Universum* 13): As autopoietic systems assemble meaningful images|concepts from an inherently meaningless multiplicity of optico-chemical irritations ("blocks of life|sensation") rather than from a "book of nature," the cognitive passage leads "[f]rom the abyss of intervals to the surface, from the most *abstract* to the seemingly concrete" (21). As Flusser notes laconically about photographs, "[t]hus it is wrong to ask *what* they mean (except one would give the meaningless answer: they mean photons). What one needs to ask is *why* do they mean what they show" (43).[38] While it is impossible for the autopoietic system to not translate irritations into information and for the observer to not transfer that information into meaning, technical apparatuses have "no problems with the point-elements: they neither want to capture them, nor to imagine them, nor to understand them" (18). For the apparatuses, the elements are nothing but a medium to which it is "structurally coupled:" "For them, the point-elements are nothing but a field of possibilities for their functioning [they] blindly translate [...] the effects of photons on molecules of silver nitrate into photographs" (18). In fact, the apparatus operates solely according to the energetic logic of the medium:

> If one looks at the technical images more closely, one sees that they are not images at all, but symptoms of chemical or electronic processes. [...] "Read" in this way, technical images are objective depictions of processes within the point-universe. (32)[39]

A further consolidation between ecology, systems theory and Deleuzian philosophy, therefore, lies in their setting up a radical distinction between medium and environment.

radical   space

In ecological terms, the "medium of the world"—the world's "plane of immanence"—consists of the multiplicity of physical and biochemical, both organic and nonorganic carriers that form a constantly shifting field defined by material|physical movements|aggregations, such as tectonic shifts and geological aggregations, biological evolutions, chemical colonizations, viral sweeps and animal nomadisms. In *The Logic of Sense*, Deleuze develops the conceit of the surface of sense to align this intensive world with the world of the observer, or to align the lived multiplicity of the medium with the observed multiplicity of the environment, by way of a "flat" ecology (in the sense that Manuel DeLanda talks of Deleuze's "flat" ontology) which radically redistributes not only the focus of many debates within the ecological, but also the way in which to radically rethink the relations between nature and culture.

In this topological conceit, Deleuze goes back to the Stoics who "distinguish between two kinds of things. First, there are bodies with their tensions, physical qualities, actions and passions, and the corresponding 'states of affairs' [...] There are no causes *and* effects among bodies. Rather, all bodies are causes" (4).[40] Causes and effects form two radically incompossible series, causes having to do with "physical qualities and properties" (4) while effects are "logical or dialectical attributes. They are not things or facts, but events" (4–5). According to Deleuze, this "new dualism of bodies or states of affairs and effects or incorporeal events entails an upheaval in philosophy" (6).

The causal realm consists of energetic|intensive operations and changes, while the effectual realm consists of the structural|computational integration and the observation of these changes. This radical logic once more defines structural and behavioral operations as two separated domains: "the realm of interactions, in which behaviour is observed, and the structural realm, in which structural changes happen stand so-to-say orthogonally to each other [...] and [create] non-intersecting fields of phenomena" (Maturana, *Erkennen* 20). Structural changes pertain to the autopoietic system's energetic constitution, while behavior pertains to its computational|observational constitution. Accordingly, Maturana differentiates between the human being as a living entity, "a purely zoological being" (*Was ist Erkennen* 95) that exists in the space of matters-of-fact, and the "human being" as *Homo sapiens sapiens* who exists "in relational space" (95).

Although these two realms are radically separated, "everything happens at the boundary between things and propositions" (Deleuze, *Logic of Sense* 8) and thus at the boundary between the material and the immaterial. As Deleuze notes, "events, differing *radically* from things, are no longer sought in the depths, but at the surface, in the faint incorporeal mist which escapes from bodies, a film without volume which envelops them, a mirror which reflects them" (9–10, emphasis added).

What Deleuzian philosophy adds both to ecology and to the radical constructivist notion of thinking the paradoxical unity of the realms of the medium and the environment and to the problem of thinking "the unity of the difference between [autopoietic] system and environment" (Luhmann, *Social Systems* 212) is his definition of the plane of immanence as a "projective plane;" a surface that becomes the site of a flat eco*logics* in which the medium and the environment, as well as the biological and the cultural, are arranged on one "radically projective" surface:

> the old depths no longer exists at all, having been reduced to the opposite side of the surface. By sliding, one passes to the other side, since the other side is nothing but the opposite direction. [...] It suffices to follow it far enough, precisely enough, and superficially enough, in order to reverse sides and to make the right side become the left or vice versa. (Deleuze, *Logic* 9)

In such a flat ecology, it is "by skirting the surface, or the border, that one passes to the other side, by virtue of the strip. The continuity between reverse and right side replaces all levels of depths; and the surface effects in one and the same Event, which would hold for all events, brings to language becoming and its paradoxes" (11).[41] In this projective topo*logics*, the sides are distributed on a plane, whose "continuity of reverse and right sides [...] permits sense [...] to be distributed to both sides at once, as the expressed which subsists in propositions and as the event which occurs in states of bodies" (125). *This unilateral, projective surface is the site of a flat ecologics.*

structural couplings   **evolution**

Ultimately, Guattari's ecosophy, Deleuze's philosophy and the theory of autopoiesis converge on the radical negotiation of human eigenorganizations and their "life;" what Maturana\Varela call the structural couplings between autopoietic systems and their medium. Most generally, Deleuzian philosophy is ecological in that it is "a philosophy of structural couplings."

As Luhmann notes, "[s]tructural couplings [...] rest on [...] a material (or energetic) continuum, into which the borders of the system do *not* inscribe themselves, which means primarily on a physically functioning world" (*Gesellschaft* 102). In this physically functioning world, structural couplings define the dynamic compatibilities between autopoietic systems and the bio-energetic, "atmospheric" irritations they receive from their medium: "[t]he continuous interactions of a structurally plastic system in an environment with recurrent perturbations will produce a continued selection of the system's structure" (Varela, *Biological Autonomy* 33). In opposition to the visible, conscious operations within the realm of observation, structural couplings tend to operate automatically on imperceptible| unconscious levels: They are machinic operations that fly below the radar of the perceptual thresholds of the respective systems of integration| consciousness above them. As Luhmann notes, "the structural coupling happens continuously and unobserved, it functions also, and especially, when one does not think or talk about it" (*Gesellschaft* 106).

From an autopoietic system's perspective, structural couplings concern the routines by which it keeps an energetic equilibrium within the set of forces, tensions and stresses that define the field of intensities that it operates in; in ecological terms, the stresses of the ecosystem|medium of which it is a part and that allows it to stay operative as a system.[42] In other words, structural couplings define both the *reflexive* and *reflective* protocols the autopoietic system has developed to maintain its structure within its ecosystem, or, put more simply, the programs it has developed to stay alive: "The result of the establishment of this dynamic structural correspondence, or *structural coupling*, is the effective spatio-temporal correspondence of changes of states of the organism with the recurrent changes of state of the medium while the organism remains autopoietic" (Maturana, "Organization" 320): As every autopoietic system is the result of an "infinity" of structural couplings between itself and the medium, structural couplings are often extremely weird and bring about what Dillard calls utterly "improbable lives" (173).

The medium, as a continuously changing set of intensities, irritates the autopoietic system on a quantitative|analog rather than a qualitative|digital level. As Ernst von Glasersfeld notes, "our sense organs announce|register always only a more or less strong impact upon an obstacle, but never convey to us the features or characteristics of what they impact upon" (quoted in Glaser 60–61),[43] which means that "[t]he systemic, interior side of the structural coupling can be designated as irritation (or disturbance, or perturbation)" (Luhmann, *Gesellschaft* 118), or, as von Foerster notes somewhat more laconically, "in none of them [cells] is the quality of the cause of activity encoded, only the quantity" (*Understanding* 233). To address first of all the "*intensity* of the stimulus" (von Foerster, *Wissen* 274), structural couplings concern the irritations that trigger digital operations that in turn cause changes in the operational architecture of the system (which an observer reads as the system's behavior) or in its structural setup (which an observer reads as the system's evolution):

The history of the structural coupling of an organism and its nervous system to a medium is thus a history of interactions, during whose course a structure is modulated by operational relations, which seem to an observer as behavior, but which are ausschliesslich structurally bedingt and realized and thus do not follow or are caused by semantic "meaning" or "function." "Meaning" and "sense" are without exception characteristics of the *description* that an *observer* constructs. (Maturana, *Erkennen* 21)

The difference between autopoietic species and individual autopoietic systems consists in that each autopoietic system has a specific range of operational and navigational responses *vis-à-vis* the set of perturbations it encounters in its medium; a range that can be measured as its "amount of agency"—both physically: "what its body can do" and psychically: "how plastic is its cognitive apparatus"—and that ultimately defines its range of evolutionary options because "the structure of the environment cannot specify its changes, but can only *trigger* them" (Maturana|Varela, *Tree* 131, emphasis added). On the other hand, even extremely rigid systems can be extremely successful in terms of evolutionary tenacity: Although Dillard at some point notes "the abysmal stupidity of insects" (67), this says nothing about their evolutionary potential.

Deleuze provides a philosophical context for the interplay of such physical and psychic operational responses when he defines the brain as an interval in the arc between excitation|stimulus and response. For Deleuze, the brain is an integrational rather than a mimetic machine that "does not manufacture representations" (*Bergsonism* 24) of objects, but "merely complicates the relationship between a received movement (excitation) and an executed movement (response). Between the two, it establishes an interval" (24). As a digital machine it "divides up excitation infinitely" (52) and " in relation to the motor cells of the core [...] leaves us to choose between several possible reactions" (53). Although this specific quote talks about a *human* brain ("us"), according to Deleuze, such intervals reach down to the most microscopic levels of autopoietic operations and while "[n]ot every organism has a brain, and not all life is organic, [...] everywhere there are forces that constitute microbrains, or an inorganic life of things" (Deleuze|Guattari, *Philosophy* 213). As with Guattari's concept of a virtual chaos, the smaller the interval, the faster the response, which is why on the plane of virtual chaos things happened infinitely fast: "Even at the level of the most elementary living beings one would have to imagine micro-intervals. Smaller and smaller intervals between more and more rapid movements" (Deleuze, *Cinema 1* 63). In every eigenorganization, the scale of analog|digital transfers goes from the level of the infinitely large (the human being as part of the cosmos) to the infinitely small (the cell as a constitutive part of the human being), with the "infinitely analog" functioning as the ideal of a pure unconscious.

As complex systems' reactions to the medium rely on their capacity for internal operational responses, their range of agency concerns their internal plasticity, their capacity for internal modulations and their specific perceptual thresholds, which determine the amount of choices they have to respond to structural changes. At the same time, every level of "options of response" separates systems further from the medium because it implies a further level of mediation. To "live" ecologically, therefore, implies both becoming more perceptive *and* becoming more unconscious and imperceptible by opening up one's perceptual thresholds to more and more anonymous levels. As Varela notes,

> knowledge appears more and more as built from small domains, that is, microworlds and microidentities [...]. The very heart of this autonomy, the rapidity of the agent's behaviour selection, is forever lost to the cognitive system itself. Thus, what we traditionally call the "irrational" and the "nonconscious" does not contradict what appears as rational and purposeful: it is its very underpinning. (quoted in McMurry 80; also see Varela et al., *Embodied Mind* 336)

According to this filtering process, "[a]n atom...perceives infinitely more than we do and, at the limit perceives the whole universe" (Deleuze, *Cinema 1* 63–64) or, as Dillard notes, "Donald E. Carr points out that the sense impressions of one-celled animals are *not* edited for the brain: This is philosophically interesting in a rather mournful way, since it means that only the simplest animals perceive the universe as it is" (19). As the levels of conscious perception are much more mediated and rough-scale in complicated aggregates, Deleuze's project is to find ways of making these levels finer and to open up the aggregate to unconscious levels of perception. In *Pure Immanence*, his final, posthumously published text, Deleuze celebrates the processes of developing the range of pick-ups for intensities as movements of becoming unconscious or becoming imperceptible that have as their perspective point (in topological terms: as their "point at infinity") a "powerful non-organic life" that "carries with it the events or singularities that are merely actualized in subjects and objects" (29). Deleuze's exquisitely ironic and at the same time eminently ecological project is to raise, rather than to lower, quality to quantity: "to raise lived perceptions to the percept and lived affections to the affect" (Deleuze|Guattari, *Philosophy* 170) with affects being *"precisely these nonhuman becomings of man*, just as percepts [. . .] are *nonhuman landscapes of nature"* (169).

In *Monte Grande, What is Life?*, the documentary about his life and death, Varela refers to a preorganic, anonymous field that is similar to the one Deleuze is interested in; a field he relates to the Buddhist idea of "another level of consciousness," a "subtle consciousness that in fact is the foundation for the other types of consciousness." This consciousness sometimes shines through in "very special dreams or very intense moments of special meditation practices or the moment when you die." After the individual's death it flows through the world, in order to manifest itself anew, maybe as the core of a new, different consciousness. "The interesting view there is this kind of . . . the idea of a flow of consciousness which has moments in which it manifests as in a more layered consciousness including mental phenomena, cognition; and then after death it continues like a flow and it comes up again."

This notion of "a consciousness, in other words, an awareness that is aware of itself, without brain"[44] resonates directly with Deleuze's notion of a "transcendental *field*" (*Immanence* 25) that manifests itself as a "pure stream of a-subjective consciousness, a pre-reflexive impersonal consciousness, a qualitative duration of consciousness without a self" (25); a field in which sensation is "only a break within the flow of absolute consciousness" (25). At the perspective point of such a radical reversal, every part of the living world contemplates itself in a communal, anonymous meditation: "But we shouldn't enclose life in the single moment when individual life confronts universal death. *A* life is everywhere, in all the moments that a given living subject goes through and that are measured by given lived objects" (29).

While many structural couplings are related to the purely biological, in the case of the human, they are also related to the framing processes that are important for the construction of the observer's "finite" environment. In fact, Deleuze sees philosophy, in opposition to science, as retaining in its concepts the infinity of structural couplings.[45] The philosophical concept keeps the medium "in" the environment, although the environment is, as a result of the reductive character of perceptual operations, only a subset of the medium:

> Through structural couplings, a system can be docked to highly complex environmental conditions, without having to compute or reconstruct their complexity. [...] As one can see with the physically narrow spectrum of eyes and ears, structural couplings always perceive an extremely limited segment of the environment. (Luhmann, *Gesellschaft* 107)[46]

Structural couplings, however, do not only define the relationship between the system and its medium in the sense of the "ecological milieu" within which it operates; they are also operative between several autopoietic systems: "This means that two (or more) autopoietic unities can undergo coupled ontogenies when their interactions take on a *recurrent* or more stable nature" (Maturana|Varela, *Tree* 75): radical parasitism.[47]

One reason why Guattari had proposed to broaden the autopoietic reference to sociopolitical and linguistic systems, in fact, had been that language, as the medium of human communication, brings about the structural coupling of its users on the level of a shared linguistic code. "[W]e are in language [...], we 'language' only when through a reflexive action we make a linguistic distinction of a linguistic distinction. Therefore, to operate in language is to operate in a domain of congruent, co-ontogenetic structural coupling" (Maturana|Varela, *Tree* 210). In this way, language, which emerges in between the multiplicity of the structural couplings that define the overall autopoietic system, is radically linked to the biological phylum:

> The moment and the context of the emergence of a communicative behaviour are not chance events, they are caused by the structural isomorphism of the communicating organisms and similarly by their structural coupling to the medium in which they exist. (Maturana, *Erkennen* 291)

The notion of the human eigenorganization as a recursively arranged aggregate that involves elements that make distinctions, and elements that make distinctions of distinctions, implies two levels of critique: one that concerns the idea of language as "always already" cultural (although this is radically true, language is at the same time an attribute of real energetics) and one that concerns the reduction of information transfers to the linguistic level of the observer, as in Lacanian psychoanalysis. As Serres had shown, the multiplicity of languages and codes that are operative in the human aggregate can only be addressed by a scaled logic. Deleuze|Guattari, therefore, speak of "'natural' codings operating without signs" (*Thousand Plateaus* 117) and of levels on which "writing [...] functions on the same level as the real, and the real materially writes" (141); a realm Jacques Monod relates to the "microscopic cybernetics" of enzymes. According to such a notion of writing, human language is pervaded, at every moment, by an infinite number of material writings and "signaletic material [*matière signalétique*]" (Deleuze, *Cinema 2* 33); an intensive, productive and emergent writing whose inscriptions often do not congeal into a linguistic sign.

pour   **down**

Let me pour these theoretical observations into the ecological artistic practice of Robert Smithson's projects *Asphalt Rundown* (Rome 1969) [*Figure 2*: detail], which consists of the event of pouring 1,000 tons of asphalt over a ledge, *Glue Pour* (Vancouver 1970) [*Figure 3*: detail] and *Concrete Pour* (Chicago 1969) [*Figure 4*: detail]. These "flow works," in which materials (asphalt, glue and concrete respectively) are poured down an inclined slope,

Figure 2   Figure 3   Figure 4

can be read as a direct commentary on a radically evolutionary and eco*logical* dynamics.

In radical constructivist terms, nonhuman evolution is the result of an ecology of structural couplings between autopoietic systems and their medium on both energetic and operational|informational levels. In the case of the human, evolutionary parameters include the level of the

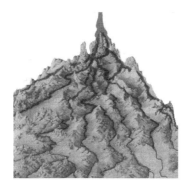

Figure 5

environment and the observer.[48] Because of the nonlinearity of evolutionary processes, radical constructivism defines evolution as "a *natural drift*, a product of the conversation of autopoiesis and adaptation" (Maturana|Varela, *Tree* 117). As Emmeche notes, "for these kinds of complex systems there is no other way to predict their evolution than by tracking it" (87). To define the dynamics of such a "pachinko evolution," Maturana|Varela use the image of a drop flowing down an evolutionary mountain [*Figure 5*: "The Natural Drift of Living Beings Seen as a Metaphor of the Water Drop." Maturana|Varela 111, detail], and Maturana that of a skier going down an entropic slope:

> How does the living system find the right course? Very simply—by itself. It glides—as you slalom when skiing [...]. This dance-like slalom through life I call "drifting," and as a game with structures it lasts as long as the organization and the adaptation are maintained. The only way this game can end, therefore, is in death. (Maturana, *Was ist Erkennen* 81)

Edward Lorenz uses the same image in his description of nonlinear processes that are "dependent on initial conditions." His subtraction of the skier, however, reduces the operation to a purely physical one: "For an

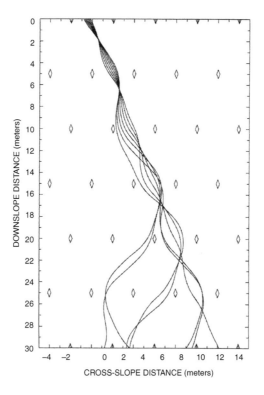

Figure 6

everyday system that […] can […] serve to illustrate many of the basic properties of chaotic behavior, […] let us consider one that still bears some resemblance to the pinball machine. The new system will again be a slope, with a ball or some other object rolling or sliding down it" (25) [*Figure 6*: "Diagram of Ski-Slope." Lorenz 33]. Deleuze's "ecological philosophy" is based on a similar operation of moving *with* and *in* the medium rather than moving against it:

> We got by for a long time with an energetic conception of motion, where there's a point of contact, or we are the source of movement. Running, putting the shot, and so on: effort, resistance, with a starting point, a lever. But nowadays we see movement defined less and less in relation to a point of leverage. All the new sports—surfing, windsurfing, hang-gliding—take the form of an entering into an existing wave. There's no longer an origin as starting point, but a sort of putting into orbit. The key thing is how to get taken up in the motion of a big wave, a column of rising air, to "get into something" instead of being the origin of an effort. (*Negotiations* 119)

Smithson's flow works embody these medial movements and their radical logic almost literally: Like radical constructivism, Smithson differentiates between two separate levels. The "matter-of-fact," physical action of pouring concerns what Smithson calls the "site" aspect of the work, while the various forms of its portable documentation, such as photographs or films, concern its "non-site aspect." As the registers that define the "site" aspect of the work are purely physical ones, this "side" of the work is "not involved with nature, in the classical sense. There's no anthropomorphic reference to environment" (Smithson, "Interview" 204). The non-site aspects, on the other side, are fully "observational." The physical, matter-of-fact action of pouring is, at least in *Concrete Pour*, literally "concrete" in that the actual pour-down operation is simply to do with the propagations of masses and intensities, such as weights, velocities, frictions. Symptomatically, Smithson's drawings of *Asphalt Pour* recapitulate almost literally the diagrams of evolutionary drift and of the sensitive dependence on initial conditions that defines nonlinear systems [*Figure 7*: "Asphalt on Eroded Cliff;" *Figure 8*: "1000 Tons of Asphalt"]. The glue, the

Figure 7 Figure 8

asphalt and the concrete compute|contemplate the best way to flow, similarly to the way the sand does in *Walden*:

> The material was sand of every degree of fineness and of various rich colors, commonly mixed with a little clay. When the frost comes out in the spring, and even in a thawing day in the winter, the sand begins to flow down the slopes like lava, sometimes bursting out through the snow and overflowing it where no sand was to be seen before. Innumerable little streams overlap and interlace one with another, exhibiting a sort of hybrid product, which obeys half way the law of currents and half way that of vegetation. As it flows it takes the forms of sappy leaves or vines, making heaps of pulpy sprays a foot or more in depth, and resembling, as you look down on them, the laciniated lobed and inbricated thalluses of some lichens; or you are reminded of coral, of leopard's paws or birds' feet, of brains or lungs or bowels, and excrements of all kinds [...]. You thus find in the very sands an anticipation of the vegetable leaf. (565)

The non-site aspects of the operation address it as a relational, observed event. (See also Mark Tansey's painting "Purity Test" (1982) [*Figure 9*], which is a direct comment on the question of "observation.")

Figure 9

Although these non-site aspects are radically different from the initial, matter-of-fact operation, they emerge directly from the material level of the propagation of forces|movements and they remain their attribute.

The flow works are radical in that they operate at the interface|threshold of the world of material matters-of-fact and immaterial events. Creating a "projective space" of sites and non-sites, Smithson's work becomes what might be called "radical art." Symptomatically, Smithson eschews the aesthetics of conceptual art for an aesthetic of the radical separation of material and conceptual registers:

> I think that conceptual art which depends completely on written data is only half the story; it only deals with the mind and it has to deal with the material too. Sometimes it is nothing more than a gesture. I find a lot of that written work fascinating. I do a lot of it myself, but only as one side of my work. My work is impure; it is clogged with matter. I'm for a weighty, ponderous art. There is no escape from matter. There is no escape from the physical nor is there any escape from the mind. The two are in a constant collision course. You might say that my work is like an artistic disaster. It is a quiet catastrophe of mind and matter. (Smithson, "Fragments" 194)

If the flow works address the notion of entropy because they involve an inclined slope ["a stronger tendency towards the inorganic than to the organic. The organic is closer to the idea of nature: I'm more interested in denaturalization or in artifice than I am in any kind of naturalism" (Smithson, "Interview" 204)], they also address the notion of evolution. In fact, Smithson's work combines a maximum amount of deceleration with an infinitely slow becoming: "The world is slowly destroying itself," he notes. Instead of a sudden cataclysm, he prefers "the lava, the cinders that are completely cold and entropically cooled off. They've been resting in a state of delayed motion. It takes something like a millennium to move them. That's enough action for me. Actually that's enough to knock me out" (205). When the interviewer interjects: "[a] millennium of gradual flow..." (205), Smithson's geological answer evokes precisely this almost infinitely slow becoming: "You know, one pebble moving one foot in two million years is enough action to keep me really excited..." (205).

conclusion   vampyroteuthis infernalis

To return the discussion to a radical haecology: The functional parameters of a "radical haecology" are: As a didactics, it aims at making the operations of "structural couplings" conscious (Guattari). As a science, it aims to generate knowledge about structural couplings and to develop routines that allow for regulated responses to structural disequilibria (von Foerster, Maturana|Varela). As a practice, it aims at implementing parameters within which the equilibrium between the autopoietic systems and their media can be kept, both locally and globally (Bateson). As an artform, it provides "blocks of sensation" (Smithson). As a philosophy, it creates concepts that radically link immaterial concepts and material movements (Deleuze).

Structural couplings link autopoietic systems that share a history of biological and|or communicative coevolution.[49] As Varela notes, "[en]vironmental regularities are the result of a conjoint history, a congruence that unfolds from a long history of codetermination" (*Embodied Mind* 198). The life-form Flusser describes in *Vampyroteuthis Infernalis*—a species of squid that lives in deep-sea regions—seems to share nothing with the human except for the fact that it is in all aspects its diametrically opposed "other." In both biological and philosophical terms, it embodies a world of pure intensity: Its operational parameters are defined by the very terms that designate the human's relation to intensity as "conatus," "appetitus" or "libido".

Already in terms of its optical perception, the difference between the human and the vampyroteuthic world is a categorical one. While the human world is enlightened, the vampyroteuthis lives in "dark weather." His world comes into being only as the result of the illumination it provides by itself, because "only his light-producing organs create the appearances, the phenomena" (Flusser, *Vampyroteuthis* 40). Not only operationally, but quite *literally*, his optical world is radically constructed by him, because the vampyroteuthis sends "light cones into the world," using "his tentacles to tear blocks of information from these cones and paralyz[ing] them into data" (46).[50]

In all senses, his relation to the world is based on irritation rather than integration. If humans operate by integrating and stratifying the flow of molecular microperceptions and molar perceptions into stable concepts and histories, the vampyroteuthis encounters pure fluxes and intensities|imprints: "the world as perceived by vampyroteuthis is a fluid, centripetal vortex [...] We have 'problems', it has 'impressions'. Its mode of perception is impressionistic" (*Vampyroteuthis* 39–40). Whereas humans project themselves into their medium through physical and psychical movement—movement being one of their most important spectra of evolutionary agency—vampyroteuthis is mostly stationary, sucking in a moving medium that literally flows into it: "vampyroteuthis sucks in the world instead of treating it. Vampyroteuthis turned into a passionate, erotic subject, whose objective counterpart is active, rushes towards him in order to be passionately enjoyed by vampyroteuthis" (*Vampyroteuthis* 41).

Its aim in life, therefore, is not so much integration as intensification. Where humans sublimate, it produces, so that "our wakeful consciousness is its unconscious" (*Vampyroteuthis* 42). While human knowledge|philosophy aims at complete sublimation, vampyroteuthis thinks orgasmically: "what measuring and cutting reason is for human thought, the apprehending coitus and orgasm is for vampyroteuthis" (46).[51] Whereas the human world is geometrical, computed and contemplative, its world is dynamic, irritating and libidinal:

> Space for it is a coiled tension that is filled with energy [...]. This is why for it, geometry is what we call dynamics [...] the world that rushes towards it constantly surprises it with its mutability and the plasticity of its impacts. Its theory, its philosophical gaze is orgiastic rather than contemplative. (43)

Its unconscious is what in the human functions as the agency of the observer. "In the end vampyroteuthis will be forced to dive [...] deeply into repressed regions that are analogous to those in which we practice analytic geometry" (47). In the ethics of its libidinous philosophy, the highest achievement is not to reach the level of a universal spirit but that of a universal intensity.

Accordingly, "for us, its murderous and suicidal anarchy is a hellish society, but for it, it is the unattainable heaven of freedom" (58).

My point in this juxtaposition is that only a truly radical eco*logics* can adequately "deal with" this life-form; an eco*logics* in which metaphysics is born from physics and in which psychology is not "a field of knowledge that can be separated from biology, but that branch of biology that deals with specific complexities" (29). Like Deleuze|Guattari and Maturana|Varela, Flusser traces the genesis of autopoietic systems from the virtual chaos of the plane of immanence and from the multiplicity of physical forces, pressures and repressions that move over|in it. In fact, organisms are first of all nothing but "accumulation[s] of repressed pressures" (29). An eco*logical* analysis calls for a material psychology in which the ratio of the deceleration of the virtual chaos and the accompanying loss of elasticity and plasticity are direct results of the processes of subjectivation. As Flusser notes, "the more rigid the cramp, the stronger the personality" (29).

If the pure plane of immanence is a world|medium of single cells that are pushed and pulled into varying formations, singular life on this plane of immanence is an energetic arrangement compressed into the complexity of eigenorganizations; a process in which cells "lose their individuality and live merely as functions of society" (53); a process by which they are subsumed in "hierarchies of human apparatuses of administration" that make up an "organism" (53). Symptomatically, it is only within complex formations|organizations that the cellular, "teeming, immortal, mosaic life" (52) becomes "mortal." What Deleuze calls "life" is for Flusser a multiplicity of "germ cells" that "traverse the organism with contempt and have remained immortal" (53). In fact, if individual life is the compression of "a life" into eigenorganized complication, its decompression is individual death: "the organism is an energy that has been contracted into life, which explodes when the cramp, which is the organism, relaxes" (30).

Vampyroteuthis embodies quite literally the explosive force of such an anonymous life as captured by and caught in an organism, which is why his ethics is one of dissolution and orgasm; orga*nis*m. While humans fear individual death, vampyroteuthis literally enjoys (in) it, which is why "[t]he animals tend towards suicide and cannibalism. They devour their 'arms' even in tanks that are filled with food, even when it is teeming with crabs around them" (*Vampyroteuthis* 25).

Ultimately, a radical eco*logics* and a radical philosophy ask us to consider this intensive life, which is always inclined towards a dissolution into anonymity (the respect for *haecceities* of all kinds, for the world living "a" life, anonymous and intensive) not only as one more aspect of an endangered biodiversity, but as a life that operates on the same projective surface as "our" life. Vampyroteuthis embodies the "other side" of our life, which we can reach without crossing a threshold, because our world and its world are positioned on a projective plane. A radical eco*logics* asks us to embrace this life as the intensive, wild and anonymous life we have emerged from and to which we are still immanent. Deleuze provides the radical philosophy for this embrace.

Notes

1. Cf. Žižek; Badiou.
2. The English translations oscillate between "plane of consistence" and "plane of consistency." "The plane of consistency is the abolition of all metaphor; all that consists is Real. These are electrons in person, veritable black holes, actual organites, authentic sign sequences. It's just that they have been uprooted from their strata, destratified, decoded, deterritorialized, and that is what makes their proximity and interpenetration in the plane of consistency possible. A silent dance" (Deleuze|Guattari, *Thousand Plateaus* 69).
3. Michael E. Zimmerman notes "three major branches of radical ecology: deep ecology, social ecology, and ecofeminism" (1).
4. Luhmann, *Ecological Communication*.
5. Radical constructivism is a research field assembled from the work of, among others: Ludwig von Bertalanffy, John von Neumann, Gregory Bateson, Norbert Wiener, Ernst von Glasersfeld, René Thom, Erik Christopher Zeeman, David Ruelle, Edward Lorenz, Mitchell Feigenbaum, Steve Smale, James A. Yorke, John H. Holland, Murray Gell-Mann, Stuart Kauffman, Ilya Prigogine, Francisco J. Varela, and Humberto R. Maturana.
6. See also Guattari, "Space & Corporeity:" "It is fitting to specify them [machinic components] as autopoietic systems as qualifies them Francisco Varella [sic] who by the way assimilates this type of system to machines."
7. The term "eigenorganization" derives from "eigenvalue," which was introduced into linear algebra, together with that of "eigenfunction," by German mathematician David Hilbert in his 1904 essay "Grundzüge einer allgemeinen Theorie der linearen Integralgleichungen." The mathematical context was the modeling of transformations or operations by way of (1) the identification of those vectors within the transformation|operation that remain invariant and (2) the measurement of the scalar changes of these vectors, such as those brought about by operations of stretching or compression. In such transformations|operations, Hilbert called a preserved direction an "eigenvector" (the eigenvectors of a linear operator are non-zero vectors which, when operated on, result in a scalar multiple of themselves) and the associated amount by which it has been scaled its eigenvalue (if there is a vector (object x) that has been preserved by H apart from submitting it to a scalar multiplier k, x is called an eigenvector of H with the eigenvalue k). A transformation that does not affect the direction of a vector, Hilbert called an eigenvalue equation (*Eigenwertgleichung*). *The notion of eigenvalue, then, concerns the orientation and the scale of vectors within a mathematical transformation|operation.* The subsequent use of the term in other contexts is based on a generalization from "vectorial change" to "systemic change." As eigenvalues|eigenvectors can come to define invariants not only within mathematical, but also within physical or biological systems while these undergo changes, they can be used to model these systems in relation to the changes that they undergo. Or, if one stresses the processual rather than the systemic side, as eigenvalues|eigenvectors define invariants within processes, they can be used to isolate "systems" within these processes: if to identify an eigenvalue|eigenvector means to define a system as a cluster of invariant characteristics within a specific process, then a system can be defined, *mutatis mutandis*, within a specific process as "something undergoing change." Any system x, therefore, might be defined as "the invariance in undergoing the transformational process y." As instances of "eigenimages," for instance, "eigenfaces," which are used for computerized face recognition, are composed of sets of statistically generated individual face ingredients, or facial

eigenvectors in the field of physics, eigenvalues have found a wide range of application, from quantum mechanics to resonance theory, which deals with questions of how to model "systems in process" in terms of wave functions and of their invariant "eigenfrequencies."

8. Thus every organic body of a living being is a kind of divine machine or natural automaton, which infinitely surpasses any artificial automaton, because a man-made machine is not a machine in every one of its parts. For example, the tooth of a brass cog-wheel has parts or fragments which to us are no longer anything artificial, and which no longer have anything which relates them to the use for which the cog was intended, and thereby marks them out as parts of a machine. But nature's machines—living bodies, that is—are machines even in their smallest parts, right down to infinity. That is what makes the difference between nature and art, that is, between the divine art and our own (Leibniz 277). On Gottfried W. Leibniz, see also Deleuze, *The Fold* "Leibniz discovers that the monad as absolute interiority, as an inner surface with only one side, nonetheless has another side, or a minimum of outside, a strictly complementary form of outside. Can topology resolve the apparent contradiction? The latter effectively disappears if we recall that the 'unilaterality' of the monad implies as its condition of closure *a torsion of the world, an infinite fold*, that can be unwrapped in conformity with the condition only by uncovering the other side, not as exterior to the monad, but as the exterior or outside *of* its own interiority." (111, emphasis added)

9. While it is true that according to the theory of autopoiesis "[t]he organization of a machine thus has no connection with its materiality" (Guattari, *Chaosmosis* 39), it does not follow that, as Guattari notes, autopoiesis "lacks characteristics essential to living organisms, like the fact that they are born, die and survive through genetic phylums" (39).

10. Cf. Deleuze|Guattari, *Thousand Plateaus* 266. While the concept of the plane of immanence denotes the field of energetic multiplicity that the human being is materially part of (a set of pure quantity|intensity), the "plane of consistency|composition" denotes the plane of the material|perceptual contractions|modifications that make up the autopoietic assemblages (an eigenorganized set of quantity|quality). The "plane of transcendence" (which is a set of pure quality|metaphor) is the plane from which the observer "surveys" these two planes. Deleuze differentiates, however, between the observer, who is always in a position of "*n*+1," and a mode of survey that operates from within the field that is being surveyed and is therefore in a position of "*n*−1." In his conceptualization of the latter form of survey, Deleuze relies on Ruyer's *Neo-Finalisme*. The important point is that a philosophical concept is, according to Deleuze, in a position of "*n*−1" because it is immanent to the problematics it negotiates: "the concept is in a state of *survey* [survol] in relation to its components, endlessly traversing them according to an order without distance" (Deleuze|Guattari, *Philosophy* 20). As such, it is "a *multiplicity*, an absolute surface or volume, self-referents, made up of a certain number of inseparable intensive variations according to an order of neighborhood, and traversed by a point in a state of survey. The concept is the contour, the configuration, the constellation of an event to come" (32–33). Because of this, "[t]he philosophical concept does not refer to the lived, by way of compensation, but consists, through its own creation, in setting up an event that surveys the whole of the lived no less than every state of affairs" (33–34).

11. Most postmodern ecologies derive their arguments from "social constructivism." Andrew McMurray draws on "*self-organizing, self-referential, complex social systems*" (15) and on "the indirect, *mediated* character of our encounter with it [nature]" (27). As he notes, "nature exists as something either systematized by society (becoming the focus of internally significant communications) or externalized by society as uncommunicative environment" (5). Although he is aware of that paradox, he leaves it curiously unexplored: "[N]ature as ultimate biological horizon poses a practical challenge to the theoretical posture of total constructivism: [...] As the material limits of the world assert themselves in the form of global warming, changes in weather patterns, etc., we are reminded that nature does not dissolve into discourse. [...] It must be understood, then, that when we speak of nature we are talking about real limits, forces and entities—but ones that are always mediated through culture" (46). James Grier Miller is also "primarily concerned" with "conscious and social" (62) systems. For a critique inspired by Deleuze see William E. Connolly: "To escape the curse of reductive biology, many cultural theorists reduce body-politics to studies of how the body is *represented* in cultural politics. They do not appreciate the *compositional* dimension of body-brain-culture relays" (xiii). As such "they lapse into a reductionism that ignores how biology is mixed into thinking and culture and how other aspects of nature are folded into both" (3).
12. "Habit draws something new from repetition—namely difference [...]. In essence, habit is contraction" (Deleuze, *Difference* 73).
13. As a belief is a habit—"a deliberate, or self-controlled, habit is precisely a belief" (Peirce, "Survey" 330)—the oscillation between belief and doubt is the one between habitualization and de-habitualization. "Belief is not a momentary mode of consciousness; it is a habit of mind essentially enduring for some time, and mostly (at least) unconscious; and like other habits, it is (until it meets with some other surprise that begins its dissolution) perfectly self-satisfied. Doubt is an altogether contrary genus. It is not a habit, but the privation of a habit" (Peirce, "What Pragmatism Is" 279).
14. "May not the laws of physics be habits gradually acquired by systems" (Peirce, "Design and Chance" 553). See also: "The laws of Nature are nothing but the immutable habits which the different elementary sorts of matter follow in their actions and reactions upon each other. In the organic world, however, the habits are more variable than this" (James, *Psychology* 104), and "[n]ature exhibits only changes, which habitually coincide with one another so that their habits are discernible in simple 'laws'" (James, *Radical Empiricism* 74). Cf. Massumi 237, 11.
15. "The principle is the habit of contracting habits. [...] Habit is the root of reason, and indeed the principle from which reason stems as an effect" (Deleuze, *Empiricism* 66), or, "as Bergson said, habits are not themselves natural, but what is natural is the habit to take up habits" (44). In fact, as the result of natural|habitual operations, reason is also "an affection of the mind. In this sense, reason will be called instinct, habit, or nature" (30).
16. Cf. Cramer.
17. "As soon as we find that a belief shows symptoms of being instinctive, although it may seem to be indubitable, we must suspect that experiment would show that it is not really so" (Peirce, "Issues" 297).
18. "The passive synthesis of habit constituted time as the contraction of instants under the condition of the present, the active synthesis of memory constitutes it as the interlocking (emboîtement) of the presents themselves" (Deleuze, *Difference* 81, revised translation).

19. The term eigenorganizations has become, albeit in a symptomatically nontheoretical version, an important reference in the ecological as the "intrinsic value" of a tree, landscape or animal as opposed to its "use-value."

20. "By retaining the infinite, philosophy gives consistency to the virtual through concepts; by relinquishing the infinite, science gives reference to the virtual, which actualises it through functions. Philosophy proceeds with a plane of immanence or consistency; science with a plane of reference. In the case of science it is like a freeze-frame" (Deleuze|Guattari, *Philosophy* 118) while art creates "a plane of composition that is able to restore the infinite" (203); ["*plane of immanence of philosophy, plane of composition of art, plane of reference or coordination of science; form of concept, force of sensation, function of knowledge; concepts and conceptual personae, sensations and aesthetic figures, figures and partial observers*" (216)]. If "the philosopher brings back from chaos [...] *variations* that are still infinite but that have become inseparable on the absolute surfaces" (202), the scientist "brings back from the chaos *variables* that have become independent by slowing down [...] finite coordinates on a secant plane of reference" (202) while the artist "brings back from the chaos *varieties* that no longer constitute a reproduction of the sensory in the organ but set up a being of the sensory [...] on an anorganic plane of composition that is able to restore the infinite" (202–203).

21. Cf. Thom.

22. Cf. Naess, "Simple in Means": "*Sophy* comes from the Greek term *Sophia*, 'wisdom,' which relates to ethics, norms, rules, and practice. Ecosophy, or deep ecology, then, involves a shift from science to wisdom" (183).

23. I am not quite sure why the geometrical coordinates should be those of "supposedly closed systems." Maybe Deleuze|Guattari want to stress the impossibility of language|thought to refer directly to the realm of matters-of-fact. In actual fact, these systems are neither open nor closed; they are quite simply "what they are."

24. Deleuze is especially concerned with modes of how to reach back from within representational, active syntheses to passive syntheses. In relation to time|memory, the question is how, given that passive syntheses are subrepresentative, to reach back to them "despite" the representational operations of memory. In other words, "whether or not we can penetrate into [*penetrer dans*] the passive synthesis of memory; whether we can in some sense live the being in itself of the past in the same way that we live the passive synthesis of habit" (*Difference* 84, revised translation).

25. This notion of wilderness shares aspects of what Jean-François Lyotard calls "matter" as opposed to "nature."

26. "On the plane of consistency, *a body is defined only by a longitude and a latitude*: in other words the sum total of the material elements belonging to it under given relations of movement and rest, speed and slowness (longitude); the sum total of the intensive affects it is capable of at a given power or degree of potential (latitude). Nothing but affects and local movements, differential speeds (Deleuze|Guattari, *Thousand Plateaus* 260). "We call longitude of a body the set of relations of speed and slowness, of motion and rest, between particles that compose it [...] We call latitude the set of affects that occupy a body at each moment, that is, the intensive states of an *anonymous force* (force for existing, capacity for being affected)" (Deleuze, "Ethology" 629). "The plane of consistency contains only haecceities, along intersecting lines. Forms and subjects are not of that world" (Deleuze|Guattari, *Thousand Plateaus* 263).

27. Deleuze differentiates very carefully between what Spinoza calls affect [affectus] and affection [affectio], the first one of which is related to analog duration [*duree*] and intensity, the second to digital chronology [*temps*] and representation| observation: "There is a difference in nature between the affect and the affection. The affect is not something dependent on the affection, it is enveloped by the affection [...]. What does my affection, that is the image of the thing and the effect of this image on me, what does it envelop? [...] It is not a comparison of the mind in two states, it is a passage or transition enveloped by the affection, by every affection. Every instantaneous affection envelops [...] a lived passage, a lived transition, which obviously doesn't mean conscious. [...] There is a specificity of the transition, and it is precisely this that we call duration [...]. You have [...] two states which could be very close together in time. [...] They are very close together. I am saying: there is a passage from one to the other, so fast that it may even be unconscious [...] The affect is what? It is the passage. The affection is the dark state and the lighted state. Two successive affections, in cuts. The passage is the lived transition from one to the other. Notice that in this case here there is no physical transition, there is a biological transition, it is your body which makes the transition." ("Les Cours de Gilles Deleuze")

28. An epistemology has to be "of such a kind [...], that it explains itself, or in Hilbert's language, that it is an eigentheory. [...] Experience is the cause | The world is the effect|The epistemology is the rule of transformation" (von Foerster, Wissen 368–369).

29. Cf. "I shall act always so as to increase the total number of choices" (von Foerster, *Understanding* 282). See also Naess, "Deep Ecological Movement." As Arne Naess notes, human and nonhuman life forms have "values in themselves" (197). His ethical project is also to "[m]aximize symbiosis" and to "[m]aximize (long-range, universal) diversity" (209).

30. Radical constructivism argues that "organization exists or it does not exist [...]. It is 'digital' and not 'analog': Between 'yes' and 'no' there is, one might say, a quantum jump, without transition" (Maturana, *Was ist Erkennen* 78). For Deleuze|Guattari, however, qualities emerge from quantities through a topological process not of *cutting* but of *contraction*: "What, in fact, is a sensation? It is the operation of contracting trillions of vibrations onto a receptive surface. Ouality emerges from this, quality that is nothing other than contracted quantity" (Deleuze, *Bergsonism* 74).

31. "A unity exists in a medium determined by its properties as the domain in which it operates as a unity. Anything that a unity may encounter as a unity is part of its medium. Anything from which a unity may become operationally cleaved through its operation as a unity, is part of its medium." The important thing for a radical ecology is that "the environment overlaps with the medium but is not necessarily included in it" and that "the medium of a unity may include features that the observer does not see as parts of the unity's environment, but which, to him, appear to be inside the boundaries of the unity, even though they are operationally necessarily outside them" (Maturana, "Cognition").

32. According to the media theory of German systems theorist Niklas Luhmann, this potentiality rests on the fact that the medium and "its" formations stand in a relation of reciprocal independence.

33. See also the concept of "media ecology" in the work of Marshall McLuhan.

34. "[D]igital distinctions introduce GAPS into continuums [...] whereas analog differences, such as presence and absence, FILL continuums" (Wilden 186).

35. On the relation between Bateson and Guattari, see Conley.
36. In cybernetic terms, "any digital message MUST at some level be a metacommunication about an analog relation" (Wilden 171).
37. Cf. "[C]ognition [*Erkenntnis*] is different from the environment, because the environment does not contain differentiations, it merely is as it is" (Luhmann, *Gesellschaft* 223).
38. "[T]he technical images are the expression of the attempt to gather [*raffen*] the point elements around us and in our consciousness into surfaces, in order to fill up [*stopfen*] the intervals between them; of the attempt to set into images [*in Bilder zu setzen*] on the one hand elements such as photons and electrons and on the other bits of information" (Flusser, *Universum* 17). "[T]o turn point-elements into two-diomensional images. To emerge from the zero-dimensionality into the image-dimensionality. From the abyss of intervals to the surface, from the most *abstract* to the seemingly concrete; seemingly, because in actual fact it is impossible to contract points into surfaces. Because every surface consists of an infinite number of points, one would have to contract an infinite number of points in order to create actual surfaces. This is why the gesture of the imaginators [*Einbildner*] can create only apparent images. Surfaces that are in fact full of intervals, grid-like surfaces." (21)
39. "A photograph shows the chemist what kind of reactions specific photons have triggered in specific molecules of silver compounds. A television image shows the physicist what paths specific electrons have run within a tube" (Flusser, *Universum* 32).
40. "The only time of bodies and states of affairs is the present" (Deleuze, *Logic of Sense* 4), a "lived present" (chronos), while the time of events is the "experienced present" that includes the past and the future (aion): "the becoming which divides itself infinitely in past and future and always eludes the present" (5). The result of such a chrono*logics* is a simultaneous reading of time: "as the living present in bodies which act and are acted upon. Second, it must be grasped entirely as an entity infinitely divisible into past and future, and into the incorporeal effects which result from bodies, their actions and their passions" (5).
41. Cf. *The Fold*, in which Deleuze describes the topology of the world as a "torsion that constitutes the fold of the world and of the soul" (26).
42. As McLuhan states, "[t]he function of the body, as a group of sustaining and protective organs for the central nervous system, is to act as a buffer against sudden variations of stimulus in the physical and social environment" (*Understanding Media* 53).
43. "In a terminology taken from computer science one can also state that structural couplings *digitalize analog* states. As the environment and the other system in it operate simultaneously with the referential system of observation, there are initially only analog conditions (that run in parallel). From these, the involved systems cannot draw information, because this implies digitalization. Structural couplings, therefore, first need to transform analog into digital states, if the environment should influence the system through them" (Luhmann, *Gesellschaft* 101).
44. Cf. Hayward|Varela.
45. "[P]hilosophical concepts have events for consistency whereas scientific functions have states of affairs or mixtures for reference: through concepts, philosophy continually extracts a consistent event from states of affairs—a smile without the cat, as it were—whereas through functions, science continually actualizes the

event in a state of affairs, thing, or body that can be referred to" (Deleuze|Guattari, *Philosophy* 126).

46.   "Instead of searching for mechanisms in the environment that turn organisms into trivial machines, we have to find the mechanisms within the organisms that enable them to turn their environment into a trivial machine" (von Foerster, *Understanding* 152, also see 164).

47.   "[T]he structural coupling of two *independent* structurally plastic unities is a necessary consequence of their interactions, and is greater the more interactions take place. If one of the plastic systems is an organism and the other its medium, the result is ontogenic adaptation of the organism to its medium: the changes of state of the organism correspond to the changes of state of the medium. If the two plastic systems are organisms, the result of the ontogenic structural coupling is a consensual domain, that is, a domain of behaviour in which the structurally determined changes of state of the coupled organisms correspond to each other in interlocked sequences" (Maturana, "Organization" 326, emphasis added). See also: "We call social phenomena those phenomena that arise in the spontaneous constitution of third-order couplings, and social systems the third-order unities that are thus constituted" (Maturana|Varela, *Tree* 193).

48.   On Deleuze and Smithson see also O'Sullivan.

49.   "The interactions within a consensual realm [can] be described as communicative interactions" (Maturana, *Erkennen* 290).

50.   From the human perspective, this is a history "in reverse" in which molar perceptions are translated into molecular perceptions. "When they have reached the central nervous system, these data are processed, compared to data that are already stored and sent to other vampyroteuthes by way of intraspecific codes, to be stored in their memories. This brings about a developing dialogue between the vampyroteuthes, with the help of which the sum of available information steadily increases. This is the history vampyroteuthis." (Flusser 51)

51.   "Thinking is the process by which reason (nous) reaches behind phenomena (phainomena) in order to contemplate them [...]. Within this process, reason functions like a knife: it cuts appearances into definable blocks. [...] Ultimately, therefore, human thought is a manipulation by knife, and the stone knives of the Paleolithic Age, the oldest human instruments, prove, at what time we began to think." (Flusser, *Vampyroteuthis* 45)

   The result of this integrative procedure is "a back and forth between cut appearance and empty concept" (46).

Works cited

Badiou, Alain. *Deleuze: The Clamor of Being.* Trans. Louise Burchill. Minneapolis: University of Minnesota Press, 1999.

Bateson, Gregory. *Mind and Nature: A Necessary Unity.* Cresskill: Hampton Press, 2002.

——. *A Sacred Unity: Further Steps to an Ecology of Mind.* Rodney E. Donaldson, ed. New York: HarperCollins, 1991.

"Close up of Dune Sand in Wind Tunnel." http://www.weru.ksu.edu/new_weru/multimedia/movies/dust003.mpg. Last accessed September 9, 2008.

Conley, Verena Andermatt. *Ecopolitics: The Environment in Poststructuralist Thought.* London: Routledge, 1997.

Connolly, William E. *Neuropolitics: Thinking, Culture, Speed.* Minneapolis: University of Minnesota Press, 2002.

Cramer, Friedrich. *Symphonie des Lebendigen: Versuch einer allgemeinen Resonanztheorie.* Frankfurt a.M.: Insel, 1998.

Deleuze, Gilles. *Bergsonism.* Trans. Hugh Tomlinson and Barbara Habberjam. New York: Zone Books, 1988.

———. *Cinema 1: The Movement-Image.* Trans. Hugh Tomlinson and Barbara Habberjam. Minneapolis: University of Minnesota Press, 2003.

———. *Cinema 2: The Time-Image.* Trans. Hugh Tomlinson and Robert Galeta. Minneapolis: University of Minnesota Press, 2003.

———. *Difference and Repetition.* Trans. Paul Patton. New York: Columbia University Press, 1994.

———. *Empiricism and Subjectivity: An Essay on Hume's Theory of Human Nature.* Trans. Constantin Boundas. New York: Columbia University Press, 1991.

———. "Ethology: Spinoza and Us." Trans. Robert Hurley. *Incorporations.* Jonathan Crary and Sanford Kwinter, eds. New York: Zone Books, 1992, 625–632.

———. *The Fold: Leibniz and the Baroque.* Trans. Tom Conley. Minneapolis: University of Minnesota Press, 2001.

———. *The Logic of Sense.* Trans. Mark Lester with Charles Stivale. New York: Columbia University Press, 1990.

———. *Negotiations 1972–1990.* Trans. Martin Joughin. New York: Columbia University Press, 1995.

———. *Pure Immanence: Essays on a Life.* Trans. Anne Boyman. New York: Zone Books, 2001.

———. "Les Cours de Gilles Deleuze." http://www.webDeleuze.com/php/texte. php?cle= 191&groupe=Spinoza&langue=2. Last accessed April 1, 2008.

Deleuze, Gilles, and Félix Guattari. *Anti-Oedipus: Capitalism and Schizophrenia 1.* Trans. Robert Hurley, Mark Seem and Helen R. Lane. Minneapolis: University of Minnesota Press, 1983.

———. *A Thousand Plateaus: Capitalism and Schizophrenia 2.* Trans. Brian Massumi. Minneapolis: University of Minnesota Press, 2005.

———. *What Is Philosophy?* Trans. Hugh Tomlinson and Graham Burchell. New York: Columbia University Press, 1994.

Deleuze, Gilles, and Claire Parnet. *Dialogues II.* Trans. Hugh Tomlinson and Barbara Habberjam. New York: Columbia University Press, 2002.

Dillard, Annie. *Pilgrim at Tinker Creek.* New York: Harper & Row, 1985.

Emmeche, Claus. *The Garden in the Machine: The Emerging Science of Artificial Life.* Trans. Steven Sampson. Princeton, NJ: Princeton University Press, 1996.

Flusser, Vilém. *Ins Universum der technischen Bilder.* Göttingen: European Photography, 1989.

Flusser, Vilém, and Louis Bec. *Vampyroteuthis Infernalis: Eine Abhandlung samt Befund des Institut Scientifique de Recherche Paranaturaliste.* Göttingen: Immatrix Publications, 1987.

Foerster, Heinz von. *Understanding Understanding: Essays on Cybernetics and Cognition.* New York: Springer-Verlag, 2003.

———. *Wissen und Gewissen: Versuch einer Brücke.* Frankfurt a.M.: Suhrkamp, 1993.

Glaser, Eckehard. *Wissen verpflichtet: Eine Einführung in den Radikalen Konstruktivismus.* München: Herbert Utz Verlag, 1999.

Guattari, Félix. *Chaosmosis: An Ethico-Aesthetic Paradigm.* Trans. Paul Bains and Julian Prefanis. Bloomington: Indiana University Press, 1995.

———. "Space & Corporeality: Nomads, City, Drawings." Trans. Heghnar Zeitlian. *Semiotext(e)/Architecture.* Hratzan Zeitlian, ed. New York: Semiotext(e), 1992.

Guattari, Félix. *The Three Ecologies*. Trans. Ian Pindar and Paul Sutton. London: Athlone Press, 2000.

Hayward, Jeremy W. and Francisco J. Varela. *Gentle Bridges: Conversations with the Dalai Lama on the Sciences of Mind*. Boston: Shambhala, 1992.

Hilbert, David. "Grundzüge einer allgemeinen Theorie der linearen Integralgleichungen (Erste Mitteilung)." *Nachrichten von der Gesellschaft der Wissenschaften zu Göttingen, Mathematisch-Physikalische Klasse, 1.-6. Note* (1904): 49–91.

James, William. *The Principles of Psychology, Vol. 1*. New York: Dover Publications, 1980.

——. *The Works of William James, Vol. 3: Essays in Radical Empiricism*. Frederick H. Burkhardt, Fredson Bowers and Ignas K. Skrupskelis, eds. Cambridge: Harvard University Press, 1976.

Leibniz, Gottfried Wilhelm. "Monadology." *G. W. Leibniz: Philosophical Texts*. Roger S. Woolhouse and Richards Franks, eds. Oxford: Oxford University Press, 1998, 267–281.

Lorenz, Edward N. *The Essence of Chaos*. Seattle: University of Washington University Press, 1993.

Luhmann, Niklas. *Ecological Communication*. Trans. John Bednarz. Chicago: University of Chicago Press, 1989.

——. *Die Gesellschaft der Gesellschaft*. Frankfurt/Main: Suhrkamp, 1997.

——. *Social Systems*. Trans. John Bednarz, Jr. with Dirk Baecker. Stanford: Stanford University Press, 1995.

Lyotard, Jean-François. *The Inhuman: Reflections on Time*. Trans. Geoffrey Bennington and Rachel Bowlby. Stanford: Stanford University Press, 1992.

Massumi, Brian. *Parables for the Virtual: Movement, Affect, Sensation*. Durham, NC: Duke Press, 2002.

Maturana, Humberto R. *Erkennen: Die Organisation und Verkörperung von Wirklichkeit*. Trans. Wolfram K. Köck. Braunschweig: Vieweg, 1982.

——. "Cognition." http://www.enolagaia.com/M78bCog.html. Last accessed April 1, 2008.

——. "The Organization of the Living: A Theory of the Living Organization." *International Journal of Man-Machine Studies 7* (1975): 313–332.

——. *Was ist Erkennen?* Trans. Hans Günter Holl. München: Piper, 1994.

Maturana, Humberto, and Francisco J. Varela. *Autopoiesis and Cognition: The Realization of the Living*. London: D. Reidel Publishing Company, 1980.

——. *The Tree of Knowledge: The Biological Roots of Human Understanding*. Trans. Robert Paolucci. Boston: Shambhala, 1998.

McLuhan, Marshall. *Understanding Media. The Extensions of Man*. London: Abacus, 1974.

McMurry, Andrew. *Environmental Renaissance: Emerson, Thoreau and the System of Nature*. Athens: University of Georgia Press, 2003.

Miller, James Grier. *Living Systems*. Colorado: University Press of Colorado, 1995.

Naess, Arne. "Simple in Means, Rich in Ends." *Environmental Philosophy: From Animal Rights to Radical Ecology*. Michael E. Zimmerman et al., eds. Englewood Cliffs: Prentice Hall, 1993, 182–192.

——. "The Deep Ecological Movement: Some Philosophical Aspects." *Environmental Philosophy: From Animal Rights to Radical Ecology*. Michael E. Zimmerman et al., eds. Englewood Cliffs: Prentice Hall, 1993, 193–212.

——. "The Shallow and the Deep." http://www.alamut.com/subj/ideologies/pessimism/Naess_deepEcology.html. Last accessed April 1, 2008.

O'Sullivan, Simon. *Art Encounters Deleuze and Guattari: Thought Beyond Representation*. London: Palgrave, 2006.

Peirce, Charles Sanders. "Design and Chance." *Writings of Charles S. Peirce: A Chronological Edition, Vol. 4*, 1879–1884. Christian J.W. Kloesel, ed. Bloomington: Indiana University Press, 1986, 544–554.

——. "Issues of Pragmaticism." *Collected Papers of Charles Sanders Peirce, Vol. V: Pragmatism and Pragmaticism*. Charles Hartshorne and Paul Weiss, eds. Cambridge, MA: Belknap Press, 1965, 293–316.

——. "A Survey of Pragmaticism." *Collected Papers of Charles Sanders Peirce, Vol. V: Pragmatism and Pragmaticism*. Charles Hartshorne and Paul Weiss, eds. Cambridge, MA: Belknap Press, 1965, 317–345.

——. "What Pragmatism Is." *Collected Papers of Charles Sanders Peirce, Vol. V: Pragmatism and Pragmaticism*. Charles Hartshorne and Paul Weiss, eds. Cambridge, MA: Belknap Press, 1965, 272–292.

Pynchon, Thomas. *Against the Day*. New York: Penguin, 2006.

Reichle, Franz, dir. *Monte Grande: What is Life*. Switzerland: t&c film, 2004.

Ruyer, Raymond. *Neo-Finalisme*. Paris: Presses Universitaires de France, 1952.

Segal, Lynn. *The Dream of Reality: Heinz von Foerster's Constructivism*. New York: Springer-Verlag, 2001.

Serres, Michel. "The Origin of Language: Biology, Information Theory, & Thermodynamics." *Hermes: Literature, Science, Philosophy*. Josué V. Harari and David F. Bell, eds. Baltimore: Johns Hopkins University Press, 1982, 71–83.

Smithson, Robert. "Asphalt on Eroded Cliff (1969)." *Robert Smithson: The Collected Writings*. Jack Flam, ed. Berkley: University of California Press, 1996, 267.

——. "Asphalt Rundown, Rome (1969)." *Robert Smithson*. Eugenie Tsai and Cornelia Butler, eds. Berkley: University of California Press, 2005, 183.

——. "Concrete Pour, Chicago (1969)." *Robert Smithson*. Eugenie Tsai and Cornelia Butler, eds. Berkley: University of California Press, 2005, 224.

——. "Fragments of an Interview with Patsy Norvell (1969)." *Robert Smithson: The Collected Writings*. Jack Flam, ed. University of California Press, 1996.

——. "Glue Pour (1970)." *Robert Smithson: A Retrospective View*. Robert Hobbs. Ithaca, New York: Cornell University Press, no year, 104.

——. "Michael Heizer, Dennis Oppenheim, Robert Smithson: Interview with *Avalanche*." *Land and Environmental Art*. Jeffrey Kastner, ed. London: Phaidon Press, 1998, 202–205.

——. "1000 Tons of Asphalt (1969)." *Robert Smithson*. Eugenie Tsai and Cornelia Butler, eds. Berkley: University of California Press, 2005, 182.

Spencer-Brown, George. *Laws of Form*. London: Allen and Unwin, 1969.

Tansey, Mark. "Purity Test (1982)." *Mark Tansey: Visions and Revisions*. Arthur Danto, ed. New York: Abrams Incorporated, 1992, 69.

Thom, René. *Structural Stability and Morphogenesis: An Outline of a General Theory of Models*. Reading, MA: Addison-Wesley, 1989.

Thoreau, Henry David. *Walden; or, Life in the Woods*. New York: Dover Publications, 1995.

Varela, Francisco J. *Principles of Biological Autonomy*. New York: Noth Holland, 1979.

Varela, Francisco J. Evan Thompson and Eleanor Rosch. *The Embodied Mind: Cognitive Science and Human Experience*. Cambridge, Mass.: MIT Press, 1993.

Wilden, Anthony. *System and Structure: Essays in Communication and Exchange*. London: Tavistock, 1972.

Zimmerman, Michael E. *Contesting Earth's Future: Radical Ecology and Postmodernity*. Berkeley: University of California Press, 1994.

Žižek, Slavoj. *Organs Without Bodies: Deleuze and Consequences*. New York: Routledge, 2004.

5
Subjectivity and Art in Guattari's *The Three Ecologies*

Gary Genosko

Introduction: how many ecologies?

I want to begin with a deceptively simple question. Why did Guattari settle on a fixed number in his book *The Three Ecologies*? How many ecologies are there, anyway? Well, if we believe Guattari, there are at least three. I do not know precisely how he arrived at this number. In lieu of what might count as solid autobiographic, explanatory evidence, then, or even other kinds of justifications based on precise theoretical or practical considerations, let us consider this number as a way of critically appreciating the tasks to which Guattari set these ecologies. But the questions pile up: how do these ecologies cohere? What are their constraints and combinational possibilities? Clearly, to speak of three is not so simple. The number of ecologies answers to Guattari's delineation of the ethical, esthetic and political foundations of transdisciplinary knowledge. In fact, he brings together ecology and disciplinarity in order to pose the problem of what the passage from inter- to transdisciplinary knowledge looks like (Guattari, "Les fondements").

Ecology is, in its plural, heterogeneous forms, not a conjuration (supernatural) but a conjugation (as in an interchange of closely related elements) with important qualifications and implications. Ecology in the plural does not offer a magical solution to the question of how disparate knowledges and practices cohere. It is not some sort of general pedagogy, a training course for which one might sign up. Guattari clearly rejected such a notion in order to exclude any pretenders to such specialized educational expertise ("Les fondements" 15). Indeed, to invoke in the same introduction transdisciplinarity and ecology is not to "solve" anything. It may even appear tautological if by transdisciplinary one means more than a juxtaposition of many (multidisciplinary) and greater than a coordinated interconnection (interdisciplinary); thus, it is a somewhat rebellious and always critical kind of ecology of knowledge. When transdisciplinarity is used as a buzzword, an abracadabra word that merely anoints a project in the eyes of potential funders, as Guattari once put it, it changes nothing because nothing really changes at the level of process ("De la pluridisciplinarité" 6). Ecology served Guattari as an example of how to pose the question of transdisciplinarity on a large and stratified scale. In this he is not very original. However, within its plural form, ecology has for Guattari a macroscopic referent (Guattari, *La Philosophie* 47); THINK and ACT GLOBALLY (Conley 651). Yet the macro is not divorced from the micro. The prospects of transdisciplinarity are thought through in terms of three (and maybe more) ecologies, the articulations of which demonstrate the difficulties, potentialities and stakes of knowledge and action in the face of global challenges. Although transdisciplinary ecology goes beyond the multidisciplinary and interdisciplinary pretenders, it is not a higher-level synthesis or overriding solution.

The winding road to nontranscendent ecology

In terms of intellectual biography, Guattari shifted his exploration of transdisciplinary projects from groups and institutions to that of ecology later in his life, which is say, during his fifties. He spent most of his career experimenting with combinations of knowledge; indeed, from the time of his youth forward he lived a life of transdisciplinary implementation. Guattari created numerous experimental groups and publications, and participated in many other projects and institutions whose purpose was the exploration of organization through flexible participatory organograms, such as the evolving schedule of job rotation at Clinique de la Borde, where he worked as a psychoanalyst for almost 40 years (see Genosko, "Félix Guattari"). This system was an analytic instrument by means of which individual and collective affects could be articulated with institutional demands (material, social, bureaucratic, therapeutic tasks) towards the goals of enriching social relations, promoting the assumption of responsibility, participating in collective inventions (local jargon development), not only for patients, but for doctors and support staff, as well. In short, he envisaged the transformation of those involved in the extraordinarily complex negotiations and interactions (progressive and regressive) entailed within the Clinique on daily, weekly, and monthly bases, and in longer cycles. Like the institution thus reconceived, ecology is a similar kind of hypercomplex operator, a catalyst of change, a caretaker of concern, with great and grave stakes. But the most important stake was the development of a new kind of subjectivity.

Guattari came to ecology in the latter half of the 1980s as an antidote to what he dubbed the "winter years" of the first half of that decade, which saw the rise of many conservatisms and neoliberal economic policies, and the ascendancy of "vague vogues" like postmodernism, which he despised (Chesnaux and Gentis 544). The grayness of a political field perfused with sabre-rattling and the exponential expansion of what he called the unidimensionalizing forces of Integrated World Capitalism (what is now called

Empire) on the back of rapid technological change yoked to reactionary social archaisms almost led to his permanent relocation from France to Brazil. In this respect Guattari's encounter in 1982 with Lula da Silva, presently Brazil's president, but then radical leader of the recently legalized Workers' Party, allowed him to find some much needed energy in a revised, open unionism consonant with that of Solidarity in Poland (see Genosko, *The Party Without Bosses*).[1] After the mid-1980s Guattari struggled to gain some distance from the dirty tricks of the French state, such as the sinking of Greenpeace's ship *Rainbow Warrior* in Auckland harbor. He visited Japan during this period on numerous occasions and became fascinated by the strange and singular turns that capitalist subjectivity was taking with the unprecedented wealth of the bubble economy. The year 1989, original publication date for *Les trois écologies*, was not propitious for the environment: recall the massive oil spill in Prince William Sound off the Alaska coast caused when the Exxon tanker Valdez ran aground, creating long-standing ecosystemic damage. But by the same token, 1989 was a good year for thinking about the potential for collective social reinvention as the old Communism came crashing down with the Berlin Wall, inaugurating a whole new set of relationships with an emergent eastern Europe.

What sort of ecologies do we get when a practicing analyst and political radical schooled in far left social movements and transdisciplinary experimentation gets hold of them? What we get are ecologies that can be represented by *écosophes* such as Franz Kafka and Samuel Beckett (Guattari, *La Philosophie* 42).[2] Kafka or Beckett will not be found in the *Dictionary of the Environment*. I cannot do justice in this presentation to Guattari's far-reaching critique of psychotherapeutic methodologies; suffice to say that his vision of ecology's psychological dimension owed less to the standard-bearers Freud, Lacan and Klein than to "ecologists of the phantasm" like Marquis de Sade, or master of the refrain, Marcel Proust, to both of whom Guattari regularly turned when he wanted to tackle the most intractable problems of the psyche.

There are, then, arts and logics of the eco; but also praxes of large-scale change linking micro and macro; varieties of ecology that encompass the environment at the macro-level in terms of the extent of ecocatastrophes (Chernobyl, global warming); social relations at the intermediary level; and mental ecology at the micro or molecular level (Guattari, *La Philosophie* 47). The threesome is multi-leveled, but the hierarchy is misleading if it leads one to think that the macro holds the greatest value. It is, for Guattari, precisely the opposite. It is perhaps not very surprising for a psychoanalyst to gravitate towards the psychical realm or for a seasoned thinker of collectivities to situate the social in between. Yet even this is oversimplifying, since all the levels are intimately connected and solutions at one level entail changes at the others: earthly spheres, social tissues and worlds of ideas are not compartmentalized.

In an unpublished manuscript concerning the "Great Ecological Fear," Guattari wrote of an iceberg: the tip, above water and visible, represented environmental disasters and menaces; down below the water line was the bulk of the worry, that is, the degeneration of social relations, like the rise of organized crime organizations in the detritus of Stalinism and as parasites of hypercapitalist growth, and mental pollution caused by media infantilization, passivity-inducing postpolitical cynicism, to which may be added the traumas of globalization and (anti)terrorism. The iceberg represented a continuum of material encompassing the fabric of everyday life, large-scale crises, and habits of thought (Guattari, "La grand-peur écologique").

My task is to interrogate Guattari's three ecologies in terms of the leading questions I have posed with careful attention to their prospects for a transformative thought and action that have been and will continue to be transmissible by artists and through the arts in general as connectors of micro- and macro-dimensions. This is not meant to imply either that artists should subordinate themselves to ecological imperatives, say, the amelioration of esthetic degradation, or that artistic practices can be adequately judged with reference to a transcendent concept such as ecosystemic balance. Ecology is not art's prop; neither is art ecology's secret weapon. These restraints are helpful if only to underline that the interrogation of the three involves transits across, transformative powers, and nothing less portentous than, as Guattari claimed, "the production of human existence itself in new historical contexts" (*Three Ecologies* 34).

Three ecologies

For Guattari, there are three fundamental types of ecology: environmental, social, and mental (*Three Ecologies* 35). These types—biospherical, social relations, human subjectivity—are also figured as registers and "multipolar issues" (*Three Ecologies* 29) whose ethicopolitical articulation, as opposed to technocratic solution, is the proper concern of ecosophy (*Three Ecologies* 28). What makes this articulation superior to technocratic solution-mongering, for instance, the American model of emissions trading which displaces industrial pollution instead of reducing it, is that it will effect a revolution of the subject formations and social groupings charged with tackling ecological issues. In the absence of such profound change at the level of mentalities, that is, of real existential mutations, even the proffering of technocratic solutions lacks the resolve for their authentic deployment. The will, in short, is just not there. This is not a wholesale rejection of technology and international environmental bureaucracies, or environmental science, for that matter, which are often subsumed by and reduced to oft-repeated and poorly understood slogans—Rio Declaration on sustainable development, Kyoto Protocols on climate change, to cite two examples from the 1990s. To the extent that Guattari tried to get beyond tired left–right, east–west, socialist–capitalist, science–anti-science distinctions, international initiatives hold some promise in terms of the contribution they can make to the complexification of the contexts in which ecological issues are understood. During his lifetime Guattari was signatory to hundreds of good causes. Still, he sought to regain human values against an "unbelievable scientistic myopia" that sometimes infects international conferences. Guattari had in mind the Heidelberg Appeal first presented at the Rio Earth Summit in 1992, and subsequently signed by thousands of scientists. This Appeal constituted a discrediting in the name of technoscientific elites of all naysayers as "irrational" romantics (Guattari, "Remaking" 264). What fails in such positioning is that the connection between the material and immaterial is not made, that is, the circle that includes the mutual need for change in material, social, and environmental conditions and in mentalities. This is what Guattari foregrounds in his conception of a generalized ecology, an ecosophy (Guattari, "Remaking" 264). There is no easy trade-off in Guattari's work between a rejection of science and its replacement by art. On the contrary, he would reject any "unequivocal ideology" because it leads to profound impasses and implosions. Guattari wrote with a kind of desperation about the need for biomedical success in the fight against AIDS, for example. But such success would need to be channeled by ethical motivations "in less absurd, less dead-ended directions" than those dictated by interests based solely on profit, property, scarcity, and restricted distribution (Guattari, "Remaking" 268).

By the same token, Guattari's ecosophic perspective cannot be unified by the simple sloganeering of eco-revivalists, yet the three ecologies are "complementary headings" (*Three Ecologies* 41) and "points of view" that are, in effect, like "interchangeable lenses" (*Three Ecologies* 42). Levels, types, views, visions, lenses—Guattari shifts his descriptors throughout his book. His goal is to elucidate the "common principle" of the three ecologies in the conception of subjectivity. This is his most original contribution to the theorization of ecology. Guattari's concern with the quality of subjectivity is what holds together art and ecology.

The Guattarian subject is not an individual, an individuated person, thinking and thus being; no climax of philosophical striptease in originary intuition; no ego shipwrecked from real territories of existence, as he underlined (*Three Ecologies* 35). Rather, the Guattarian subject is an entangled assemblage of many components, a collective (heterogeneous, multiple) articulation of such components before and beyond the individual; the individual is like a transit station for changes, crossings, and switches (*Three Ecologies* 36). In the development of Guattari's conceptual language, assemblage came to replace group. This is not to deny the existence of core elements; on the contrary, there are nuclei or especially dense crossing points where interiority is found and from which energy can be extracted for further differentiation, complexification and enrichment. Such nuclei replaced for Guattari the prevailing psychoanalytic languages of complex, system, and structure, making subjectivity irreducible to a universal syntax, mathemes, imagos, mythemes, etc. This subject is also polyphonic—of many relatively independent parts—because it assembles components in order to posit itself in terms of some points of reference (body, social clusters, etc.), in an existential territory, a field in which it is incarnated, but out of which it also ventures. For this productive self-positing is relational, subsuming both autonomous affects of the prepersonal and preverbal world and multitudinous social constructions. Emergent and processual, producing and produced by mutual self-engenderings, the subject emerges as it finds a certain existential consistency, without getting tied down to an identity once and for all, in the crossing points of components, in their intra- and interassemblage relations, sometimes deflating into involutions, blockages, and encystments; at other times taking off through transformations (potential consistencies). Open and full of potential, this subject is truly a work in progress|process outflanking both essentialist and constructionist postulates. Radically creative and at times aggravating in its abstractness, "subjectivity still gets a bad press," Guattari admitted (*Three Ecologies* 36). This is hardly determinative. The stakes are high and this abstruseness is a cost, I think, of escaping from takeovers and annexations by fixed and single and exterior coordinates—psychoanalytic, structural, or postmodern "plinths" upon which the subject may be mounted like a botched taxidermic specimen (see Baker 63).

For Guattari, the three ecologies point the way toward emancipatory praxes whose "major objective [is] to target the modes of production of subjectivity, that is, knowledge, culture, sensibility and sociability... " (*Three Ecologies* 49). Ecology's business is to attend to the regimes by means of which subjectivity is produced and to intervene in them. Ecology is readied for this task by Guattari insofar as he shifts into the delineation of the dynamics of eco-logic—how the three ecologies communicate (the terms are affective intensities rather than delimited sets like stages, complexes, linear phaseal developments, or universal structural coordinates). Although Guattari abandoned typical psychoanalytic psychogenetic stages for the sake of a heterogenetic becoming (giving to singularization a constancy), he still needed to retain some sense of a self's prospective unfolding without slavishly adhering to a developmental model punctuated by decisive events and sticking points. Only an emergent self would suffice; and the phases of such an emergent organization, while at work over time from childhood through adulthood, would also be available in parallel at different degrees and in a variety of combinations over a lifetime.

One of the ways in which Guattari translated this insight into practical criticism may be seen in his observations on the American "artificial realist" painter George Condo. Guattari observed that critics of Condo's work experienced an acute disorientation before his paintings of figures (many with comic, contorted heads) in landscapes. The cataloguing of countless modern masters as seminal influences and the proliferation of reference points which a single Condo painting seems to visit, often subsuming several periods of a given painter's works, led Guattari to suggest the following to Condo about the polyphonic character of his work: "all your periods coexisted—blue, clown, linear, volumical, monochrome, etc. It is like a symphony articulating all the levels of your 'self', simultaneously exploring and inventing it through your painting" (Guattari, "Introduction" 5).[3] The details of this example are perhaps less pertinent than the translation of the coexistence of stratifications of subjectivity (with varying degrees of formedness, capacity to be shared, what one might call degrees of fixity) onto periods of painting each with its distinctive thematics. This example of what might be called transversal criticism—a tool for the enactment of an adventurous connectivity that skirts around the abyss of a list of influences and precursors that only point towards the past and freezes Condo's work in a crowded representational space—brings the paintings flush with the engendering of a subjective territory.

Eco-logic concerns new incarnations of subjectivity in partially formed existential territories not yet yoked to normalized extrinsic pillars, whether these are certain family members, respectable academic grammars, religious fixations or esthetic styles. This lack of fixity is a fecund amodality (an abstract, intense feeling of vitality not object-oriented or attached to causes) that is ripe for the eco-logic. This is where the logic turns to praxis. The eco-praxes "scout out" (*Three Ecologies* 45) somewhat opportunistically "catalysts of existential change" that lack solid support in the assemblage yet are full of passive potential for swerving from normality, running counterclockwise, as it were, but not running completely amok, either. Of course, this is also where things can also go horribly wrong. Instead of summoning forth and assisting new traits of subjective particularity, incomparable singularities breed banal imitations, or we get another aneconomic myth of a return to Nature or similarly counterproductive manifestation (Sea Shepherd Society ships threatening Japanese whaling fleets or Paul McCartney lending his face to the global media slaughter of Newfoundland sealers). Eco-praxes are on the watch for dissident vectors, ruptures and mutations of subjectification in all walks of life and thus in all the ecologies and in any existential territory. But these have to be delicately *turned* toward productive and active ends and provided with scaffoldings and guy-ropes so they do not just twist in the wind.

Eco-logic is by definition activist, but not in a narrow sense—vigorous, yes; dedicated, certainly; but not motivated by single-issue ecopolitics, or animated by the generation of a paper trail of non-binding agreements. It is colored, perhaps a better term would be marbled, by a therapeutic ethos. Guattari does not simply exclude the political goals of "single-issue" ecological movements or "archaic attachments" to Walden Pond or stereotype of the "eco-Indian," but instead occupies different ground: "Ecology," he wrote, "must stop being associated with the image of a small nature-loving minority or with qualified specialists" (*Three Ecologies* 52).

Indeed, Guattari himself talked the talk and walked the walk since he entered into the fray of French Green politics by taking out memberships in both of the groups into which the movement had split, with the goal not of reunifying the politicos whose social ecology was the economy and the militants who went out and defended the earth, but of inhabiting a social ecology torn asunder by leadership quarrels, in order to explore the potential for collectively discovering, with the others who followed his lead, not another tired axis (left, right, or neither), but a new way of working together. And it is for this effort that his Green colleagues remembered him.

Breakaway components of subjectification must be handled with care and sobriety; even so, the gentle loosening and tutelage of such catalytic components (or segments thereof) inevitably leads for Guattari, using a chemical metaphor for precipitating change, to certain kinds of redundancies (what he calls existential refrains) in a given assemblage upon which subjectivity becomes focused or which fixes subjectivity in a way that interrupts the diversity of the components at play. Despite everything else going on, one is glued to the television set whose screen then becomes a circumscribed existential territory (Guattari, "Subjectivities" 199–200). Examples of complex refrain motifs are found by Guattari across the arts—in fiction, theatre, visual art, with a preponderance of musical examples, even in literature (the model being *la petite phrase* that captivated Swann in the salon of Madame Verdurin) (Guattari, "Ritournelles" 239ff). Refrains (recurrent beatings of time understood in relation to a milieu and its components) are established when motifs are detached from the flux of components, when an established texture is interrupted and a motif curls up without spinning around hopelessly, acquiring the ability to generate a positive process of self-reference. This is not such a rarified phenomenon, for, after all, it happens with those few notes from pop songs that occasionally come back to one as motifs detached from personal turning points or lingering, somewhat autonomous, affective qualities; but, instead of inviting us into rich universes of personal reference, reactivated in the present for the future, they find themselves hijacked and affixed to automobile tires or boxes of breakfast cereal. Subjectivity shifts but merely onto commodities (capitalistic refrains of advertising), thus restricting its potential for enhancing and enriching itself through the exploration of its own universes of value.

Eco-praxes try to nurture the ruptures and flights of catalysts of change and their productive evolutions, keeping them from turning in circles, getting "shut inside" the Nintendo universe, trapped in the compounds of reality TV programs or in the "doped voids" of classic rock revivals. Enthusiastically, but vaguely, some performance art, Guattari thought, "shoves our noses up against the genesis of beings and forms, before they get a foothold in the dominant redundancies—of styles, schools, and traditions of modernity" (*Chaosmosis* 90). Guattari draws on numerous examples from the arts because for him these are positive paradigms, though at times vertigo-inducing, of how the full implications of subjectivity's mutational forward flight from consumerism and other "steamrollers" and "contractions" can be explored by creators themselves, observers, critics, and non-experts. He even reserves art's traditional role of providing refuge for dissident vectors of expression (*Three Ecologies* 46). Still, "[t]his is not about making artists the new heroes of the revolution," Guattari insisted (*Chaosmosis* 91). It is about the esthetic dimension of eco-praxis. Ecosophic activism "resembles" the work of artists in extracting details that serve as path-breakers for subjective development and as guidance in responsibly negotiating refrains (*Three Ecologies* 52).

Circa 1989, Guattari was moved by the work of New Yorker David Wojnarowicz, especially in terms of the queer activist esthetics that this painter and writer knit into his provocative critiques of healthcare (especially redesigned money) in the United States that "trigger an existential movement, if not of revolt, at least of existential creativity" (Guattari, "David Wojnarowicz" 76–77). This entails that ecosophic artists are engaged in a form of anti-Empire critique since work on the multitude of dissident, singularizing vectors of subject-formation goes hand-in-hand with this, with the Guattarian proviso that there is no falling back on old state socialism or the welfare state, nor pre-fab dialectical solutions (*Three Ecologies* 52–53). Thus, Guattari tended to privilege so-called political art (Wojnarowicz's production of an accelerated reaction of viewers with regard to the politics of the management of AIDS as a global phenomenon), but not absolutely, since there are no guarantees that catalytic components will be engaged, and ethico-esthetic commitments will be generated. There are many possible strategies in this area. For example, Canadian photographer Ian Wallace treats the assembly of anti-logging protesters at Clayoquot Sound, British Columbia in 1993 through a style of disrupted documentation that attempts to provide some partial existential support for collective action by underplaying the highlights of protest and arrest and refusing to indulge in star-focused rally reportage (in one instance seated protesters are looking in every direction but towards a speaker with a microphone atop a van).

The artistic examples that Guattari mentioned over the course of his life-work are too numerous and diverse to constitute a definitive esthetic, except that this alone tells us of the importance of heterogeneity, yet his emphasis was on forging new value systems (not simply renewing existing traditions of militancy), short of offering a fully elaborated alternative. On this point Guattari was quite candid (*Three Ecologies* 66). Yet the ethico-esthetic contact point between art and its audiences involved for Guattari inducements precipitating the assumption of responsibility in an existential transference of singularities leading to the assembling of new constellations of components with their own intrinsic and extrinsic relations. The ability of buildings like Shin Takamatsu's signature works of the 1980s and early 1990s in Kyoto and Osaka (such as ARK, Pharaoh, Kirin Plaza, Syntax and Imanishi) to effect a profound transformation of many of the facets of their urban environments was for Guattari a key example of existential transferences brought about by contextual mutations that trigger in each person gripped by the vision of the architectural project their own taking of initiative within the new existential territories and worlds of reference thus opening up before them (Guattari, "Les machines"). For Guattari, stellar architectural projects provided evidence of changes in habits, routines, spatiotemporal coordinates, opening cracks in interpretive grids, and rearranging situatedness for those who occupy and work in them, pass through, or simply observe. Guattari was thinking less about influence than inspiration, a kind of contact with works that gets collective subjects moving towards the acquisition of the means for their own production.

Refrains can be quite precarious. They may implode psychically in deathly repetition. But Guattari's myriad artistic examples possess an open precariousness, the capacity to sustain "praxic openings-out" from an existential territory that do not remain trapped by exploitative coordinates or wrapped up in postpolitical alienation (*Three Ecologies* 53) Not only are these openings enunciated but they find a consistency that makes them habitable by politically, ethically, esthetically, and psychoanalytically engaged projects; this is certainly not the psychoanalysis that Slavoj Zizek describes as miserably giving one the freedom to enjoy as much as one wants of what one does not want![4] Being carried beyond familiar territories into alterities of all sorts permits the emergence of new valorizations, new social practices, new subjectivities. Artists can provide the means for these creative forward flights, these breakaways. For Guattari, art begins with the expressive features of a territory that become for its inhabitants flight paths beyond its borders. Art begins not with a home but with a house, not with inner-directness, but with outer-directedness; when in 1993 Rachel Whiteread set up her casting operation on Grove Road she turned what was once a home inside out into a house by filling in all the frames and planes and sections so that the work could not shelter anything, but simply point outward, the functional having become expressive and mobile (Deleuze and Guattari, *What is Philosophy?* 183–186).

Tri-ecological vision

The paths of this vision may be "tangled" (Guattari, *Three Ecologies* 67), but Guattari's call for the "permanent recreation of the world" begins by attending to a melody of nature and art that suggested "renam[ing] environmental ecology *machinic ecology*" (*Three Ecologies* 66). The unity of biosphere and mechanosphere means that biological life, including human beings, is involved in the vast techno-informatic infrastructures in the era of planetary computerization and the IT revolution. Subjectivity is thus dependent on machinic phyla (telecommunications; synthetics; new temporalities brought about by increasing processing capacities; and biogenetic engineering of life forms) and engenders itself with machinic components from ipods to iris scans (Guattari, "Regimes" 103). This did not displace the biospherical challenge of large-scale problems like ozone depletion, for instance; on the contrary, the machinic dimension of the depletion of stratospheric ozone by catalytic chain reactions initiated by imbalances introduced by CFCs and Halons is well recognized. It also meant for Guattari that eco-praxes at the environmental level on "natural equilibria" would involve more and more sophisticated interventions and transversal criticisms—like the quaintly named "hamburger connection" that linked rainforest habitat destruction for pasture with fast food but which is now more ominous as beef has become a vector for global food insecurity; or the algae bloom in the Venice lagoon which Guattari linked to the proliferation of exploitative New York real estate redevelopers (Donald Trump algae) who generate unknown levels of homelessness and despair in poisoned social urban ecologies (*Three Ecologies* 43). The reconstruction of group belonging and institutional life, driven by processes of subjectification that find in realms both intimate, even fantasmatic and more distant, perhaps objective ways of reevaluating the censoring and concealing shrouds (*Three Ecologies* 68) engulfing them, begins at the most "miniscule level" (*Three Ecologies* 69) but opens toward the global. The omnidirectional openness of subjectivity needs, Guattari also warned, to find real existential anchors and outlets that allow it to simultaneously install itself in all three ecologies, lest it fly away from lack of consistency and perspective (*Three Ecologies* 69).

Transdisciplinary ecology

Asking after the number of ecologies is not really a quantitative question at all. Readers of *The Three Ecologies* are in much the same position as Gilles Deleuze before the triptychs of Francis Bacon. Deleuze discovered the non-linear distribution of forces whose laws of rhythm make visible invisible musical temporalities across the three panels, and this led him to boldly conclude that there are only triptychs in Bacon (*Francis Bacon* 70). Guattari's three ecologies are themselves evidence of his refusal of transcendent judgment,[5] synthesized or subsumed; the three ecologies maintain the paradigm of creativity, soberly serving crossings and connections across disparate domains, running from the intimate, everyday to the planetary in scale. In theorizing this complex three, Guattari showed how to grasp the generality of ecological vision; perhaps this is firmly situated in the French tradition of transdisciplinary studies grounded in the human science of communications, but subject to occasional revisions as a "science of the event" (sociology of the present) and then various rapprochements between the life and social sciences (see Mattlelart 20–27). The three ecologies are an assemblage that shows how disparate domains constantly engage one another. There is a transference here between art's and ecology's hope in the creation of new universes of value. Looking through the lenses of the three ecologies one sees, better, one senses another world, not yet real, but unfolding itself toward actualization, and for which one is always responsible, as existential grounds are sought for the incorporeal universes brought into being.

Transdisciplinary ecology is inspired by the interdependent hypercomplexity of its object or problem—that of subjectivity. Unlike so many contemporary definitions of the term, this transdisciplinarity does not seek a transcendent, extrinsic ground or plane for the subject from which its parameters and obligations issue forth.[6] Its knowledge is engaged, ethico-esthetic and political, for it seeks nothing less than creating conditions conducive to subjectivity's self-transformation. Guattarian transdisciplinarity does not seek to transcend, it seeks to transform. It does not recoil from chaos. It does not solve the problem of chaos by positing a fixed, univocal ground, thereby vouchsaving the differentiated "subject." It is not forever on the path back from chaos.[7] It does not retrieve a unity from the "massive and immediate ensemble of contextual diversity" (*Chaosmosis* 80), thus turning around the alleged degradations of flux, but in the very act of assisting subjectivity in finding a position, a "node" around which a territory can be built and universes of meaning find affirmation, chaos in all its discomfort is respected, its textures analyzed, and its tributaries explored. *Emergence* in this sense is not toward a higher level of abstract integration, but is something that must be continuously confronted and permanently reappraised. Art can provide a model for subjectivity's eco-conscious heterogeneous explorations, without betraying their singular textures and crushing their freedoms. Of course, artists have to grapple with their own social ecologies of the schools, gallery system, art market, fickle arts councils, and fashions of criticism in which they place themselves, and whose dictates do not make a turn toward singular Guattarian theories all that obvious!

More and more Guattari tended to refer simply to "aesthetic perception" while eschewing stylistic categorizations. Through this perception one may be oriented towards ecosophic activism of the highest order by exploring refrains and extracting innovative segments of components in collective processes of subjectification, the existential impacts of which are never decided in advance. Near the end of his final book *Chaosmosis*, Guattari stated: "Perhaps artists today constitute the final lines along which primordial existential questions are folded?" (133). This sober "perhaps" simply tells us that while artists do contribute important components to the tri-ecological vision, there are no guarantees: it is no easy task to throw off certain self-satisfactions of creativity, deal with anti-intellectualism, and economic marginality, expand one's world so as to take responsibility for matters that were once conveniently outside one's purview, and adapt one's means of working in accordance with the demand to subtract that is required in contributing to collective processes of subjectification—an inspiration and object that is, frankly, as much a matter of the self-transformation of artists themselves through the artistry of living and aging as it concerns the renewals and deviations initiated by works of art.

Notes

1. See "Summit of Radicals" in my *The Party without Bosses: Lessons on Anti-Capitalism from Félix Guattari and Luís Inácio 'Lula' da Silva*. Semaphore Series. Winnipeg: Arbeiter Ring, 2003.
2. See also Guattari, "L'intervention institutionnelle," Fonds Félix Guattari (IMEC), typescript of an interview. ET 09-26 (1980), 146.
3. See also Guattari, *Chaosmosis*. Trans. P. Bains and J. Pefanis. Bloomington: Indiana University Press, 1995, 6–7, and Genosko, *Félix Guattari: An Aberrant Introduction*. London: Continuum, 2002, 49ff.
4. Hear his remarks in Ben Wright's film, *Slavoj Zizek: The Reality of the Virtual* (2004).
5. "I refuse transcendent judgements." Guattari, "Entretien," by E. Videcoq and J. -Y. Sparel, *Chimères* 28 (Printemps-été 1996): 22.
6. Several participants in the colloquium on transdisciplinarity at L'Abbaye de Royaumont call for a "transcendent language" or "transcendent explanatory power." In *Transdisciplinarity: reCreating Intergrated Knowledge*. Eds M. A. Somerville and D. J. Rapport. Oxford: EOLSS, 2000.
7. Famously, E. O. Wilson wrote about the "path back from chaos" and this is picked up as a definition of transdisciplinarity by one of the editors (Rapport) of the volume *Transdisciplinarity* previously cited, p. 135.

Works cited

Baker, S. *The Postmodern Animal*. London: Reaktion Books, 2000.

Chesnaux, J. and Gentis, R. "Félix, Our Friend." Trans. M. McMahon. *Deleuze and Guattari: Critical Assessments of Leading Philosophers*. Ed. G. Genosko. Volume 2. London: Routledge, 2001, 542–545.

Conley, V. "New Ecological Territories." *Deleuze and Guattari: Critical Assessments of Leading Philosophers*. Ed. G. Genosko. Volume 2. London: Routledge, 2001, 645–664.

Deleuze, Gilles. *Francis Bacon: The Logic of Sensation*. Trans. D. W. Smith. Minneapolis: University of Minnesota Press, 2003.

Deleuze, G. and Guattari, F. *What is Philosophy?* Trans. H. Tomlinson and G. Burchell. New York: Columbia University Press, 1994.

Genosko, G. "Félix Guattari: Towards a transdisciplinary metamethodology." *Angelaki: Journal of the Theoretical Humanities* 8/1 (April 2003): 129–140.

——. *The Party without Bosses: Lessons on Anti-Capitalism from Félix Guattari and Luís Inácio 'Lula' da Silva*. Semaphore Series. Winnipeg: Arbeiter Ring, 2003.

——. *Félix Guattari: An Aberrant Introduction*. London: Continuum, 2002.

Guattari, Félix. *La Philosophie est essentielle à l'existence humaine* [Philosophy is Essential to Human Existence], entretien avec Antoine Spire, Michel Field and Emmanuel Hirsch. Paris: Editions Aube, 2002.

——. *The Three Ecologies.* Trans. Ian Pindar and Paul Sutton. London: The Athlone Press, 2000.

——. "La grille." *Chimères* 34 (1998): 7–20.

——. "Remaking Social Practices." *The Guattari Reader.* Ed. G. Genosko. Oxford: Blackwell, 1996, 262–272.

——. "Entretien." Interviewed by E. Videcoq and J. -Y. Sparel. *Chimères* 28 (Spring–Summer 1996): 19–32.

——. *Chaosmosis.* Trans. P. Bains and J. Pefanis. Bloomington: Indiana University Press, 1995.

——. "Les machines architecturales de Shin Takamatsu." *Chimères* 21 (Winter 1994): 127–141.

——. "De la pluridisciplinarité à la transdisciplinarité." Written with Sergio Vilar, Barcelona–Paris, September 1992. Fond Felix Guattari (IMEC): ET 05-13.

——. "Les fondements éthico-politique de l'interdisciplinarité. Handwritten text, April 1991. Fonds Félix Guattari (IMEC): ET 10-24.

——. "David Wojnarowicz." *Rethinking Marxism* 3/1 (1990): 76–77.

——. "Ritournelles et Affects existentiels." *Cartographies schizoanalytiques* [Schizoanalytic Cartographies]. Paris: Galilée, 1989, 251–267.

——. *L'inconscient machinique.* Fontenay-sous-Bois: Recherches, 1979.

——. "Introduction." *George Condo.* Paris: Daniel Templon, 1990, 5–8.

——. *Les trois écologies.* Paris: Galilée, 1989.

——. "L'intervention institutionnelle." Typescript of an interview (1980). Fonds Félix Guattari (IMEC): ET 09-26.

——. "La grand-peur écologique. Handwritten ms (n.d.). Fonds Félix Guattari (IMEC): ET 10-03.

——. "Regimes, Pathways, Subjects." *The Guattari Reader,* 95–106.

——. "Subjectivities: For Better and for Worse." *The Guattari Reader,* 199–200.

Mattlelart, A. and M. *Rethinking Media Theory.* Trans. J. A. Choen and M. Urquidi. Minneapolis: University of Minnesota Press, 1992.

Somerville, M. A. and Rapport, D. J. (eds) *Transdisciplinarity: reCreating Integrated Knowledge.* Oxford: EOLSS, 2000.

Wright, B. *Slavoj Zizek: The Reality of the Virtual,* 2004.

6
Artists or "Little Soldiers?" Félix Guattari's Ecological Paradigms

Verena Andermatt Conley

In their copious writings, Gilles Deleuze and Félix Guattari often have recourse to terms, such as world, earth, geology and *ecumenon*, that may strike the reader as having environmental implications. It is Félix Guattari, however, who writes openly about ecology in what may well be his most widely read essay, *The Three Ecologies* (1989), which has become his testament of sorts, as well as in his latest and even more challenging *Chaosmosis* (1992). In both texts, Guattari addresses the environment somewhat surprisingly by way of ecology. In 1992, as if to prolong his ecological pronouncements with concrete politics, Guattari also ran (unsuccessfully) for office as candidate positioned between France's two green parties.

In the pages to follow I propose to visit the short, concise essay almost 20 years after its publication and juxtapose it to some passages from his final text, *Chaosmosis*, in order to see whether, and how, an alliance with the inventor of schizoanalysis is indeed possible. I will first show that Guattari is indebted to the anthropologist, Gregory Bateson, whose ideas of ecology he adapts to his own ends when distinguishing between mental, social and environmental ecologies before reviewing some of Guattari's main ecological arguments. The latter concern singular and collective processes of subjectivation, esthetic and ethical paradigms, emphasis on enunciation and the place of militantism. Lastly, I will see whether and how we can use Guattari's environmental paradigms or, in the psychoanalyst's words, what kind of a dynamism we can lift from them, to help us with environmental dilemmas confronting us today.

Guattari is indebted to Gregory Bateson, whose work both he and Deleuze quote repeatedly. The very notion of "plateau" as well as that of the "double bind" introduced in *A Thousand Plateaus* from the title to the chapter on geology (40) are derived from the British anthropologist. In fact, Guattari's essay begins with an epigraph from Bateson to the effect that "there is an ecology of bad ideas, just as there is an ecology of weeds" (*Three Ecologies* 27).

In his revolutionary *Steps to an Ecology of Mind*, Bateson distinguishes between physics and information theory. With recourse to the latter, he makes several important points. He draws attention to the mind by showing how information is transmitted as differences that move along circuits and pathways in the brain in discrete and discontinuous fashion (see Conley 1997). In addition, transmission of information takes place between individuals but goes also, at the same time, through the environment. An individual does not "end" at his or her skin but is part of a general circulation of information that links him or her to, and with, the environment. As the anthropologist puts it: organism + organism + environment (*Steps* 491). Bateson argues too that, with the environment, it is often not the impact itself but the response—or feedback—that can be delayed or take place far away that will be of importance. The proverb to which Bateson gave birth is that a butterfly, flapping its wings somewhere on the globe, can produce a hurricane several thousand miles away. Over time, a system loses what Bateson calls its flexibility and becomes overcoded. Like blood in clogged arteries, circulation or information in society slows down and becomes fixed around centers of power.

Guattari adapts several of Bateson's ideas. His choice of the term "ecology" in the title of the essay rather than that of "environment" is significant. Ecology studies a generalized interaction of living organisms and their habitat while the term environment usually distinguishes between humans and their surroundings. For Guattari, ecology, the law of the *oikos*, studies how humans and other forms of life exchange on, and with, the earth. As he puts it: "The root 'eco' is used here in its original Greek sense of *oikos*, that is, 'house, domestic property, habitat, natural milieu'" (*Three Ecologies* 91n52). At the heart of Guattari's writings is a critique of present forms of exchange and the importance of resisting them by inventing new ones. These new ways of domesticating will inform what he calls an "ecosophy," that is, a wisdom of dealing with one's *oikos* that includes the environment. This ecosophy is a form of schizoanalysis (*Chaosmosis* 127). Guattari focuses especially on Bateson's ecology of mind and the circulation of ideas that he renames "mental ecology." He is most interested in the reconstruction of the subject—or what he calls processes of subjectification—and of the social. Human interaction with the environment is of importance, yet for him the latter is also populated with machines and technologies. Nature and culture can be less separated than ever (*Three Ecologies* 43) now that humans live in second nature. The least developed point in Guattari's writing is that of the natural environment itself.

Guattari published the French version of his essay in 1989 at the time of the disintegration of the Soviet Union and the apparent triumph of the "West," especially of the United States and of the market economy. He raised two major points: first, the development of the techno-sciences has not brought about a concurrent development of humans and of social relations (*The Three*

Ecologies 27) second, with the disappearance of what he calls the partially phantasmatic East–West antagonism, a space can be opened from where to address the imbalances between North and South (30). These changes, he claims in a rather universal statement, are rendered more difficult by the sorry state of people who have been infantilized by the media to the point where they no longer think. The replacement of Marxist emphasis on social experimentations by that of capitalism's sole valorization of profit and self-interest is creating a world from where all forms of metaphoric and real life are rapidly disappearing. Zombie-like subjects, degraded social relations, global misery, ongoing violence and environmental catastrophes are the sad truth of this kind of economy. Bad ideas have taken hold. Global catastrophes such as AIDS and Chernobyl—an explosion at a Ukrainian nuclear plant in 1986 that affected the surrounding areas but also much of Western Europe and even the Eastern part of the United States—show the limits of techno-sciences.

Guattari makes it clear that in order to bring about change new paradigms are needed. He demarcates himself from "folkloristic" environmentalists obsessed with bygone ways of living. With reference to Paul Virilio, Guattari (*The Three Ecologies* 42) shows that we have to analyze the world in, and from, today's conditions, that is, with a sizable increase in demographics and huge urban agglomerations as well as with technologies. What holds for human geography is true for environmental questions as well. To illustrate his point, Guattari gives the example of a presenter on television who takes an octopus from polluted water and puts it in a tank with clean water where it dies immediately (42). Guattari insists that we need to think "transversally" about interactions between ecosystems, the mechanosphere and individual and social universes of reference. Just as mutating algae invade the lagoon of Venice, so also degenerate images invade television screens. Another kind of algae, he adds, proliferate in the domain of social ecology where people like Donald Trump proliferate freely and unchecked. They evict thousands of people who are like the dead fish of environmentally polluted waters. The latter analogy gives pause today in view of Trump's recent successful show on NBC TV, *The Apprentice*, where, during a contest, the tycoon-turned-show-host fired participants who did not execute their task with the prerequisite competence and raw sense of competition. The show is seen to teach rules of business people can apply to their own work lives—and succeed. It would appear that the task of changing sensibilities, in Guattari's parlance, has become even more onerous.

The real culprit, writes Guattari, is global capitalism or what he prefers to call "integrated capitalism," which equates goods, nature and people and puts everything under the sign of profit. The only way out is through changes in and even mutations of value. As in his previous writings, Guattari urges his readers to think "transversally," that is, across economic, scientific, subjective and other regimes. One cannot think the environment separate from the way humans think and act in other areas. We cannot rely solely on technocratic adjustments. To produce change, we have to begin, Guattari argues, by reconstructing singular and collective processes of subjectification. Such a reconstruction will help the individual and the social. It will enable new ways of thinking away from the values of global capitalism and will open possibilities for focusing on the environment. In order to do this, we cannot be concerned exclusively with visible relations of force on a grand scale, but will have to take into account "molecular domains of sensibility, intelligence and desire" (28).

Guattari invites humans to resist not so much technologies or new media as the official "media" and their marketing devices that, under the sway of a powerful elite, exploit people by turning them into consumers and by preventing them from thinking. Guattari reaffirms the necessity for a creative resistance, be it singular or collective, and stresses the importance of an aesthetic and ethical politics. To create is to resist. It is to open a space as a condition for other ways of thinking. (The latter can be good or bad!) Guattari's esthetic paradigm emphasizes singularity and minimizes power takeovers. People have to detach themselves continuously from the present order so as to differ from difference and to become revolutionary. Revolutions, including the Marxist revolution, quickly became stale, or molarized in his words, and marred by power takeovers. Guattari wants to prevent stratification through the ongoing, creative opening of molecular spaces.

To prevent rigidification such as the one he sees underlying scientific objectivity, Guattari joins those who privilege the artist for being part of an avant-garde or, in Freud's words, ahead of common mortals. Guattari credits artists with a power of invention that escapes those who are closer to science and whose focus on objectivity condemns them to repetition, loss of flexibility and stasis. "Besides," he writes,

> are not the best cartographies of the psyche, or, if you like, the best psychoanalyses, those of Goethe, Proust, Joyce, Artaud and Beckett, rather than Freud, Jung and Lacan? In fact, it is the *literary* component in the works of the latter that best survives (for example, Freud's *The Interpretation of Dreams*, can perhaps be regarded as an extraordinary modern novel!). (*The Three Ecologies* 37, emphasis added)

Schizoanalysis, elaborated in *A Thousand Plateaus* as an alternative to the official psychoanalysis and mobilized in *The Three Ecologies* under the guise of an *ecosophy*, will lead to transformations in mental ecology that will have repercussions on the ways humans interact with the environment.

Though he is well versed in philosophy, Guattari was trained as a psychoanalyst and specialized in alternative forms of therapy at the La Borde Clinique. The main part of his essay aims to transform singular and collective mental and social ecologies. As he claims with Gilles Deleuze in *Anti-Oedipus*, it is the *institutionalization* of psychoanalysis that imprisons and dooms the patient. The introduction of the Oedipal complex in the unconscious kills life by insisting on an obligatory passage for all humans and the generalization of a model predicated on repetition and the death drive. As an analyst, he wants to turn the discipline toward life and creativity. Just as it was not psychoanalysis but the way it had been institutionalized, so today it is not technologies but the way they are adapted or taken over by the media and marketing that produces zombie-like subjects. Unlike many post-68 French theorists, Guattari does not use a Heideggerian blueprint. He advocates the construction of new subjectivities with technologies. With Deleuze, he writes of a "becoming-radio, becoming-electronic, becoming-molecular" (*A Thousand Plateaus* 473) and of computer-assisted subjectivities. Alliances between people and technologies are part of a "machinic heterogenesis." The latter favors rhizomatic alliances, mutations and becomings that undo traditional genealogies associated with a fixed growth that is likened to the shape of a tree.

Guattari wants to turn technosciences toward humans. In a digital world that valorizes skills and objectivity, to create an opening of space is of even greater importance. To make his point, he mobilizes Ilya Prigogine and Isabelle Stengers, two physicists who in their book, *Entre le temps et l'éternité*, focus on the possibility of a temporal element in science (*Three Ecologies* 40). He also (mis)quotes in French a lengthy passage from Walter Benjamin about the difference between information and storytelling: "When information supplants the old form, storytelling, and when it itself gives way to sensation, this double process reflects an imaginary degradation of experience" (67). Though they are both offshoots of storytelling, he adds, information or the report tries to convey the pure essence of things. Real storytelling sinks the thing into the life of the storyteller. To produce changes in mental ecologies, transformations of space|time coordinates are necessary.

When appealing to esthetics for a recomposition of processes of subjectification that would counter striation and stratification creating rigid territories, Guattari argues for gentle forms of singular or collective deterritorialization that will, eventually, lead to bifurcations and new forms of composition while avoiding the violent confrontation inherent in the oedipal scenario (*The Three Ecologies* 45). Guattari is close to Gianni Vattimo's concept of *Il pensiero debole* or weak thought and also again to Ilya Prigogine and Isabelle Stenger who, in their book *Order out of Chaos: Man's Dialogue with Nature*, note that the world is never centered but always in movement and far from equilibrium until a bifurcation brings about a new order. Again, the new order can be good or bad, as in the case of cancer. The subject, for Guattari, is "autopoietic" and capable of producing change from within. As he puts it, "Something is detached and starts to work for itself, just as it can work for you if you can 'agglomerate' yourself to such a process" (*Chaosmosis* 132–133). The process creates new affects, percepts or concepts. Transformations also happen by connecting with the outside, by making new alliances that can lead to bifurcations and even mutations.

An ecological politics based on esthetics has to be complemented by ethics. Analyzing the contemporary world, Guattari argues that the Marxist division between infrastructure and superstructure no longer holds. In today's world of signs and abstraction, only interchangeable semiotic regimes prevail. He distinguishes four regimes: economic, juridical, techno-scientific semiotics and semiotics of subjectification (*Three Ecologies* 48). While again asking his readers to think transversally, that is, across different regimes, and to remember how the economic semiotics have infiltrated those pertaining to subjectification, he nonetheless makes it clear that one should work from one's own discipline. He addresses all those in a position of intervening in people's psyches to put active pressure on their students, patients, even customers and constituents, to enable them to open spaces for transformation.

Not only psychiatrists, who have to work only slightly with singular and collective assemblages to produce new cartographies of the psyche (*The Three Ecologies* 40), but all those who deal directly or indirectly with regimes of subjectification are responsible for their work and cannot seek refuge behind a so-called neutrality. It is necessary for them to *be engagés*, in Sartrean parlance, that is, to be committed. Included here are all those who deal with education, healthcare, culture, art, the media, sports or fashion (*The Three Ecologies* 39). They are all in a position to produce affect, to transform consumers into producers. They have an ethico-political responsibility to bring about transformations in mental and social ecology, in new ways of thinking and exchanging that will open to new processes of individual and collective subjectification. In turn, the latter will make possible the rethinking of the natural environment. Only under these conditions can humans "repair the Amazonian 'lung'" or "bring vegetation back to the Sahara" (66). Guattari has faith in technology for these repairs.

In a world of signs and abstractions, aesthetic and ethical politics rely increasingly on performative enunciations as creative resistances. Every enunciation brings about an ever-so-slight change. Performative enunciations, singular or collective, that cannot be reduced simply to information are most necessary to make humankind ever more diverse. Changes in mental ecology that include processes of subjectification and in social ecology that bring forth the creative power of the social are the prerequisites to improving environmental ecology. Like Deleuze, Guattari has faith in the creative abilities of the social before and outside institutions. While recourse to militantism is necessary from time to time, it is subordinated to other, more creative ways of thinking. In Guattari's words, one has to be "analytically militant." Theory and practice always go together.

In the domain of social ecology there will be times of struggle in which everyone will feel impelled to decide on common objectives and to act *"like little soldiers,"* by which I mean like good activists. But there will simultaneously be periods in which individual and collective subjectivities will "pull out" without a thought for collective aims, and in which creative expression as such will take precedence. (*The Three Ecologies* 52, emphasis added)

The social continually recreates itself and institutions are marginalized.

How, we can now ask, is it possible to use the philosopher–analyst's pronouncements for our environmental dilemmas today? As Guattari claims, revisionism is the norm and theories have to be continually adjusted. Almost two decades after Guattari wrote his essay, it is safe to say that the world has taken a turn for the worse or, at least, that the scales have been magnified. To AIDS and Chernobyl, there have been added more famines and wars in Africa, extreme violence in the Middle East, global terrorism, accelerated extinction of numerous species, and, recently, clear signs that climate change with all its predictable and unpredictable consequences is well upon us. How can Guattari help us think through "eco-subjects," be they singular or collective, and how can we have recourse to his texts to help put pressure on environmental problems? Guattari writes of the necessity of an ongoing deterritorialization. On the one hand, the world is by definition always off-equilibrium, ready to bifurcate. On the other, active deterritorialization also produces change that can lead to rupture. We cannot, therefore, remain fixated on *The Three Ecologies* but have to deterritorialize ourselves in the way Guattari himself did in his own writings. What then continues to be productive in Guattari's texts for thinking about the environment today and where are we most in the process of detaching ourselves from his essay?

We cannot but agree with Guattari that capitalism has grown extensively and intensively. It has left a void in the psyche of many people by depriving them of existential territories. This may be in part why, as Guattari says, we witness a return to conservative values that are the outcome of a desperate move to find new existential territories. Of continued importance, therefore, is the philosopher–psychiatrist's plea to make humans think transversally and to make them see how economic regimes have invaded the environment and the human psyche. Guattari does not outline a program but rather indicates new ways of thinking and novel ways of domesticating our *oikos* that would enable conditions for change. In the remaining pages, I will probe the importance of his essay today by focusing on the four main points discussed so far: (1) transversal thinking and singular and collective creative resistance; (2) the ethical obligation of those who can intervene in people's psyches; (3) performative enunciation as helping the evolution of environmental practices; and (4) the respective merits of creativity, militantism, and institutions.

1. The necessity of transversal thinking and the reorientation of individual and collective subjectivity by means of an affirmative creative resistance

are of value. In many ways, we can say that Guattari's lessons have been well heeded. We can only agree with him on the question of infantilization of the psyche by many media shows all over the world, though the latter are by far not as uniform as he makes them out to be. It can be argued, however, that, with the proliferation of new media, creativity has become more accessible to many people, including minorities, by way of tapes, CDs, DVDs, webcams or videocams. Poets, musicians, painters, filmmakers flourish in greater numbers than ever all over the world, including Africa. Websites from YouTube to MySpace enable further the circulation of much creativity that can be good or bad. Bands proliferate. Painters have more venues for exhibiting their work on line (such as on the Saatchi gallery website) and around the world as museum space increases globally.

In a recent conversation at the Harvard Humanities Center, Glenn D. Lowry, Director of the Museum of Modern Art (MoMa) in New York, argued against some critical voices that, today, museums are open to all and the public has changed. It no longer consists of a select elite but of the mainstream (A Conversation with Glenn D Lowry, the Harvard Humanities Center, October 25, 2006). Many people turn to collecting art as a form of investment. Guattari dismisses most of such activity as contaminated by the system (*Chaosmosis* 132). True, the febrile and generalized creativity is often not entirely divorced from a notion of profit that, today, is no longer quite as suspect as it was for Guattari. When the state does not take care of its citizens, as might still be the case in France but not in places like the United States, people have to make at least a "sustainable" profit. The sheer number of products shows that esthetic creativity is thriving and that the media does not only manipulate people and turn them into vitamin-fed consumers.

If the media are not all infantilizing, and more diverse than Guattari makes them out to be, listeners or viewers are far from simply being co-opted. Technologies and the media help uncover manipulations by governments and industries at unprecedented rates. Scandals including media cover-ups and censorship are readily exposed. Congressional hearings denounce government censorship about findings concerning the environment (C-Span, January 30, 2007). Not only do regularly scheduled talk-show hosts (from Chris Matthews on NBC to Anderson Cooper on CNN) ask hard questions and uncover scandals, it is especially private citizens who, through websites (Moveon.org or Peace and Justice, and many more), are able to rally many more people than previously at the grass-roots level. They begin to form what is becoming a still controversial public sphere. While Guattari's caution against the sleep-inducing quality of many media is to the point, creativity proliferates as never before and people often resist going to sleep as ordered by the media through recourse to other media.

Creativity in literature, film and the arts is heeding Guattari's call. Even if the impact and the seduction of the economic discourse and of money are greater today than in 1989, on the creative front, a battle between quality and quantity is ongoing. The battle is continuously fought by creators who oppose the sheer valorization of quantity and profit. It is never won but never quite lost. There are always those who fight for the creation of existential territories with new forms of subjectification or singular and collective visions that can have implications for thinking the environment. As a Marxist, or perhaps as a French person with state support, Guattari continues to refuse any association of creativity with money.

2. The psychoanalyst's call for renewed responsibility of those who are in a position to intervene in others' psyches in the most varied domains and who are also to varying degrees caught up in capitalism's emphasis on profit is of continued importance. Guattari exposes the false neutrality of these professionals and urges them to show active commitment. Though the degree of commitment is uneven and the economic regime has invaded many of these disciplines, there are many more signs of renewed attention to the relation between subjectification and the environment, be it in the arts, or even more so in architecture or urbanism. Even fashion is paying more attention to the production and labor abuse, and also to the impact of textiles on the soil. The food industry from producers to growers, grocery chains and restaurants, is greening. More attention is paid to the survival of species and to methods of production not only in France, with the, at times parochial, resistance to biotechnologies, but also in the United States, with attention ranging from fishing practices to soil contamination.

3. To produce change in a world that functions increasingly according to signs and abstractions, Guattari emphasizes the importance of singular and collective enunciation as a performance of sorts. Each enunciation creates a slight evolution, which can be good or bad (*Three Ecologies* 40). Subjects are engaged in ongoing negotiations—*pourparlers*—with one another. This emphasis can be even further developed. In today's multipolar world where the balances of power are constantly shifting and where new actors arrive on the scene daily, these performative enunciations also have to be acts of translation. In order to deterritorialize themselves these actors will deviate from a dominant economic discourse or, in the words of another French thinker, Etienne Balibar, reformulate "burning questions" in other ways. Such an active reformulation is a kind of resistance. To construct a global world-space or a world in common, deterritorialization and ongoing negotiations are more than ever necessary, but cannot exist without acts of translation. The environment today is the only concern that is simultaneously global *and* local. As Guattari says, it concerns life on, and the future of, the planet.

Almost 20 years after the publication of *The Three Ecologies*, the environment is suddenly on everyone's agenda, including that of politicians and business people. Al Gore's *An Inconvenient Truth* was on the New York Times bestseller list for several months and the film version won its director, Davis Guggenheim, numerous awards all over the world, including an Oscar for best documentary (February 24, 2007). In a somewhat unlikely fashion, Guattari's call for a revolution was recently echoed rather unexpectedly by Jacques Chirac, who, as President of the French Republic, declared upon hearing the report from the United Nations Intergovernmental Panel on Climate Change in Paris that "It is time for a revolution." We need, he is quoted as saying, a "triple revolution to save the earth: a revolution of consciousness, an economic revolution and a revolution in political action" (www.lemonde.fr, February 2, 2007). While many of these calls may be reactive or even cynically speculative in the context of political elections, they are nonetheless symptomatic of a transformation. There are slight indications that a threshold may have been reached where a changing sensibility, a new intelligence, and even a broad desire of people to rethink the environment have become noticeable. Indeed, to address problems of the environment a desire is needed as well as an understanding to make possible new forms of cooperation and replace what Guattari repeatedly denounced as current forms of "savage competition." To bring about these changes, pressure is needed for political action. This pressure can come from the social or from institutions.

4. Guattari advocates execution of pressure by people who are engaged in fields that help create subjectivities. Indeed, it is often those professionals whom Guattari enumerated who are more likely to have retrained their own sensibility before intervening in that of others. Yet Guattari is suspicious of militantism and cautions against going back to forms of blind allegiance to a party, such as those he remembers perhaps being practiced by the communists, whose excesses are now denounced everywhere. If militantism past and present can lead to abuse, environmental issues—even more than others—cannot be addressed without a modicum of singular and collective militantism, though the latter has undergone changes. As Balibar—echoing Guattari—puts it, we cannot go back to a former style of blind allegiance and militantism. However, he points out, there is currently no organized resistance against global capitalism. It is not, for Balibar, the slightly folkloristic World Social Forum—the self-proclaimed rival to the World Economic Forum—that will put sufficient pressure on capitalism to produce change (*L'Europe, l'Amérique, la guerre* 168). Pressure has to come from above and from below. Some militantism is necessary as long as its own claims are constantly evolving. It has to come from individuals and the social, from grass-root organizations *and* from institutions, those of government or other, specialized, including

non-governmental organizations. In a rapidly changing context, subjects no longer simply oppose the state as they did in Europe a few decades ago. The state and institutions are taking on renewed importance for environmental action. Guattari rightly says that there has to be a *desire* among the people to tackle dilemmas of anthropological differences but also those of the environment. This desire would hopefully be affirmative and not just based on reaction and panic. Yet Balibar is critical of what he calls Deleuze and Guattari's formulations of nomadism, "of open spaces with free-floating individuals," presumably—though there is no footnote to back it in his text—with reference to *A Thousand Plateaus*. Balibar thinks that subjects, though always changing, have to be contextualized. Artistic resistance has to be complemented by new forms of militantism, grass-roots organization by institutions. What Balibar calls Deleuze and Guattari's "free-floating individuals" in nomad space are too easily co-opted by the very media discourse that the psychoanalyst and the philosopher decried.

The Three Ecologies, however, does acknowledge, if not institutions, at least a geopolitical context. Guattari asks for the recomposition of individual and collective subjectivities in the *context* of new techno-scientific *and* geopolitical coordinates. The essay, somewhat surprisingly, addresses the question of the "subject" that, Guattari writes, "is not a straightforward matter" (*The Three Ecologies* 35). In an interview with Gilles Deleuze, Toni Negri had noted that the philosopher had deterritorialized himself over the years and seemed to have become more attentive to problems of subjectification (*Negotiations* 176). The same, we can say, holds for Guattari, who, when dealing with ecological problems in 1989, speaks of a more specific *context* than in his earlier texts. In a global world, he calls for addressing the imbalances between what he calls the North and the South and the problems of the Third World (*Three Ecologies* 30). Though he contextualizes, he remains vague and continues to refer mainly to singular and collective subjects and a vague socius. When problems of the environment are addressed, the question of institutions cannot simply be dismissed. At some point, grass-roots organizations connect with institutions that are continually being transformed from the inside.

In conclusion, we can say that Guattari sees well the necessity to singularize *and* to act collectively on issues of anthropological differences *and* on those of the environment. As Balibar reminds his readers, to act collectively or to militate does not mean to go back in time or to be identitarian. In the wake of Deleuze and Guattari, Balibar notes that one has to travel in one's identities and even dis-identify (*Droit de cité* 129). The world has accelerated manifold since Guattari wrote *The Three Ecologies*. Today, instantaneity is ubiquitous and, indeed, the distance between esthetics, ethics and militantism is reduced. Of ever greater importance is to reduce further the distance between thinking and acting, theory and practice.

An enunciation, like the flap of a butterfly's wing mentioned by Bateson, produces a transformation, perhaps even a storm. The composition of a common world can only be done through performative enunciation, ongoing negotiations and acts of translation. Continued deterritorialization is of importance to prevent rigidification, yet it has to be coupled with new forms of militantism *and* appeal to institutions. As Deleuze argues, no more is there a voice to be discovered that would become the bearer of a collective consciousness, and marketing has reduced the possibilities of invention. However, the opening of new spaces, he concludes with Guattari, remains possible even in the most striated of all worlds. Every entity is leaking and people are continually deterritorializing. Any active intervention in the environment today has to be done from the grass-roots level *and* from that of institutions. There is a need to reorient intelligence, sensibility and, especially desire. People have to *want* to create a common world-space that focuses on the environment. This common space is predicated on ongoing translation and negotiation in a world that is electronically connected but less unified. Since the 1960s in the Western world, many paradigms have changed. Students are taught how to make money instead of how to think. In the same context, they are taught skills in a digital world based on the binary digit rather than the fraying and opening of passages. Guattari esthetico-ethical paradigms will have to be modified to help change this trend.

The Three Ecologies continues to be a valid treatise and, as a result, we can make a qualified alliance with its author while being aware of new contexts and other dynamisms in a multipolar world. Guattari's essay cannot become a fixed bible among environmentalists. The latter have to heed Guattari's injunction and deterritorialize in ongoing fashion. The invention of new coordinates of time and space, such as duration and the realization that not only human but natural phenomena cannot be reduced entirely to the speed of electronic transmission and are not just a question of skills, is of continued importance (*Chaosmosis* 127). So is Guattari's insistence on desire, sensibility and intelligence. In the words of the psychoanalyst, one has to be "analytically militant," combine theory and practice. Yet, like science, theory has to evolve, Guattari keeps reminding his readers. Today, the balance between the artistic and ethical paradigms on the one hand and the militantism and institutional negotiations on the other will have to be redrawn. With some qualifications we can espouse and transform Guattari's thinking and militate analytically for problems that have become far more pressing and complex than they were when he wrote about them with such foresight.

Works cited

Balibar, Etienne. *Droit de cite: culture et politique en démocratie*. La Tour d'Aigues (Vaucluse): Ed. De l'Aube, 1998.

——. *L'Europe, l'Amérique, la guerre: réflexions sur la médiation européenne*. Paris: Editions de la Découverte, 2003.

Bateson, Gregory. *Steps to an Ecology of Mind*. New York: Ballantine, 1972. French edition, Paris: Seuil, 1980.

Benjamin, Walter. *Illuminations*. Trans. Harry Zohn. Ed. Hannah Arendt. London: Fonatana, 1992 [1973].

Conley, Verena Andermatt. *Ecopolitics: The Environment in Poststructuralist Thought*. London, New York: Routledge, 1997.

Deleuze, Gilles. *Negotiations, 1972–1990*. Trans. Martin Joughin. New York: Columbia University Press, 1995, 169–182.

Deleuze, Gilles and Félix Guattari. *Mille Plateaux*. Paris: Minuit, 1980. Trans. Brian Massumi. *A Thousand Plateaus*. Minneapolis: University of Minnesota Press, 1988.

Gore, Al. *An Inconvenient Truth*. Emmaus: Rodale Press, 2006.

Guattari, Félix. *Les trois écologies*. Paris: Galilée, 1989. Trans. Gary Genosko. *The Three Ecologies*. London: Athlone Press, 2000.

——. *Chaosmose*. Paris: Galilée, 1992. Trans. Paul Bains and Julian Pefanis. *Chaosmosis: An Ethico-aesthetic Paradigm*. Bloomington: Indiana University Press, 1995.

Priogine, Ilya and Isabelle Stengers. *Order out of Chaos: Man's Dialogue with Nature*. New York: Bantam, 1984.

——. *Entre temps et éternité*. Paris: Fayard, 1988.

Vattimo, Gianni and Piero Aldo Rovatti (eds). *Il pensiero debole*. Milan: Feltrinelli, 1988.

7
Subjectivity, Desire, and the Problem of Consumption

Jonathan Maskit

> Destiny is not inscribed in an infrastructure. Capitalist societies secrete a society, a subjectivity which is in no way natural, in no way necessary. One could very well do something else. What I refuse is the idea of an inevitable and necessary program.
>
> (Guattari, *Soft Subversions* 277)

Introduction

Environmentalism can be put in rather simple terms: there are too many people using up too much stuff too quickly for it to regenerate itself.[1] As Juliet Lichtenberg put it (describing the 1992 Rio Earth Summit): "the North accused the South of overpopulation, and the South accused the North of overconsumption" (155). One can imagine several possible strategies to address this problem, but all will consist of one or more of the following: reducing the human population, reducing the average level of human consumption, or finding some way to regenerate material resources more quickly. While it is clear to me that reductions in human population are necessary, they are not my concern here. I also take it as a given that while some resources are substitutable (energy sources—oil, coal, wind, solar—are generally substitutable while air and water are not), the basic processes of nature—photosynthesis, fermentation, etc.—are nonsubstitutable and cannot be, in general, made to work significantly faster or more efficiently than they have heretofore (and that such technical manipulations are only achievable within a social context that may be, for reasons suggested below, objectionable). This leaves the second category: consumption. Yet, while almost all would agree that questions having to do with consumption must be addressed, environmental philosophers have generally resisted taking them up, focusing instead on questions having to do with the value of nature, the idea of ethical responsibility, and, in recent years, environmental policy.[2]

When environmental philosophers have thought about consumption, they have done so almost entirely without addressing questions having to do with subjectivity. Their strategies have usually been of two types. Some have asserted, as Arne Naess has done, that the knowledge that it is ecologically undesirable to consume more, or even as much as one does, will lead one simply to want to consume less (*Ecology* 88). An alternative version of this is found in the moralizing mantra "reduce, re-use, recycle," which, coupled with a set of social practices oriented as much towards shame and embarrassment as anything else, sends the clear message that environmentalism is primarily about individual responsibility, i.e., the curbing of individual desire. The second strategy has been to treat consumption as a problem for social or governmental policy (Maskit, "Deep Ecology and Desire"). If it is generally agreed that people consume too much, policies need to be put in place that function as incentives or constraints in order to bring individual behavior in line with the desired outcome. We can look at such policies as simply an externalization of the internal control that the first model seeks. Thus, rather than asking that individuals be responsible for themselves, the policy model despairs of such a possibility and instead finds ways to coerce people into behaving as if they were capable of being responsible. In this paper I take a different tack, drawing on the work of Félix Guattari, whose insightful investigations of subjectivity and desire have much to offer in this arena.

What is refreshing about Guattari's approach to this problem is that he begins with an earlier question, which then makes questions about consumption not (necessarily) about limits but about the structure of subjectivity itself. Questions about subjectivity arise for him in the context of what he terms the "three ecologies": the ecology of nature, the ecology of society, and the ecology of the subject. This complex context allows Guattari at least to begin thinking about environmental issues in a far richer way than do many others. We might frame these three ecologies as follows. How will nature be (how will we conceive nature)? What will society be like? What sorts of subjects will we be? This framework brings into play many issues that more traditional environmental philosophical approaches leave untouched. While questions about what nature is like are common, they are usually not framed as having to do with how we would like nature to be.[3] Such prescriptive normativity is somewhat more common when questions about social organization (political, economic, ethical, etc.) are raised.[4] Questions about subjectivity often either are absent from environmental philosophy or tend to offer an account of some "natural" form of subjectivity that is somehow being violated by the status quo. What is unique and powerful about Guattari's position is that he treats all three types of ecology as sites of contestation. That is, we cannot appeal to the natural order of nature to make sense of whether our actions are good or bad. We cannot appeal to what we "know" to be the best form of social organization in order to then see how compatible that form is with environmentalism. And, finally, we cannot appeal to how we "know" subjects to be in order to defend or criticize the status quo. For Guattari, such criticism is essential, but must be carried out not in the name of reestablishing some nature, society, or subjectivity that has been violated but in the name of opening up a space for investigating what nature, society, and we ourselves can be like.

I focus in what follows on the third question, as I find it to be the most searching and overlooked of the three. However, I do so against the background of Guattari's claim that all three ecologies need to be addressed. The paper has three sections. The first briefly surveys the extant literature on Deleuze|Guattari and the environment and then investigates how environmental philosophers up till now have treated desire, consumption, and subjectivity. The second is an elucidation of Guattari's position. The third goes beyond Guattari to take a more substantive look at the sorts of issues, problems, and questions that need to be addressed if we really are to rethink subjectivity.

What's wrong with consumerism?

Since the publication of the first works on Deleuze and Guattari and the environment, by Patrick Hayden, Verena Andermatt Conley, and myself, there has been a slow trickle that is, perhaps, ready to become a stream (Hayden, "Gilles Deleuze and Naturalism" and *Multiplicity and Becoming*; Conley, *Ecopolitics*; Maskit, "Something Wild?"). Nevertheless, this area of research remains rather small. One striking thing about this literature is that, oddly, Guattari, who, unlike Deleuze, wrote explicitly about environmental issues, is often ignored or effaced, subsumed under the name of Deleuze.[5] Interestingly, it is only amongst the French that Guattari seems to have been taken seriously on environmental issues in his own right (Antonioli, especially 225–249; Marange; A. Querrien).

In this paper I take seriously Guattari's contribution. My goal here is to return to an issue I addressed several years back in a paper on Arne Naess (Maskit, "Deep Ecology and Desire"). In that essay I criticized Naess for his claim that simply knowing that it is neither good for oneself nor for the environment to purchase, own, or have a particular product (his example is a new camera), I simply will not want to have that thing. I argued that desire is far more complex than Naess was willing to admit and suggested a neo-Kantian solution to the problem, to wit, one where a democratic state would coerce us to act as we would act were we truly rational beings. Since the time of the publication of that essay, several, although by no means many, environmental philosophers have taken seriously the problem of desire and consumption. I turn now to a brief review of some of this recent literature.

Issues concerning consumption and economics, while clearly central to environmentalism as practiced, have not played a central role in the environmental philosophical literature to date. Insofar as philosophers have discussed consumption (often in dialogue with economists), such discussion has often been a debate as to whether or not material consumption is something worth worrying about. This debate has generally not considered how nature, society, and subjectivity are conceived within economic theory. For my purposes, it is these unposed questions that are of primary import. If we do not take such issues seriously, then, even if we do believe there is reason to consider consumption to be a problem, we are likely to find ourselves treating it as a technical problem requiring a modification of the types of things consumed (although not necessarily the amount), a modification in the productive processes used to produce those things, or a reduction in the amount and|or type of things consumed achieved through governmental policies. The first modification might entail replacing conventional automobiles with hybrids or other more efficient cars. The second addresses consumption only indirectly by focusing on production, calling, for example, for the use of less energy or water.[6] It is only the third that involves an actual reduction in consumption, although this reduction is to be achieved either through voluntary measures or through, for example, tax structures that make certain sorts of consumption relatively more expensive, e.g., London's tax on automobiles that enter The City. These approaches often find themselves asserting the paradoxical position of being in favor of a reduction in consumption while also conceding the "necessity" of continued economic growth.

An alternative approach to the issue of consumption treats it as a devastatingly serious problem that can only be addressed if we look at the structure of subjectivity. I am interested in this latter approach because it raises the possibility of embracing environmentalism not as an ascetic practice that requires denial but as an opportunity to rethink who we are as subjects.

The literature on consumption and subjectivity often evidences an a-historical sense of human nature. While apologists for modern, consumer culture often praise that culture for its ability to satisfy human desires whose origin is taken to be natural, it might surprise us to find even critics of this societal form such as Hana Librova naturalizing (or quasi-naturalizing) desire and subjectivity, thus making a reduction in consumption contingent upon a certain violence to subjectivity ("The Disparate Roots of voluntary Modesty").

Juliet B. Schor reminds us that economics did not always bracket questions about subjectivity and desire, that ours was not always a consumer culture, and that the desires that shape us as consumers were ones that had to be constructed ("A New Economic Critique"). For all that, when she comes to offering some sort of critical response, it is almost as if she forgets her own analysis and ends up simply reminding us that moral arguments to consume less will be ineffective because of the intransigence of the very subjectivity she has just shown is contingent.

Of those who have gotten so far as to engage with subjectivity when addressing questions of consumption, several stand out. Philip Cafaro and Peter Wenz give arguments against consumption primarily rooted in virtue ethics. Such an argument, however, requires that we accept, in place of the economist's a-historical notion of subjectivity, a virtue ethicist's a-historical account of subjectivity, which can be seen in two ways. First, Cafaro argues that consumer culture does not really produce happiness or human flourishing. Second, our failure to live up to our environmentalist commitments is taken as evidence of character flaws ("Less is More" and "Gluttony"). Wenz, despite giving an insightful analysis of *consumerism* ("treating consumption as good in itself" (198)), ends up arguing that the real problem with consumerism is that it celebrates and depends upon revaluing traditional vices (for example, pride, envy, etc.) to the detriment of traditional virtues such as frugality, temperance, or generosity ("Synergistic Environmental Virtues"). From a slightly different perspective Laurie Michaelis too argues that a reinvigorated notion of the good life is needed to combat modern consumerism's attack on traditional virtues ("Ethics of Consumption").

Rogene A. Buchholz argues that it was the collapse of the Protestant ethic of working and saving that led to our contemporary culture of consumption. To counter this collapse she calls for a new environmental ethic that will help us to temper our desires ("The Ethics of Consumption"). Buchholz and Sandra B. Rosenthal take this argument further, arguing that consumption is bad for people. As a response they offer a reinvigoration of a pragmatist notion of subjectivity and community ("Towards an Ethics of Consumption"). As with Cafaro and Wenz, I cannot accept an argument that holds that there is some correct form of subjectivity that is currently being affronted. While Judith Lichtenberg treats desire not as a given, but as produced by social forces, she does so against the background of a subject who stands fast ("Consuming Because Others Consume"). Paul Wachtel too argues that consumer society is damaging to subjectivity, calling for changes in individual life and social structure, but again against the background of a fixed notion of subjectivity ("Alternatives to the Consumer Society").

Roger J. H. King is the only recent philosopher who has tried to take seriously the links between desire, consumption, *and* subjectivity, wondering how one becomes someone who wants to consume less.[7] As King puts it, "how will citizens change their understanding of self, their sense of identity, in order to be able to live environmentally responsible lives, not just sacrifice environmentally destructive habits?" ("Playing with Boundaries" 174). King points to three strategies for "breaking the hold of consumerist and other environmentally dysfunctional beliefs and practices" (175). First, technological changes to industrial processes without changing consumerism. Second, an ethical strategy that criticizes consumerist ideology head-on.[8] And finally, civic environmentalism, a term borrowed from William Shutkin and others, which would foster "support groups" that take responsibility for social and environmental impacts of a community's way of life (*The Land That Could Be*; see also Rubin, "Civic Environmentalism"). Since the first two strategies have already been discussed and rejected above (as they are provisionally rejected by King), it is the third that requires consideration. While King sees civic environmentalism as potentially valuable, insofar as it may give individuals communal strength to resist their "inner drives" (178), he worries that the power of community will not necessarily lead to environmentalism, that the abyss of cultural difference will not be so easily crossed, and that modernity's pervasive mobility makes the building of "true" community impossible (182). In the end, despite having opened a window onto the possibility (indeed necessity) of rethinking both subjectivity and community, King rejects civic environmentalism too and puts his eggs in the basket of ethical change. For my part I would like to pursue a bit further the path King does not follow, albeit from a different theoretical perspective.

Guattari on capitalism and subjectivity

As is well known, subjectivity, at least since Heidegger, has come in for a lot of criticism. We might sum up the post-Heideggerian critique of the subject as focusing on the subject's substantiality (Cadava, Connor, and Nancy, *Who Comes After the Subject?*). While this critique, whether offered on Heideggerian, Foucauldian, Derridian, Lyotardian, or other grounds, is often taken as a rejection of the very idea of subjectivity, the rejection seems too quick. Guattari, for his part, sometimes seems defensive in his insistence on discussing subjectivity, yet if we understand subjectivity as nonsubstantial, historically rooted, and culturally shaped, it seems that we may be able to keep the benefits of subjectivity—the possibility of some notion of autonomy and its concordance with grammar not least among them—without subscribing to a substantialist metaphysics we may find impossible to defend.

In *The Three Ecologies*, Guattari argues, albeit too briefly, for the consideration of the ecologies of "the environment, social relations, and human subjectivity" in response to contemporary environmental problems (*Trois Écologies* 12–13; *Three Ecologies* 28).[9] He contrasts this approach with one that focuses narrowly on technological manipulations intended to address industrial pollution. He characterizes his approach as follows: "the only true response to the ecological crisis is on a global scale, provided that it brings about an authentic political, social and cultural revolution, reshaping [*réorientant*] the objectives of the production of both material and immaterial goods" (*Trois Écologies* 13–14/28). The question is, what would it mean to bring about a revolution in subjectivity? Guattari elucidates as follows:

at every level, individual or collective, in everyday life as well as in the reinvention of democracy (concerning town planning, artistic creation, sport, etc.) it is a question in each instance of looking into what would be the dispositives of the production of subjectivity, which tends towards an individual and|or collective resingularization, rather than that of mass-media manufacture, which is synonymous with distress and despair. (*Trois Écologies* 21/33–34)

Guattari's analysis here, focusing on the structures (dispositives) of subjectivity, relates to that pursued in both volumes of *Anti-Oedipus*, although here he seeks to be more concrete. He thus seeks to outline what he calls the project of mental ecosophy (his term for an ecology broadened to include mental and social life in addition to nature) not as a (re) discovery of some essence of the subject, but as a reinventing of a whole network of different types of relations in which subjects find themselves and which are, at least partially, constitutive of subjectivity itself. Guattari is clear that there are both relations that need to be reinvigorated and those that need to be fought against. In the former group, we find "the relation of the subject to the body, to phantasm, to the passage of time, to the 'mysteries' of life and death." In the latter group, in need of "antidotes," are "mass-media and telematic uniformization, the conformism of fashion, the manipulation of opinion by advertising, surveys, etc." (*Trois Écologies* 22–23/35).[10] There are here clear echoes of both influential predecessors (chiefly Marx) and contemporaries (Deleuze (of course), but also Foucault, and Derrida). What we need to elucidate is how the second group of relations currently poisons the first (thus requiring an antidote). For example, how is it that processes of uniformization affect the subject's relation to the body? Much work in this area has been undertaken by scholars such as Susan Bordo and Kathryn Pauly Morgan, who have argued convincingly that the epidemic of eating disorders and voluntary plastic surgery in our culture are traceable, at least in part, to the presentation of a type of feminine body image, in print, on television, in films, etc. (Bordo, "Reading the Slender Body" and Morgan, "Women and the Knife"). From a rather different perspective, we could investigate how the automobile and other transportation technologies shape our relations to our bodies (and to each other). The modern relation to time began, as Lewis Mumford has shown, not in factories but in medieval monasteries ("The Monastery and the Clock"). Despite these perhaps surprising origins, the modern relationship to time, in which one is expected to be "on time" and in which events are coordinated in strict adherence to a shared temporal framework, did not become socially pervasive until the industrial era. It is now so deeply entrenched that we do not even reflect on how thoroughly this temporal framework has become part of who and what we are. As a corollary to this point, one might think of our insistence on efficiency as a virtue as an instance of the temporalization of our lives.

What is important for Guattari is that such investigations take place *not* in the name of the recovery of some damaged form of subjectivity (for who truly wants to return to a preindustrial form of subjectivity, wrapped up as it is in racism, sexism, classism, etc.?) but in the name of possibilities that will be individualizing (subjectivizing, we might say) rather than uniformizing.[11] "It would be absurd," Guattari writes, "to want to return to the past in order to reconstruct former ways of living. After the data-processing and robotics revolutions, the rapid development of genetic engineering and the globalization of markets, neither human labour nor the natural habitat will ever be what they once were" (*Trois Écologies* 33/42).

For Guattari (as for Deleuze), what is essential is to think here without recourse to any sort of reification or substantialization. Rather, our focus must be on *process*, which "strives to capture existence in the very act of its constitution, definition, and deterritorialization" (*Trois Écologies* 36/44; the reflexive character of the French verbs is lost in the English.). The deterritorialization (glossed by Bonta and Protevi as "the process of leaving home, of altering your habits, of learning new tricks" (78)) is important, for it makes clear that these are processes that are not run through only once but that are always evolving. That is, society, nature, and subjectivity—the terms of the three ecologies—are all here to be thought not as things but as collections of processes that are partially, but never fully, subject to some sort of control. Rather, these processes develop according to what Guattari calls "a-signifying rupture[s]" (*Trois Écologies* 37/45). These ruptures are events or developments which cannot be made sense of from within current forms of discourse ("the assemblage of enunciation"). While such a-signifying ruptures are most easily seen in the arts, where radical developments—cubism, abstract expressionism, or the development of rock 'n roll—both resist current practices of signification and, through that resistance, call for a resignifying change, they are by no means limited to that domain. Take Critical Mass as an example. It is not even clear what to call Critical Mass, since "Critical Mass is not an organization, [sic] it's an unorganized coincidence."[12] Critical Mass is perhaps best called a "movement," although one with no leaders and no followers. Critical Mass exists as a set of monthly events in which bicyclists agree to meet at a particular place and time (usually a Friday afternoon) for a mass ride. The event takes place at the margins of the law as there are usually sufficient riders to occupy at least one complete lane of traffic. But what to call such bicyclists? They are neither out for a pleasure ride nor using their bicycles as a form of transportation. They are asserting, rather, a form of subjectivity and community that sets itself up in opposition to automobile culture. These rides routinely draw notice of the police, who seem intent on keeping the streets clear for "legitimate" users, that is, drivers. Critical Mass riders seek to challenge that very legitimacy, thus seeking to establish (albeit on a small scale) what Guattari calls an "existential Territory," that is, the linked assemblage of a place and a way of being in that place (*Trois Écologies* 38–9/46).

In contrast to my positive characterization of these a-signifying ruptures, Guattari makes clear that such ruptures are a pervasive and often negative characteristic of modern culture. He points to the "huge subjective void engendered by the proliferating production of material and immaterial goods [that threatens] the consistency of both individual and group existential Territories" (*Trois Écologies* 38–9/46). We could characterize this subjective void as the emptiness of desire produced by modern consumer capitalism, for which the object of desire is, in a sense, unimportant, so long as that desire can be only temporarily fulfilled and then always only through economic means, that is, through products and services. In addition, consumerist subjectivity is one of perpetually re-produced desire. Put otherwise, this subjectivity is one whose desires are always reestablished following any (partial) fulfillment of them. While the destructiveness of this form of subjectivity is recognized by cultural conservatives, in part because they see that it leads to "an irreversible erosion of the traditional mechanisms of social regulation," their "solution"—to attempt to reinvigorate those older forms of subjectivity and of culture—is, as suggested above, unacceptable (*Trois Écologies* 39/46–7).[13]

What is thus required is a "rebuilding [of] human relations at every level of the social body [*socius*]" (*Trois Écologies* 43/49). This task manifests itself, for Guattari, as a confrontation with what he terms integrated world capitalism (IWC). In a passage reminiscent of the work of Horkheimer and Adorno, Guattari argues that we "should never lose sight of the fact that capitalist power has delocalized and deterritorialized itself, both in extension, by extending its influence over the whole social, economic and cultural life of the planet and in 'intension', by infiltrating the most unconscious subjective strata" (*Trois Écologies* 43–44/49–50). It is this latter point that is of greatest import for both Guattari and myself. For the preceding passage has outlined the focal points of *social* ecology. When he turns to *mental* ecology, he stresses similar points, writing that "it is equally imperative to confront capitalism's effects in the domain of the mental ecology at the heart of one's individual, domestic, marital, communal [*de voisinage*], creative and personal-ethical daily life" (*Trois Écologies* 44/50). Indeed, for Guattari it is particularly in this register that work needs to be done and why his work, even if it does not fully carry through on the project it sketches, is so important. While Guattari resists characterizing an alternative form of subjectivity, he does give a compelling characterization of contemporary subjectivity, which I quote here at length:

> Capitalistic subjectivity, which is engendered through operators of all types and sizes, turns out to be manufactured so as to protect existence from any intrusion of events that might disturb or disrupt [public] opinion. For it every singularity must be either evaded or crushed [*passer sous la coupe*] in specialist apparatuses and frames of reference. Therefore,

it endeavours to manage the worlds of childhood, love, art, as well as everything in the order of anxiety, madness, pain, death, or a feeling of being lost in the cosmos... IWC forms massive subjective aggregates from the most personal... existential givens, which it connects with race, nation, the professional workforce, athletic competition, a dominating masculinity [*virilité*], mass-media celebrity.... By assuring itself of the power over the maximum number of existential refrains [*ritournelles*] in order to control and neutralize them, capitalistic subjectivity intoxicates, even anaesthetizes, itself in a collective feeling of pseudo-eternity. (*Trois Écologies* 44–45/50)

Capitalistic subjectivity could be characterized as that form of subjectivity that makes one not a cog in a productive machine (although that too may be part of it) but rather a willing participant in a form of life that is more and more structured to produce desires and pleasures that are concordant with the products and services that can be provided through the marketplace. Thus, for example, shopping becomes a form of entertainment: the day at the mall, what one does while on vacation. One's leisure time—at least if one is a member of the North's privileged class—is taken up by television, sports, eating out, remodeling one's home, and travel, all activities that are not merely economic in character, but which are shaped by and shaping of our form of subjectivity. All of our most private and personal ways of being—how we relate to our friends, our mates, and our children; how we act; how we make love; how we think of ourselves—are shaped by a media culture that exists, first and foremost, to maintain an economic and political order and to make of us the types of subjects who will fit into and support that order.

Guattari describes this state of affairs as a type of "internalization [*introjection*] of repressive power by the oppressed" (*Trois Écologies* 42/48–9). This internalization, which Deleuze|Guattari characterize elsewhere as the problem of desire desiring its own repression, is particularly pernicious because it means that even those who would like to overcome the forms of oppression found in their current situation end up working against themselves (*Mille Plateaux* 262; *Thousand Plateaus* 215). Yet there is an additional difficulty here as well in the form of two quite serious possible objections to this characterization. Either one could reject the characterization as accurate or one could accept it as accurate but deny that it is oppressive. I address these concerns in order.

It is surely true that the apologists for the status quo do not see it in these terms. Yet the assertion that our current economic system is productive of happiness and well-being is not borne out by social scientific research. Peter Wenz cites research showing that even as income in the United States doubled (from 1957 to 2001), the number of people who described themselves as "very happy" decreased from 35% to 30% of the population ("Synergistic Environmental Virtues," 205; see also Alan Durning,

"The Dubious Rewards of Consumption"). One does not have to look far to find the disaffection, dissatisfaction, and ennui amongst the inhabitants of the developed world. That we spend our leisure time pursuing individualizing pursuits that cater to our emotions suggests that there is a problem here. This brings me to the second objection.

One could easily accept everything that has been said up until now and deny that the situation as I have described it is oppressive. That is, one could ask what grounds the normativity of Guattari's position. This problem is not unique to Guattari or Deleuze. Indeed, Foucault and Derrida have both had similar objections raised against them. Guattari, however, may have an easier time answering this objection. One who accepts the characterization here given yet denies its oppressive character must not merely be ready to deny that the status quo is oppressive of human agents, but must also deny that it is socially and environmentally destructive. These critics, therefore, would have to account for either how there actually is no environmental damage associated with capitalism or why that damage is either not worth our concern or an acceptable price to be paid.

Part of the appeal of the model that I have described is that it allows for, even requires, the three ecologies to be addressed as a whole. This means that what nature, culture, or subjectivity will be like can only be addressed together. To allow one aspect of one of the three (the economic aspect of culture) to so thoroughly determine the rest means to sacrifice nature and|or subjectivity (and the non-economic aspects of culture as well) in the name of the economy. In addition, if we believe that autonomy is a value (even if it too must be thoroughly rethought in this context), I do not see how we can accept the picture I have drawn, where the very form of subjectivity is heteronomously determined, but reject my characterization of it as oppressive. Thus, we might see the constructive project before us as follows: how could we produce a form of subjectivity that fosters autonomy?

The American Dream? A call for being otherwise

How can we become the sorts of people who can not only make do with less, but want to do so? The problem, as I have argued so far, is one of restructuring subjectivity. But just how is that to be done? Guattari, I think wisely, does not say much as to what a restructured subjectivity would be like, and this for several reasons. First, he is no prophet, who can declare what the future will be (or ought to be) like. Second, his analysis is not proscriptive but therapeutic. That is, his goal here is to point to a set of problems and then to suggest some points at which we could begin to try to solve those problems. Finally, because of his commitments to process and to community, that is, to democracy, it would be strange indeed (and presumptuous) for him to declare just how it is that others should be. Rather, his work is, like Nietzsche's, intended to push us to *think* otherwise so that we might *be* otherwise. It is in that spirit that I offer the following. Its form is that of immanent critique, that is, it tries to show some of the internal tensions within our current form of politicosocial–economic organization.

Here is a (partial) list of the "values" enshrined in that organizational form: freedom of choice as to where one works and what one consumes, efficiency in the use of materials and time, growth in the size of the economic "pie."

As anyone who has ever looked for a job knows, while one is free to apply or not apply for any particular position, the general terms of one's employment are non-negotiable. One cannot determine when one works, for how long one works, what one will do while working, etc. One's freedom in this sphere is rather limited: will I take what is on offer or won't I?

As anyone who buys things knows, goods and services fall into two categories: those that are simply unavailable and those that are available in a plethora of sizes, styles, etc. Our shopping lives thus become expenditures of time either trying to find the product we want or trying to decide between the myriad of options available to us. To decide *which* phone, waffle iron, computer monitor, pair of shoes, etc., to buy can be paralyzing.

If one adds in a desire to be a "socially" or "environmentally responsible" consumer, one quickly finds that a number of things that appeared to be in the second category (plethora of options) move into the first (not available). There are, for example, no waffle irons made anywhere but China.

As anyone who tries to manage their own time knows, figuring out how best to do so is itself an exercise in time management. Of course, since time management takes place against the background of the givens of work and consumer life, one has little choice but to embrace efficiency as a personal virtue.

As anyone who watches, reads, or listens to the news knows, the economy (like political life) has now become a sort of spectator sport. Even if one's retirement funds are invested in stocks and bonds, must one really be informed on a daily basis of which indices are up and which are down, how housing starts are doing, etc.? Must one be subjectively invested in the state of the market?

This brief sketch of some perhaps obvious facts of modern life illustrates some, but by no means all, of the ways in which our form of life and our form of subjectivity are productive of activities, desires, temporalities, forms of embodiment, etc., that are often to the detriment of ourselves, our society, and the natural world. To say that there are downsides to modern life is surely not novel. What is new here is the suggestion that our addressing these concerns will require not merely technical, political, or policy suggestions, but a rethinking of what it means to be human. How could one at least begin to reshape subjectivity? Here are some ideas:

Don't watch television. Question all assertions that a practice is impossible. Know the people who produce your food. Figure out how to get from point A to point B without driving or flying.

What is interesting about this list is that some of these things look like ascetic practices. And maybe they are. But they are practices oriented not towards being the way we always could have been but towards being a way that we did not know we could be.

Notes

An earlier version of this paper was read at the Philosophy Department at LeMoyne College (Syracuse, NY), February 23, 2007. I thank the members of the Department for their helpful feedback and questions. I also thank Barbara Fultner for her invaluable feedback on an earlier draft.

1. It should immediately be noted that there is a huge body of literature debating whether this description is accurate. Conservative apologists for the status quo growth economy, such as Julian Simon, insist that scarcity is always only relative and that ingenuity driven by market prices will always allow us to find substitutes (*Ultimate Resource* and "Scarcity or Abundance?"). William Rees has argued, however, that current rates of economic activity are unsustainable because of their disruption of basic ecological processes, processes for which there can be no substitutes ("Achieving sustainability").
2. Social scientists, however, have spent considerable time trying to figure out consumption.
3. Steve Vogel is a notable exception here ("After the End of Nature" and *Against Nature*). Randall Honold argues that "the best of nature is yet to come."
4. The journal *Capitalism, Nature, Socialism* has tended to foster critiques of capitalism, but in the name of socialism, which has historically shared some key assumptions with capitalism, most importantly, that economic growth is a value to be preserved. As for liberalism, see Avner de-Shalit, "Is Liberalism Environment Friendly?"
5. The following all, to greater or lesser degrees, discuss Deleuze as if he were the sole author of works by Deleuze and Guattari together: Hayden, "Gilles Deleuze and Naturalism," 191–195; Bonta and Protevi, *Deleuze and Geophilosophy*. Authors who have credited Guattari for his important contributions here include Ed Casey, Mark Halsey, Robert Mugerauer, and David Wood.
6. Germany's law requiring that companies take responsibility for the final recycling or disposal of their durable products (cars, washing machines, etc.) might be seen as a policy of this type.
7. There are interesting and related contributions here from the social sciences. See, in particular, Mihaly Csikszentmihalyi, "The Costs and Benefits of Consuming"; Fred Hirsch, "The New Commodity Fetishism"; Tim Jackson, "Challenges for Sustainable Consumption Policy"; and Grant McCracken, "The Evocative Power of Things."
8. Such a strategy is argued for by Duane Elgin and other proponents of the simple living movement ("Living More Simply" and *Voluntary Simplicity*; see also Amitai Etzioni, "Voluntary Simplicity").
9. Further references to the French and the English (of 2000) are separated by a solidus. I have occasionally modified Pindar and Sutton's translation without indicating as such each time.

    Part of this text's apparent descent into obscurity may rest on its previously only having been available to English speakers in a partial, and not very good translation by Chris Turner (1989). Sadly, the publication of Pindar and Sutton's translation does not seem to have brought it to a much wider audience.

10. Guattari affirms this point as well in *Chaosmosis*: "Should we keep the semiotic productions of the mass media, informatics, telematics and robotics separate from psychological subjectivity? I don't think so" (*Chaosmose* 15; English translation: *Chaosmosis* 4). *Télématique* combines telecommunication and informatics. We might include under it a set of technologies not all yet known to Guattari: mobile telephones, the internet, instant messaging, etc.

11. Luc Ferry mistakenly reads Guattari as emphasizing individuality and cultural difference to the exclusion of communication. While it is true that Guattari is critical of the norm of consensus, it is clear that he is not opposed to communication (*The New Ecological Order*, 112–114).

12. From what is clearly marked as *not* "the official Critical Mass web page, because there is no official Critical Mass web page" at critical-mass.org (as accessed February 15, 2007).

13. I am thinking here of movements opposing gay rights, defending marriage as only between a man and a woman, etc. Such movements, if Guattari's analysis is right, are, it is worth remarking, *practically* misguided, since they have tended to align themselves with exactly the social, economic, and political forces that have produced the very erosion of subjectivity and culture that they so rue.

Works cited

Antonioli, Manola. *Géophilosophie de Deleuze et Guattari.* Paris: Harmattan, 2003.

Bonta, Mark and John Protevi. *Deleuze and Geophilosophy: A Guide and Glossary.* Edinburgh: Edinburgh University Press, 2004.

Bordo, Susan. "Reading the Slender Body" in *Body and Flesh: A Philosophical Reader.* Ed. Donn Welton. Malden, MA and Oxford: Blackwell, 1998, 291–304.

Buchholz, Rogene A. and Rosenthal, Sandra B. "Toward an Ethics of Consumption: Rethinking the Nature of Growth" in *The Business of Consumption: Environmental Ethics and the Global Economy.* Eds L. Westra and P. H. Werhane. Lanham, MD: Rowman & Littlefield, 1998, 221–234.

Buchholz, Rogene. A. "The Ethics of Consumption: A Future Paradigm?" *Journal of Business Ethics* 17 (1998): 871–882.

Cadava, Eduardo, Peter Connor and Jean-Luc Nancy. *Who Comes After the Subject?* New York and London: Routledge, 1991.

Cafaro, Philip. "Gluttony, Arrogance, Greed, and Apathy: An Exploration of Environmental Vice" in *Environmental Virtue Ethics.* Eds R. Sandler and P. Cafaro. Lanham, MD: Rowman & Littlefield, 2005, 135–158.

——. "Less is More: Economic Consumption and the Good Life." *Philosophy Today* 42 (1998): 26–39.

Casey, Edward. "Taking a Glance at the Environment: Prolegomena to an Ethics of the Environment." *Research in Phenomenology* 31 (2001): 1–21.

Conley, Verena A. *Ecopolitics: The Environment in Poststructuralist Thought.* New York: Routledge, 1997. www.critical-mass.org (as accessed February 15, 2007).

Csikszentmihalyi, Mihaly. "The Costs and Benefits of Consuming" in *The Earthscan Reader in Sustainable Consumption.* Ed. Tim Jackson. London and Sterling, VA: Earthscan, 2006, 357–366.

Deleuze, Gilles and Félix Guattari. *A Thousand Plateaus: Capitalism and Schizophrenia.* Trans. Brian Massumi. Minneapolis: University of Minnesota Press, 1987.

——. *Mille plateaux: Capitalisme et schizophrénie 2.* Paris: Les Éditions de Minuit, 1980.

de-Shalit, Avner. "Is Liberalism Environment Friendly?" *Social Theory and Practice* 21 (1995): 287–314.

Durning, Alan. "The Dubious Rewards of Consumption." in *The Earthscan Reader in Sustainable Consumption*. Ed. Tim Jackson. London and Sterling, VA: Earthscan, 2006, 129–135.

Elgin, Duane. *Voluntary Simplicity*. New York: William Morrow, 1981.

——. "Living More Simply" in *The Earthscan Reader in Sustainable Consumption*. Ed. Tim Jackson. London and Sterling, VA: Earthscan, 2006, 151–158.

Etzioni, Amitai. "Voluntary Simplicity" in *The Earthscan Reader in Sustainable Consumption*. Ed. Tim Jackson. London and Sterling, VA: Earthscan, 2006, 159–177.

Ferry, Luc. *The New Ecological Order*. Trans. Carol Volk. Chicago and London: University of Chicago Press, 1995.

Guattari, Félix. *Les Trois Écologies*. Paris: Éditions Galilée, 1989.

——. "The Three Ecologies." Trans. Chris Turner. *New Formations* 8 (1989) 131–147.

——. *Chaosmose*. Paris: Galilée, 1992.

——. *Chaosmosis: An Ethico-Aesthetic Paradigm*. Trans P. Bains and J. Pefanis. Indianapolis and Bloomington: Indiana University Press, 1995.

——. *Soft Subversions*. Ed. Sylvère Lotringer. NY: Semiotext(e), 1996.

——. *The Three Ecologies*. Trans I. Pindar and P. Sutton. London and New Brunswick, NJ: Athlone Press, 2000.

Halsey, Mark. *Deleuze and Environmental Damage: Violence of the Text*. Aldershot, England and Burlington, VT: Ashgate, 2006.

Hayden, Patrick. "Gilles Deleuze and Naturalism: A Convergence with Ecological Theory and Politics." *Environmental Ethics* 19 (1997): 185–204.

——. *Multiplicity and Becoming: The Pluralist Empiricism of Gilles Deleuze*. New York: Lang, 1998.

Hirsch, Fred. "The New Commodity Fetishism" in *The Earthscan Reader in Sustainable Consumption*. Ed. Tim Jackson. London and Sterling, VA: Earthscan, 2006, 136–145.

Honold, Randall. "New Nature Photography and the Future of Nature." Presented at the *International Association for Environmental Philosophy*, Salt Lake City, October 2005.

Jackson, Tim. "Challenges for Sustainable Consumption Policy" in *The Earthscan Reader in Sustainable Consumption*. Ed. Tim Jackson. London and Sterling, VA: Earthscan, 2006, 109–126.

King, Roger J. H. "Playing with Boundaries: Critical Reflections on Strategies for an Environmental Culture and the Promise of Civic Environmentalism." *Ethics, Place, and Environment* 9 (2006) 173–186.

Librova, Hana. "The Disparate Roots of Voluntary Modesty." *Environmental Values* 8 (1999): 369–380.

Lichtenberg, Judith. "Consuming Because Others Consume" in *Ethics of Consumption: The Good Life, Justice, and Global Stewardship*. Eds D. A. Crocker and T. Linden. Lanham, MD: Rowman & Littlefield, 1998, 155–175.

Marange, Valérie. "La petite machine écosophique." *Chimères* 28 (Spring–Summer 1996): no page numbers.

Maskit, Jonathan. "Deep Ecology and Desire: On Naess and the Problem of Consumption" in *Beneath the Surface: Critical Essays in the Philosophy of Deep Ecology*. Eds E. Katz, A. Light and D. Rothenberg. Cambridge, MA: MIT Press, 2000, 215–230.

——. "Something Wild? Deleuze and Guattari and the Impossibility of Wilderness." *Philosophy & Geography* 3 (1998): 265–283.

McCracken, Grant. "The Evocative Power of Things" in *The Earthscan Reader in Sustainable Consumption*. Ed. Tim Jackson. London and Sterling, VA: Earthscan, 2006, 263–277.

Michaelis, Laurie. "Ethics of Consumption" in *The Earthscan Reader in Sustainable Consumption*. Ed. Tim Jackson. London and Sterling, VA: Earthscan, 2006, 328–345.

Morgan, Kathryn Pauly. "Women and the Knife" in *Body and Flesh: A Philosophical Reader*. Ed. Donn Welton. Malden, MA and Oxford: Blackwell, 1998, 325–347.

Mugerauer, Robert. "Deleuze and Guattari's Return to Science as a Basis for Environmental Philosophy" in *Rethinking Nature: Essays in Environmental Philosophy*. Eds B. V. Foltz and R. Frodeman. Bloomington and Indianapolis: University of Indiana Press, 2004, 180–204.

Mumford, Lewis. "The Monastery and the Clock" in *Technology as a Human Affair*. Ed. Larry Hickman. New York: McGraw-Hill, 1990, 208–212.

Naess, Arne. *Ecology, Community and Lifestyle: Outline of an Ecosophy*. Trans. and ed. David Rothenberg. Cambridge: Cambridge University Press, 1989.

Querrien, Anne. "Broderies sur *Les Trois Écologies* de Félix Guattari." *Chimères* 28 (Spring–Summer 1996), no page numbers.

Rees, William E. "Achieving Sustainability: Reform or Transformation?" in *The Earthscan Reader in Sustainable Cities*. Ed. D. Satterthwaite. London and Sterling, VA: Earthscan Publications, 2001 [1999], 22–52.

Rubin, Charles T. "Civic Environmentalism" in *Democracy and the Claims of Nature: Critical Perspectives for a New Century*. Eds B. A. Minteer and B. P. Taylor. Lanham, MD: Rowman & Littlefield, 2002, 335–351.

Schor, Juliet B. "A New Economic Critique of Consumer Society" in *Ethics of Consumption: The Good Life, Justice, and Global Stewardship*. Eds D. A. Crocker and T. Linden. Lanham, MD: Rowman & Littlefield, 1998, 131–138.

Shutkin, William A. *The Land that Could Be: Environmentalism and Democracy in the Twenty-First Century*. Cambridge, MA: MIT Press, 2000.

Simon, Julian L. "Scarcity or Abundance?" in *The Business of Consumption: Environmental Ethics and the Global Economy*. Eds L. Westra and P. H. Werhane. Lanham, MD: Rowman & Littlefield, 1998, 237–245.

——. *The Ultimate Resource*. Princeton: Princeton University Press, 1981.

Vogel, Steven. "Environmental Philosophy After the End of Nature." *Environmental Ethics* 24 (2002) 23–39.

——. *Against Nature: The Concept of Nature in Critical Theory*. Albany: SUNY Press, 1996.

Wachtel, Paul L. "Alternatives to the Consumer Society" in *Ethics of Consumption: The Good Life, Justice, and Global Stewardship*. Eds D. A. Crocker and T. Linden. Lanham, MD: Rowman & Littlefield, 1998, 198–217.

Wenz, Peter. "Synergistic Environmental Virtues: Consumerism and Human Flourishing" in *Environmental Virtue Ethics*. Eds. R. Sandler and P. Cafaro. Lanham, MD: Rowman & Littlefield, 2005, 197–213.

Wood, David. "Trees and Truth (or, Why We Are Really All Druids)" in *Rethinking Nature: Essays in Environmental Philosophy*. Eds B. V. Foltz and R. Frodeman. Bloomington and Indianapolis: University of Indiana Press, 2004, 32–43.

8
Political *Science* and the Culture of Extinction

Dorothea Olkowski

The constitution

According to the science studies theorist, Bruno Latour, although a certain amount of intellectual effort has been devoted to the study of the opposition between science and discourse, virtually none has been directed to that between science and politics. Surprisingly, Latour maintains, the two may be defined by a common text, that of the *constitution*. Just as politicians draft a constitution for states, so scientists draft one for nature, but each constitution has been drafted with the aim of deliberately excluding the power of the other (*We Have Never Been Modern* 14). The apparent exclusion of science from politics and politics from science leads Latour to write or at least describe a constitution that incorporates both science and politics in relation to what he takes to be a founding moment in the history of modern science and of politics, the debate between natural philosopher, Robert Boyle, and political philosopher, Thomas Hobbes, thereby rescuing from oblivion Boyles's political theories and bringing to light Hobbes's scientific work: "Boyle has a science and a political theory; Hobbes has a political theory and a science."[1] Strangely, Boyle and Hobbes appear to agree on almost everything, and yet they still diverge.

In Latour's account Boyle develops the vacuum by constructing facts in the installation in the laboratory. Boyle argues (like Hobbes) that God knows things because he creates them but (unlike Hobbes) that *we* humans know facts, like the vacuum, that *we* have developed in circumstances under *our* complete control. But Hobbes wanted to make the Sovereign the representative of the multitude, closing off independent access to divine transcendence. Citizens are then the guaranteed Authors of the Actor who is Sovereign, and there is only one Power and so one Knowledge. The Sovereign is the only Actor designated by the Leviathan, which remains, however, a structure of nothing more than social relations. Unlike Boyle, who champions the facts created in the laboratory, Hobbes arrives at all his scientific results through mathematical demonstration. Hobbes objects to the laboratory, a closed space over which the state has no control that operates via experiments that produce their own matters of fact. As such, apodeictic reasoning, the power out of which society unfolds, is replaced by controlled doxa (*We Have Never Been Modern* 18–22). Latour argues that, in Boyle's laboratory, inert bodies were able to show, sign, and write on the experimental instruments of the laboratory, thus to take up the role of *discourse*. But it is a transcendent *discourse* insofar as the facts *speak for themselves* within the artificial chamber of the laboratory created by humans. They are more reliable than humans, since experiments are repeated under precise conditions and witnessed by reliable experts. Opposed to this is the universal power and authority of the citizen who accepts sovereignty as it is derived through mathematical deductions but whose own rights are limited to possessing property and being represented by the artificial but perfectly rational Sovereign, for whom Power equals Knowledge. By representing his experiments as legitimate—as long as *reliable, trained witnesses* attest to their factual results—Boyle created a transcendent political discourse, that of science, from which, it is claimed, he excluded Hobbes's version of politics. Similarly, insisting on the rational rule of mathematical demonstration, Hobbes is said to constitute a pure, rational politics that excludes the possibility of experimental science. In other words, Latour argues, epistemology and political theory each make claims to a transcendent truth residing *within their sphere of immanence*, and they refuse to acknowledge one another, setting into motion the work of purification, the total separation between nature and culture (*We Have Never Been Modern* 27–28).[2]

Clearly, in choosing Hobbes and Boyle, Latour establishes, respectively, an opposition between power, originating in and justified by a rational transcendence, and power originating in and justified by the conception of a transcendent discourse whose practitioners exhibit an elitist if not aristocratic superiority over the mob, meaning those without expertise. Insofar as this oppositional structure remains fundamental throughout Latour's work, he is at pains to only provide accounts of situations that support it. Thus, he chooses appropriate examples to manifest these pure oppositions: Hobbes and Boyle, and elsewhere, Socrates and Callicles. And, in carrying out his analysis, Latour utilizes a semiotic model but it operates only within another process that he names translation.

> All reasoning is of the same form; one sentence follows another. Then a third asserts that these are identical even though they do not resemble one another.... the second is used in place of the first, and a fifth affirms that the second and the fourth are identical. (*Pasteurization* 176)

Each sentence is displaced and *translated*, all the while appearing to have stayed faithful to identity. Yet, as his many diagrams attest, there is an epistemological structure that orients translation. Clearly, not only do the elements of knowledge exist in *continuous* space-time, but the chain they form conforms to the rules of dynamical systems as defined by classical physics; it must be reversible, closed, atomistic (consisting of numerous distinct elements or actors in Latour's terminology) and, within limits, deterministic (*Darwinism Evolving* 92).[3] Once established, these rules do not change, and, like the atoms described by Newton's laws of motion, *truth* must be transportable in either direction along this chain such that "truth-value circulates here like electricity through a wire," moving continuously through a *closed* system or area (*Pandora's Hope* 70–74).[4] All statements participate in the series nature of the transformations of the immanent system: locality, particularity, materiality, multiplicity, and continuity at one end, compatibility, standardization, text, calculation, circulation, and relative universality at the other. Each stage is thereby matter for what follows and form for what precedes. The two directions represent the limits of the system, that is, it is limited by reduction and realism at one end and amplification and relativism at the other, and phenomena arise in the middle between the two limits, then pass up and down the chain of transformations (*Pandora's Hope* 158).[5]

In this scheme, existence is thereby conceived of as the "*exploration* of a two-dimensional space made by association and substitution, AND and OR. An entity gains in reality if it is associated with many others that are viewed as collaborating with it" (Copi and Cohen 692). In logic, association (in semiotics, syntagm) is an expression of logical equivalence permitting the valid regrouping of simple propositions; it governs the relations, meaning the connections between subject and predicate (Devlin 46–48). Since, in Latour's conception, subject and predicate are no longer opposed, association simply asks which actor can be connected with which other actor. Substitution (paradigm in semiotics) arises as a manifestation of the hypothetical proposition which is intrinsically commutative. Reformulating any hypothetical proposition using truth tables, one arrives at its logical equivalent: "There is *no* spontaneous generation *or* there are microbes, a proposition to which commutativity now applies: 'There are microbes *or* there is *no* spontaneous generation'" (*Pandora's Hope* 159–160). Substitution asks which actor can replace which other actor in a given association. Association and substitution work together to separate the causal order of the inference, making it possible to reorder the propositions, to *disturb their causal linearity* and to eliminate any absolute demarcation between the two phenomena such that each is relatively real *and* relatively existent. Therefore, elements are defined by their associations and exist or cease to exist according to the stability of the connections and substitutions that define them (*We Have Never Been Modern* 33–34).

Latour's aim is to reveal the extent to which opposed positions reinforce one another and have reinforced one another from their very inception. Both Boyle and Hobbes, he argues, are deep into hybrids, the combining of politics, religion, technology, morality, science, and law. But both are also deep into purification, that of transcendent scientific discourse and that of the purely rational state. And both participate in the infinite distancing of a transcendent but mostly absent Deity, a "simultaneously impotent and sovereign judge" (*We Have Never Been Modern* 37). Simultaneously creating and canceling the total separation of human beings and nonhuman things, the scientific–political constitution made the moderns invincible. Whatever claim or criticism one makes of their doctrines, the moderns can also claim the opposite. Nature transcends but can also be mobilized, humanized and socialized in the laboratory; we construct society, but its laws are necessary and absolute, as transcendent as nature. The problem is that simultaneously separating and canceling the separation of humans and nonhumans, thereby obscuring their multiple mediations between science, society, and God, the moderns are impotent in the face of things and much too powerful in society (*We Have Never Been Modern* 53).[6] These oppositions have lead to a form of ontological and epistemological dualism in which Nature and the Subject|Society are divided into soft and hard aspects. Nature is both a screen for receiving the projections of social categories and a source of undisputable causes. The Subject|Society is both an effect of *sui generis* social factors and hard forces discovered by the sciences (*We Have Never Been Modern* 78).[7]

For Latour, the solution to these contradictions appears to be close at hand. We can, he argues, simply change direction in time and become premodern again. If the moderns conceived of hybrids as (impossible) mixtures of radically disconnected pure forms, then let us newly declared premoderns deploy the middle as the point of departure from which point of view the pure extremes are but provisional and partial limits (*Order Out of Chaos* 19–22). Much of what Latour proposes is extremely seductive. And yet, it too presupposes a classical model, a model that he freely admits suffers from the inability to give a coherent account of the relationship between humans and nature. Like the pure limits Latour describes via Boyle and Hobbes, classical scientific models are an effect of isolated and purified practices as well as an effort to achieve greater and greater autonomy, leading to a strange sort of universality, an immanence that isolates their structure from any other context, whether that context consists of alternative logics, new flows of matter and energy or the evolutionary novelty (*Order Out of Chaos* xxvii). What we might wish to consider is that the reversibility of classical dynamics is a characteristic of closed systems and even more so of isolated ones. This is precisely the point made by Isabelle Stengers and Ilya Prigogine. "In the classical view, the basic processes of nature were considered to be deterministic and reversible.... Today we see everywhere, the role of irreversible processes, of fluctuations" (*Order Out of Chaos* 19–22). The reorientation from the classical to the contemporary view is, for them, equally reflected in the conflict between the natural sciences and the social sciences and humanities. One consequence of the limitations of classical science is its inability to give a coherent account of the relationship between humans and nature. Many important results were repressed or set aside insofar as they failed to conform to the classical model. In order to free itself from traditional modes of comprehending nature, science isolated and purified its practices, achieving greater and greater autonomy, conceptualizing its knowledge as universal, and isolating itself from any social context (*Order Out of Chaos* 301). Perhaps we can argue, following Latour, that politics did the same, isolating itself from the discourse of science in order to claim transcendent authority. However, bringing the two together might require more than the translations or hybrids Latour proposes, which, after all, retain the structure of an isolated or closed system. It might call for an understanding of the human and nonhuman as participating in open systems and for the acknowledgment that scientific and philosophical activity are time-oriented so that the scientist must come to see herself as part of the universe she describes.[8] Minimally, science and politics might need to accept a pluralistic world in which reversible and irreversible processes coexist, or in which the former are an approximation of the latter (Bender 257).[9] This would then proscribe a restructuring of Latour's constitution, the vehicle for bringing science and politics together. In order to develop this point, let us consider some alternative figures whose politics and science might be less pure than that of Hobbes and Boyle. The point is to bring to light the limitations of classical thinking for science and for politics and to raise the question of whether or not even a structure such as that proposed by Latour evades those limits, and if it does not then to ask what consequences follow from this. Chief among these consequences, we propose, are what can be referred to as "the culture of extinction," an effect of classical scientific structures insofar as they operate in our science and in our politics. That is, a hybrid scientific and philosophical world that requires destruction in order to create anything new (Margulis and Sagan 28).[10]

The state of nature

By the seventeenth century, the same Robert Boyle (1627–1691) who is the object of Latour's interest had begun experimenting with gases, showing that air is compressible, meaning composed of particles with space between, and that gas volume is inversely proportional to pressure and temperature (Boyle's Law) (Margulis and Sagan 31).[11] More important, however, was the conceptual break of science with strict determinism when it was recognized that the motions of individual particles of gas could not be predicted. Only their behavior in structures turned out to be somewhat predictable using methods of statistical sampling when gases were studied as aggregates of particles rather than as individual particles.[12] The significance of this, for philosophy, is hard to ignore. If the stability of human ideas and reason is mediated by nature's regularity and predictability, but nature turns out to be merely probable and not certain, then where does this leave reason? Doubly probabilistic, since for empiricist philosophy, reason is a train of ideas arising from impressions, statistically probable, no longer predictable, even in an ideal sense. Soon Nicolas Leonard Sadi Carnot (1796–1832) formulated the first and second laws of thermodynamics. By quantifying the relation between heat and work, he discovered that energy is conserved yet not all heat can be turned into work. The First Law of Thermodynamics is concerned with quantity; in a closed system the total quantity of energy, whatever its transformations, will remain unchanged. The Second Law is concerned with quality; in a closed system, high-quality energy is lost to friction in the form of heat, the effect of the fast-moving atoms. The erosion of the quality of energy implies that the universe is not symmetrical with regard to time, in other words, in addition to prediction, another key feature of deterministic systems, reversibility, is lost wherever there are complex processes. Complex processes have tendencies and directions, whereas if you are just studying particles or simple processes reversibility can generally be maintained (Margulis and Sagan 29). Thus, with every scientific advance, we find ourselves moving from the laws of motion understood as transcendental Ideas that we can put to use to obtain certain knowledge, to the Laws of Thermodynamics, transcendental Ideas that are merely regulatory principles, time-oriented and probabilistic, not reversible and predictive.

At first this seems like a remarkable turn about, probability rather than predictability, a near-delirious sense of freedom. If nature is probabilistic rather than predictable, and reason is an effect of natural principles, then reason too tends towards the probable not the certain. Otherwise stated, reason too would be no more than regulative and gives us no certain knowledge. But what do we really mean when we claim that the laws used to calculate the motions of particles are probable rather than predictably certain? And what do we mean when we say that the Ideas of reason are regulative? The Second Law of Thermodynamics states that in a *closed* system, where no new matter or energy enters, disorder must increase. So, for example, in the classical mapping of gas particles distributed in two chambers, there are far more disordered states (mixtures of particles in various states) than ordered states (mixtures of particles in a limited number of states). When we say that there are many more ways for particles to be distributed evenly (in various states) than lopsidedly, what is meant is that probability is on the side of disorder, mixing, dissipation. In other words, the probable state of particles is one in which their energy—once unconcentrated—is *of little use*. So, for example, heat is of little use relative to the sunlight that generates it. Once again, we have arrived at the so-called arrow of time and left reversibility behind along with certainty, but such disorder might well be met with disappointment if not dismay. In the world of time asymmetry, Classical Thermodynamics studies structures whose complexity decreases, machines that lose the capacity to do work—closed and isolated systems sealed to incoming matter (Margulis and Sagan 34). At equilibrium, when molecules diffuse from higher to lower concentration, in other words, when molecules are most disordered (unstructured), then everything is homogeneous and *nothing interesting, remarkable or unusual* can happen; and, what is often called *complexity*, meaning the tendency toward order, is matched equally by the tendency toward disorder—heat and entropy (Devlin 79; Husserl 53).[13]

Differential calculus and vector fields

From a philosophical perspective, Gilles Deleuze takes up these developments in the natural sciences and mathematics by proposing a probabilistic but still *deterministic* philosophy formalized by means of differential calculus and vector fields. Motion and change are particularly difficult to study mathematically, as the tools of mathematics, numbers, points, lines, equations, are themselves static. Calculus, "a collection of methods to describe and handle patterns of infinity—the infinitely large and the infinitely small" made possible the use of mathematical tools to study motion and change without falling into paradoxes (Devlin 74).

> The basic operation of differential calculus is the process known as differentiation [whose aim is]...to obtain the rate of change of some changing quantity. In order to do this, the "value" or "position" or "path" of that quantity has to be given by means of an appropriate formula. (Devlin 86)[14]

Newton and Leibniz developed the rules for differentiating complicated functions by starting from the formula for a curve and calculating the formula for the gradient (or steepness) of that curve by taking small differences in the x and y directions and computing the gradients of the resultant straight lines—the gradient function is called the derivative of the original function (Devlin 90). "The crucial step...was to shift attention from the essentially *static* situation concerning a gradient at a particular point P to the *dynamic* process of successive approximation of the gradient [of the curve] by gradients of straight lines starting at P" (Devlin 87–88). Nevertheless, it is crucial to remember that the *apparently dynamic motion can only be captured, mathematically, by a static function* and, likewise, the dynamic process of closer and closer approximation to the gradient must also be captured in a static manner known as a *limit*, a sequence of approximations (Deleuze, *Difference and Repetition* 170–182).

To utilize and expand on the possibilities offered by differential calculus, Deleuze proposes what may be called a sublime Idea, an Idea in the Kantian sense insofar as it arises from and regulates its field immanently (Deleuze, *Difference and Repetition* 177). "Already Leibniz had shown that calculus…expressed problems which could not hitherto be solved or, indeed, even posed (transcendent problems)," problems such as the complete determination of a species of curve or problems characterized by the paradox of Achilles and the tortoise (Deleuze, *Difference and Repetition* 177). But what if we wish to make determinations beyond a single curve? Is there a means to make "a complete determination with regard to the existence and distribution of… [regular and singular] points which depends upon a completely different instance," an instance characterized in terms of a *field of vectors* (Devlin 44)?[15] The goal here is to explicitly link differential equations and vector fields. A vector field is defined, by Deleuze, as the *complete determination of a problem* given in terms of the existence, number and distribution of points that are its condition. This corresponds fairly well to the more or less standard mathematical definition where a vector field is defined as associating a vector to every point in the field space. Vector fields are used in physics to model observations, such as the movement of a fluid, which include a direction for each point of the observed space. Let us try to flesh this out. If, as is claimed, it is the *condition* of a problem that it would be the object of a *synthesis of the* Idea, then what is this Idea? The problem, which is that of the existence, number and distribution of points, is the object of a certain Idea, the abstract Idea of a vector field, a model consisting of vectors (in physics, an abstract entity that has magnitude and direction in a plane or in three-dimensional space, or in a space of four or more dimensions) in vector space, from which may be projected an infinity of possible trajectories in space-time (Devlin 44).[16] In mathematics (specifically linear algebra), the rules of association, commutation and distribution define vector space without reference to magnitude or directions; thus they may be utilized in a variety of fields whose terms are not material or physical.[17] The rules of association, commutation and distribution are the least restrictive set of linear rules that remain commutative; that is, for binary operations, any order is possible (Copi and Cohen 692).[18]

These rules have their equivalents in logic where association is an expression of logical equivalence permitting the valid regrouping of simple propositions; it governs the relations, which is to say, the connections between subject and predicate. In syllogistic terms, this would be expressed as the categorical relation (Copi and Cohen 694).[19] Commutation permits the valid reordering of disjunctive (hypothetical) statements (Copi and Cohen 698).[20] The rule of generation of a vector field is commutative. In logic, this means that statements may initially be combined in any order. Given a hypothetical proposition—"If there is a perfect justice, the obstinately wicked are punished"—it consists of a relation between two propositions: "There is a perfect justice," and "The obstinately wicked are punished" and can be understood as intrinsically commutative (Kant A 73–74, B 98–99). Reformulated using truth tables, we arrive at its logical equivalent—"There is *no* perfect justice *or* the obstinately wicked are punished," a proposition to which commutativity now applies ("The obstinately wicked are punished *or* there is *no* perfect justice") (Devlin 46–48). This reformulation separates, tears apart, the causal order of the inference, making it possible to reorder the propositions, to *disturb their causal linearity.* The success of this move may be debated, but that is not the most important consideration here. A third element of vector space is distribution. In deductive arguments, distribution permits the mutual replacement of specified pairs of symbolic expressions. In categorical propositions, it permits the distribution of a term if it refers to all members of a class (Devlin 86–87). In syllogistic logic, this would be expressed as *conjunction* or *connection.*

Deleuze argues that the rules for vector space apply to nature and so *nature* is associative, commutative, and distributive, where mere association, as opposed to unity, means that the laws of nature *distribute parts* which cannot be totalized, and that nature is conjunctive, expressing itself as this *and* that, this *or* that, rather than as Being, One or Whole. Following Kant's transcendental Ideas, the claim is that we can never have knowledge of nature as a whole. *Divine* power is manifest in its diverse parts (places, species, lands and waters); each *self* is not identical to any other; and every body comprising the *world* consists of diverse matter. As such, nature is associative, commutative and distributive and our knowledge of nature is limited to its immanent, regulative functions. If nature's immanent, regulative functions are associative, commutative and distributive, each of these, as mathematical or logical operations, reflects a view of nature as the power of things to exist one by one without any possibility of them being gathered together in a unity (Deleuze, *The Logic of Sense* 266–267). Following these rules, whatever has been added together can be taken apart and reformulated. Thus, becoming seems to be everywhere, nowhere is anything gathered together into a totality. How then, if nothing is gathered together, how is it possible to have a world or a concept of a world? Insofar as we are still in the realm of classical thought, the answer may be clarified using Hume's method of association of ideas. That things exist one by one without the possibility of being gathered together is the expression of purely external relations of association within an immanent field, relations of proximity that are further characterized in terms of the chance collisions of particles. However, the key to thinking through this aspect of the model that Deleuze proposes is that, at this point, the things that exist one by one are, in fact, not gathered into a Whole or a Totality governed by a Divine or transcendental principle; rather they are gathered by means of differential relations. Not surprisingly, the rules of vector space and the construction of vector fields are connected to the functions of differential calculus. That is, given a *complete* set of functions in a given space in differential calculus, each function in that space can be expressed as a combination of that complete set of functions, and functions with the property of completeness can form a vector space under binary operations.

Referring both to Lucretius and to classical physics, Deleuze argues that although the sum of atoms (or particles in physics) is infinite, they do not form a totality, an absolute and unconditioned totality for the series. Rather, atoms, which it is said are not sensible objects but objects of thought, atoms

> fall in the void and collide with one another. The result of these collisions produces motion. In the void, all atoms fall with equal velocity.... the velocity of the atom is equal to its movement *in a unique direction in a minimum of time*... [expressing] the smallest possible term during which an atom moves in a given direction, before being able to take another direction as the result of a collision with another atom. (*The Logic of Sense* 269)

Rather than a highly structured principle of causality, Deleuze argues, this collision, insofar as it takes place in a time smaller than the minimum of continuous time, a *limit* that approaches but never reaches *0*, must be something else. He calls it a *clinamen*, the original determination of the direction of movement of the atom (*The Logic of Sense* 269). This is the idea of looking for numerical and geometric patterns in a process of successive approximation. Long after Lucretius, the notation dx and dy came to be used to refer to infinitely small quantities or "the velocities of evanescent increments," once called "fluxions," ultimately, "a dynamic process of successive approximations" (Devlin 93).[21] Differential calculus, we noted, allows one to begin with a differential relation (Dx/Dy), the expression of limits, then, taking the derivative, to *derive* the formula for a curve, and then to calculate the formula for the gradient (slope) of that curve by taking small (infinitesimal) differences in the x and y directions, and computing the gradients of the resultant straight lines (Devlin 92–93).[22]

Each *derivative* of a *differential* relation takes place each time in a minimum of time. No doubt, the ultimate, subtle form of disjunction|conjunction, which in the physical world may be taken for causality, as each collision determines the next direction of the atoms, and if we were to look at patterns, rather than individual atoms, clear patterns do, in fact, emerge:

> In general, a differential equation arises whenever you have a quantity subject to change...Strictly speaking, the changing quantity should be one that changes continuously...However, change in many real life situations consists of a large number of individual, discrete changes, that are miniscule compared with the overall scale of the problem, and in such cases there is no harm in simply assuming that the whole changes continuously. (Devlin 92–93)[23]

Nevertheless, given that the use of vector fields requires the complete determination of a problem so that "if the differentials disappear in the result, this is to the extent that the problem-instance differs in kind from the solution-instance; it is the movement by which the solutions necessarily come to conceal the problem" (Deleuze, *Difference and Repetition* 177–178).

The immanent law of the state

We began with the idea that the classical model of modern natural science and the rational social contract are hybrids. Among social contract theorists, the claim was made that following the laws put forth by physics, the law of nature governs us and *reason is that law*. What is at stake in this version of the classical model of nature? Perhaps, it is argued, we began with an abstract concept, a Leviathan, a fully formed despotic state, a cerebral ideality, a transcendent regulating Idea or principle of reflection that organized the parts and flows into a whole. "But every such ideality, thought to be a *One* or a *Whole,* is nothing but a particular perishable and corruptible object, each of which we consider arbitrarily in isolation from every other object" (Deleuze and Guattari 217–222). In the end, it appears that what seemed to be first is really last.[24] For Deleuze and Guattari, the idea of the state as an abstract unity is regulated by an Idea (Deleuze and Guattari 139–144).[25] This Idea is both an Idea of reason insofar as it regulates behavior and a scientific Idea insofar as it functions according to the rules of a mathematical and natural scientific schema; thus it has *ontological* status. It is not, of course, a transcendent law, not an Idea that dominates and subordinates an otherwise fragmented field from outside that field. Rather, the state is and must be organized in accordance with an *immanent law*, a law which exercises its functions within a field of forces whose flows it *coordinates* (Deleuze and Guattari 221). It operates inside a field whose rules produce decoded flows of money, commodities, and private property; the immanent law produces a vector field that emerges out of a complete set of functions. So, for example, when we theorize the origins of the universe itself, we conceptualize that singularity, that explosion from an immensely hot, infinitely dense point 13.5 billion years ago, whose matter spread out three light years within one second, and by three minutes, cooling, it spread out 40 light years. We postulate abstractly that matter was simply traveling on the space and that the space simply expanded. And just as living matter internalizes— with increasing variation—the cyclicity of its cosmic surroundings, we too immanentize our Idea of origins (Margulis and Sagan 19–24). The structures offered by differential relations and vector fields make this possible without slipping into the formation of another metaphysical system.

Of course, many different types of states could have arisen; there is nothing intrinsically capitalist about differential relations, vector space or vector fields. Something specific is called for. It seems to be that "capitalism is the only social machine that is constructed on the basis of decoded flows, substituting for intrinsic codes an axiomatic of abstract quantities in the form of money" (Deleuze and Guattari 139). Capitalism liberates the flows of desire, when desire is understood to be the immanent binary–linear sphere of differentiations, in other words, the sphere of differential relations. This takes place under the social conditions that define the *limit* insofar as liberated desire is posited to be one with unlimited qualitative becoming. It is entirely the manner in which desire, the unlimited becoming, is associated, conjoined and distributed that determines whether it forms human subjects, works of art or social institutions (Deleuze and Guattari 224).[26] The capitalist machine establishes itself by bringing distribution expressed as *conjunction* to the fore *in the social machine* (Deleuze and Guattari 224).[27] Henceforth desire will be distributed and conjunction will be the pattern of its distribution. As nature is distributive, *every* conjunction may be transformed by the power of nature to break apart what has been connected, transforming every "and...and...and" into "or...or...or." Conjunction, as noted above, logically reformulated, immanently, is "the power of things to exist one by one *without any possibility of them being gathered together in a unity, a whole*" (Deleuze, *The Logic of Sense* 266–267).[28] Conjunction is the if...then that is so easily torn apart, separated into distinct events, each one a limit.

In economic theory, the differential relation and differential calculus make it possible to characterize change and rates of change. The differential equation can be utilized to express "the fundamental capitalist phenomenon of *the transformation of the surplus value of code into a surplus value of flux*" (Deleuze and Guattari 228). The overcoded flows of the despotic state seem to have been deterritorialized by means of nothing more than the formal power of conjunction to separate whatever has been connected. As Foucault describes it, classical representation disappears in the same manner. "There in the midst of this *dispersion* which it is simultaneously grouping together and spreading out before us...is an essential void: the necessary disappearance of that which is its foundation" (16).[29] What is left is nothing other than *pure form*. In other words, these are formal changes and there is nothing concrete in them. Whatever is concrete is *contingent*, pure effects and nothing more. So too with the disappearance of the despotic State and the rise of capitalism. It is a *pure form*. Deterritorialization simply arises in a situation of despotic overcoding, which in turn brings the despotic state to try to recode, necessitating, once again, "dissolutions...defined by the simple decoding of flows, dissolutions that arise from within in the *form* of the death instinct" understood as the tendency toward *inertial states*, the abolition of tensions, the tendency to continue in an existing state of rest or motion (Deleuze and Guattari 223).[30] This would be the point of capital and as such, the thermodynamically derived death instinct is capital; it is wealth that produces more wealth, thus it is perfectly inertial. Concretely, what inertial states, the death instinct, result in are flows: flows of property for sale, flows of money circulating, flows of production and of the means of production, flows of workers. All these flows subject to...conjunction, nothing but conjunction tearing apart, keeping separate, deterritorializing, leaving nothing but the form of decoded flows, till nature seemed to have become the pure process of production, and nothing else. But this would also be...pure death. So we should not be surprised that under these conditions, self and non-self, inside and outside, mean nothing, for nothing now has meaning, there is only the *continuous, smooth* motion of processes of production. And likewise, everything is production, there is nothing else (Deleuze and Guattari 2–3).[31]

Perhaps capitalism arose only when *industrial capital*, the direct appropriation of production by capital, the "tighter and tighter control over production" had overtaken all other flows (Deleuze and Guattari 226).[32] But what is the direct appropriation of production by capital except the pure process of production? This is not the domain of freedom from despotic regimes, but the domain of difference.

> We are in the domain of...the differential relation as a conjunction that defines the immanent social field particular to capitalism...the differential relation Dy/Dx where Dy is derived from labor power and constitutes the fluctuation of variable capital, and where Dx derives from capital itself and constitutes the fluctuation of constant capital. (Deleuze and Guattari 227–228)

Thus from the conjunction of decoded flows arises the distribution "$x + dx$" (Deleuze and Guattari 228; Devlin 79). In other words, from the surplus-value of the original investment of 100 pounds is derived the surplus-value of 10 pounds. This means that the *limit* defining differential relations is reproduced and extended, and everything, every single thing, animal, vegetable, mineral, industrial, conceptual, everything is included in the vaster and vaster vector field, which associates a vector (an abstract entity adhering to the rules that define vector space) to every point in the field space, the field of nature as processes of production. When production as a process overtakes idealistic categories, when production is the immanent principle of desire, then there are no more metaphors. Instead, something is produced; in fact, nature is now nothing but processes of production.

In the rationally calculative frame of mind of the social contract theorist, morality, especially a sense of fairness to others, was grounded in rational self-interest, yet it was able to be standardized as social virtue and pity for one's fellow human beings. Likewise, what was seen as prudent statecraft flourished, for self-reliance was demanded of everyone and contempt heaped on those who consumed without producing. Conspicuous displays of social superiority were to be replaced by the cultivation of a well-regulated, sober, and, above all, *private* inner life. But under capitalism, what can this mean?[33] When the terms of analysis, when the organization of the field of immanence under the rule of conjunction|disjunction, address quantitative characteristics alone, rate of speed, functions measuring the relation between labor power and variable capital, direction, then what are we talking about? "The person has become 'private' in reality, insofar as he derives from abstract *quantities* and becomes concrete in the becoming-concrete of these same quantities" (Deleuze and Guattari 251). In other words, the illusions fostered by the social contract are gone. Those citizens, those civil but abstract individuals, those effects of external relations are concretely nothing but their labor capacity. Each and every one is merely an application of the field as a whole. This is the so-called disappearance of the subject, the subject whose appearance was only ever brief and ghostly.

In mathematics, facts are proved by deducing them from a set of initial assumptions, called axioms, and truth has one fundamental form, the hypothetical, the conjunction: *If A, then B*, but axioms have no deduction. They must be accepted by the community of knowledgeable mathematicians and theorists. There is a certain amount of faith or belief involved in accepting them (Devlin 52). So what do we mean when we claim capitalism is axiomatic within a structure that produces what look like subjects and objects but are really nothing but effects of the vector field?

> Capitalism is indeed *the limit of all societies*, insofar as it brings about the decoding of the flows that the other social formations coded and over-coded. But it is the relative limit of every society; it effects relative breaks because it substitutes for codes [connections] an extremely rigorous axiomatic that maintains the energy of flows in a bounded state on the body of capital as a socius that is as deterritorialized, but also a socius that is more pitiless than any other. (Deleuze and Guattari 246)[34]

Although arising contingently through history, given that nature is associative, conjunctive and distributive, capitalism has become the axiomatic law of inertia, the death instinct. How can we understand this mathematically, formally?

> Suppose that a_1, a_2, a_3, ... is an infinite sequence of real numbers that get closer and closer together ... [closer to 0]. Then, there must be a real number, call it l, such that the numbers in the sequence get closer and closer to l (in the sense that, the further along the sequence you go, the closer to 0 the differences between the numbers *an* and the number l become). The number l is the *limit* of the sequence *a*1, *a*2, *a*3, ... (Devlin 98)[35]

In other words, capitalism is one with the *completeness axiom* of mathematics (Devlin 98).[36] To call capitalism a limit in a bounded system is to say that it operates as an attractor in that system. All so-called flows approach it infinitely but, since the attractor serves as a *limit,* the rules of the system cause all flows to pass through the same series of points forever.

If this is so, if this structure of hybrids accurately describes our current state, then we have no possibility of escape. Capitalism axiomatizes with one hand what it decodes with the other. If such decoding is called schizophrenia, then it is no more than capitalism's differential. If this is so, then what difference can it possibly make if one remains at the level of large aggregates or not? Clearly, Deleuze and Guattari recognize this:

> Molecular desiring-machines are in themselves the investment of the large molar machines or of the configurations that the desiring-machines *form according to the laws of large numbers*, in either or both senses of subordination.... Desiring-machines in one sense, but organic, technical, or social machines in the other: these are *the same machines under determinate conditions.* (Deleuze and Guattari 287)[37]

Below Oedipus, we find a molecular unconscious; beneath the stable forms, functionalism; under familialism, polymorphous perversity. What is the distinction between the molecular and the molar, the microphysical and the statistical? In the most formal, mathematical sense of the term, the molecular refers to Avogadro's number, the number of atoms needed such that the number of grams of a substance equals the atomic mass of the substance. An Avogadro's number of substance is called a mole.[38] Avogadro's hypothesis was key to solving many problems in the chemical sciences in the 1800s. For chemistry, molecules and moles are a matter of physical relations, ascertaining that equal volumes of gases with the same pressure and temperature contain the same number of molecules. By analogy, these terms reveal several possibilities. First, that the distinction between vitalism and mechanism does not hold; rather, "the machine taken in its structural unity, the living taken in its specific and even personal unity, are mass phenomena or *molar aggregates*...merely two paths in the same statistical direction" (Deleuze and Guattari 286). From the point of view of molar or statistical aggregates in physical relations, there is no difference between machine and life; *molar machines* may be social machines, technical machines, or organic machines. Physically, statistically, they are machines that appear as single objects and living organisms that appear as single subjects, both obeying the laws of thermodynamics.

And what of the molecular? *Desiring-machines* are of the molecular order. What does this imply? "Desiring-machines *work according to regimes of syntheses* that have no equivalent in the large aggregates...the indifferent nature of the chemical signals, the indifference to the substrate, and the indirect character of the interactions" (Deleuze and Guattari 288). So, we must note, desiring-machines do work according to regimes of synthesis, simply not those of large aggregates. Their syntheses have functional properties—properties that affect their operations, such as heightened coherence and efficiency, so that only on the level of fundamental structure does one discern the so-called play of blind combinations. The point is that the organization of molecules occurs independently; they are not structured *from above* by the large aggregates, the molar formations. That molecular syntheses gave rise to life may be due to a chance meeting of environmental factors, giving rise to highly organized living structures. The characterization of proteins as the ultimate molecular elements in the arrangement of desiring-machines and the syntheses of desire reveals the method at work here. The method requires that one goes *beyond* the molar to the molecular. How to reach this level, the most extreme level of analysis, where all aggregates are broken down and reduced to their most fundamental components? *Destroy, destroy.* The task of schizoanalysis goes by way of destruction—a whole scouring. We now distinguish between the pious Hegelian destructions where all that is negated is preserved and the impious schizoanalytical ones. With the latter, we appear to be standing before the gates of pure functionalism, that is, reduction. Perhaps this is the ultimate privatization of the individual, the reduction of each and every one to molecules, nothing more abstract and more equal in the universe.

Perhaps caution is called for here. Philosophers like Daniel Dennett have proposed that natural selection is a mindless, automatic, step-by-step process that a computer can simulate. It is a method unsuited to predicting particular cases of evolution, but capable of predicting general tendencies over large numbers, given enough time. If any process can be an algorithm, then all that would be necessary is to discover the algorithm for each process, thereby removing any notions of the accidental, the random, or the unexpected. However thrilling this may sound, it too has limitations. Many algorithms are completely uninteresting, not only because they imply the reinterpretation of an ongoing process as a step-by-step procedure, but also because the processes they organize are too general to be interesting. Moreover, even if algorithms do have some causal efficacy, there are many ways to explain or make sense of any particular phenomenon or set of phenomena. Looking to the contemporary postclassical sciences, no one demands conformity to a single theory or method simply because it is the most abstract formulation possible. Why should we do this in philosophy?

Given the molecular view of nature, the realm of desiring-machines, we might ask, what of sexuality? Did Hitler get the fascists sexually aroused? Does the bureaucrat fondle his records? Are women the pre-molar aggregates of men? And love, is it not simply a statistical molar aggregate, a Valentine's Day card, and inevitably, is the phallus not sexuality in its entirety? If so, then what are we to do? Are we to follow the rule, which is, *destroy, destroy*? Shall we reduce sexuality to its molecular components, its purely functional operations, a thousand tiny sexes? A thousand tiny sexes, so many molecules; do they too follow principles of heightened coherence and efficiency? Is this the meaning of local and nonspecific connections? Connectivity is, after all, the most primitive of rules for vector spaces and binary logic. It allows what might be called the greatest possible freedom, but for what, since, apart from analysis, it will participate in statistical aggregates? If the slogan for the desiring revolution is to be "to each its own sexes," then what is the "each" (Deleuze and Guattari 296)? For thermodynamics, "we are a particular material pattern of energy flow with a long history and a natural function. Our essential nature has more to do with the cosmos and its laws than with Rome (or any other human society) and its rules" (Schneider and Sagan xii).[39] If that "savage flow of desire" is the singularity called the "big bang," and it is one with that ontological (cosmological) flow of desire that is posited as originating and continuing to originate all existence in our universe, then certainly we can understand how molecular becomings are declared to be cosmic or cosmogenic forces and everything is called desire. And so, the culture of extinction continues. But just so we might still ask ourselves and wonder about what else might exist, and if the ultimate rule of thought must be destroy, destroy, destroy, rather than something else, something that does not commit us eternally to the death instinct and nature as nothing but processes of production, captive to capitalism and the endless proliferation of its axioms.[40]

Notes

Frederic L. Bender. *The Culture of Extinction, Toward a Philosophy of Deep Ecology.* New York: Humanity Books, 2003, 16–17. Bender traces both Western cultural and philosophical development to argue that culturally and philosophically Western ideas promoted and justified the ecological destruction of the planet. I would like to use this term to describe destruction in general. Marek Grabowski has been instrumental in helping me to understand and clarify mathematical concepts in this essay. His expertise in theoretical physics has been crucial. Needless to say, errors are strictly mine.

1. Latour cites Steven Shapin and Simon Schaffer. *Leviathan and the Air-Pump: Hobbes, Boyle and the Experimental Life.* Princeton: Princeton University Press, 1985.
2. Had Latour chosen Newton and Darwin, rather than Boyle and Hobbes, he might have had to concoct a different picture of what it means to be modern. See David J. Depew and Bruce H. Weber. *Darwinism Evolving: Systems Dynamics and the Genealogy of Natural Selection.* Cambridge: MIT Press, 1995.
3. Of course such exacting conditions are met only in highly circumscribed situations.
4. See: http://plato.stanford.edu/entries/continuity/. In his description over these pages, Latour's onto-epistemology clearly resembles that of Gilles Deleuze.
5. A discussion of such a dynamical system can be found in Dorothea Olkowski. *The Universal (In the Realm of the Sensible).* Edinburgh and New York: Edinburgh University Press and Columbia University Press, 2007, 208–219.
6. In other words, social scientists believe neither in God nor in consumer society but they do believe in science.
7. "A mediator, however, is an original event and creates what it translates as well as the entities between which it plays the mediating role."
8. The recent appearance of stinging criticism of French poststructuralism from the physicist Alan Sokal is evidence of the problem at hand. Sokal is as critical of Physicists who do not share his classical view as he is of philosophers who attempt to make use of science in their work.

9. Bender argues that modernism, which frames the world as a meaningless mechanism and is characterized by dualistic worldviews, persuades humans that they exist outside of and superior to nature, leading to the culture of extinction. I am making a different but nonetheless related argument, that our view of the *immanent laws* regulating Nature produces this culture.

10. Margulis is a well known evolutionary biologist, Sagan is a science writer. Margulis and Sagan will take this as an argument for a universe that is open and complex.

11. Ludwig Boltzmann (1844–1906) explained this one-way conversion of energy into heat and friction.

12. Margulis and Sagan point to the amount of mess, garbage, pollution, excrement, the average animal leaves behind and they argue that local order seems to produce disorder on the edges. They conclude that, if this is the case, politicians and governments should be aware of the consequences of order (structure) in an isolated system.

13. One can see that Husserl is already utilizing this model of differential calculus to solve exactly this sort of problem.

14. This required a "rigorous theory of approximation processes...the key idea of a *limit*" (87).

15. See also http://www.vias.org/simulations/simusoft_vectorfields.html and http://mathworld.wolfram.com/VectorField.html

16. "If you have a collection of entities called vectors, an operation of addition of two vectors to give a third vector and an operation of multiplication of a vector by a number to give another vector, and if these operations have the [appropriate properties, they are associative, commutative, distributive and closed under vector addition], then the entire system is called a vector space" (44).

17. See also "vector space." Britannica Concise Encyclopedia. Encyclopædia Britannica, Inc., 2006. Answers.com January 29, 2007. http://www.answers.com/topic/vector-space/

18. For example: [(p v q) v r] may be replaced by [p v (q v r)] and [(p . q) . r] may be replaced by [p . (q . r)].

19. For example: (p v q) and (q v p) as well as (p . q) and (q . p) may replace one another.

20. For example: A (B + C) = A B + AC.

21. This required a "rigorous theory of approximation processes" and the key idea of a *limit* (87).

22. So, for example, the differential equation $dP/dt = rP$ describes uninhibited growth where $(P)t$ is the size of some population and r is a fixed growth rate.

23. "Most applications of differential equations in economics are of this nature: the actual changes brought about in an economy by single individuals and small companies are so small compared to the whole, and there are so many of them, that the whole system behaves as if it were experiencing continuous change" (93).

24. 6200–6500 years ago there existed a civilization located between the Tigris and Euphrates Rivers near the Persian Gulf. The area was known as Mesopotamia and the city was called Ur. Because of the geographic location the city had fertile soil and was the perfect place to irrigate the land and raise productive crops, as well as domesticate sheep, goats and other animals. http://www.mnsu.edu/emuseum/archaeology/sites/middle_east/ur.html

25. This argument is fully developed in Olkowski, *The Universal*, chapter 2, "Love and Hatred."

26. See also Olkowski, *The Universal*, 138–139.
27. The savage social machine seems to have been organized by connection and the barbaric social machine by disjunction.
28. The hypothetical may be conceived of as the spatial manifold out of which things are determined to exist through the mutual determination of their position.
29. For an account of Foucault's formalist orientation see Vladamir Tasic. *Mathematics and the Roots of Postmodern Thought*. Oxford: Oxford University Press, 2001, 90–96.
30. See also Sigmund Freud, *Beyond the Pleasure Principle* 64–78. Inertial states refer to the first law of motion in physics, that a body in uniform motion or rest continues in that state unless it is changed by an external force.
31. "*Production* of *productions,* of actions and passions, *productions of recording processes,* of distributions and of co-ordinates...*productions of consumptions,* of sensual pleasures, or anxieties, of pain." Thus, it may not be so much the buying and selling of production that defines capitalism as it is the tighter and tighter control over production by the death instinct, that is, inertia.
32. "All sorts of contingent factors favor these conjunctions. So many encounters for the formation of the thing, the unnamable!" (226).
33. See also Max Weber. *The Protestant Ethic and the Spirit of Capitalism*. Trans. Stephen Kalberg. London: Blackwell, 2002.
34. Translation altered. Once the system is established what may have been contingent may become axiomatic. When flows are bounded by capitalism, it serves as a limit in a mathematical dynamical system; in other words, flows will cycle around it endlessly.
35. Such a limit is also known in dynamical systems theory as an attractor. See Olkowski, *The Universal*, chapter 6.
36. "The construction of real numbers and the development of rigorous theories of limits, derivations and integrals...was the beginning of the subjects nowadays referred to as *real analysis*. These days, a fairly extensive study of real analysis is regarded as an essential component of any college education in mathematics" (98).
37. Determinate conditions are statistical forms (the larger the groups averaged, the less the variation). See note at the bottom of p. 287.
38. http://scienceworld.wolfram.com/physics/AvogadrosNumber.html The Avogadro Constant, L, is a constant number used to refer to atoms, molecules, ions and electrons. Its value is 6.023×10^{23} mol^{-1}.
39. They refer here to the poet Joseph Brodsky, who wrote that humans are closer to the big bang than to Rome (xi).
40. Alternatives exist, but not in the dualism of molecular–molar. As such what is needed are completely different scales than those proposed by Deleuze and Guattari. I have suggested alternative scales in *The Universal,* chapter 1.

Works cited

Bender, Frederic L. *The Culture of Extinction, Toward a Philosophy of Deep Ecology*. New York: Humanity Books, 2003.

Copi, Irving M. and Carl Cohen. *Introduction to Logic*. New York: Macmillian Publishing, 1990.

Depew, David J. and Bruce H. Weber. *Darwinism Evolving: Systems Dynamics and the Genealogy of Natural Selection*. Cambridge, MA: MIT Press, 1995.

Deleuze, Gilles. *Difference and Repetition*. Trans. Paul Patton. New York: Columbia University Press, 1994. Originally published in French as *Différence et Répétition*. Paris: Presses Universitaires de France, 1968.

Deleuze, Gilles. *The Logic of Sense.* Trans. Mark Lester with Charles Stivale, Constantin V. Boundas, ed. Constantin Boundas. New York: Columbia University Press, 1990. Originally published as *Logique du sens.* Paris: Les Editions de Minuit, 1969.

Deleuze, Gilles, and Félix Guattari. *Anti-Oedipus, Capitalism and Szhizophrenia, vol. 1.* Trans. Robert Hurley, Mark Seem, and Helen R. Lane. Minneapolis: University of Minnesota Press, 1987. Originally published in French as *Anti-Oedipe.* Paris: Les Editions de Minuit, 1972.

Devlin, Keith. *Mathematics: The Science of Patterns. The Search for Order in Life, Mind, and the Universe.* New York: Scientific American Library, 1994.

Foucault, Michel. *The Order of Things. An Archaeology of the Human Sciences.* New York: Vintage Books, 1973.

Freud, Sigmund. *Beyond the Pleasure Principle.* Trans. James Strachey. New York: Bantam Books. 1972.

http://mathworld.wolfram.com/Topology.html

http://mathworld.wolfram.com/VectorField.html

http://www.mnsu.edu/emuseum/archaeology/sites/middle_east/ur.html

http://plato.stanford.edu/entries/continuity/

http://scienceworld.wolfram.com/physics/AvogadrosNumber.html

http://www.vias.org/simulations/simusoft_vectorfields.html

Hume, David. *A Treatise of Human Nature.* Oxford: Oxford University Press, 1968.

Husserl, Edmund. *Ideas, General Introduction to Pure Phenomenology.* Trans. W. R. Boyce Gibson. New York: Humanities Press, 1969.

Kant, Immanuel. *The Critique of Pure Reason.* Trans. Norman Kemp Smith. New York: St. Martin's Press, 1965.

Latour, Bruno. *Pandora's Hope, Essays on the Reality of Science Studies.* Cambridge: Harvard University Press, 1999.

——. *The Pasteurization of France.* Trans. Alan Sheridan and John Law. Cambridge: Harvard University Press, 1988.

——. *We Have Never Been Modern.* Trans. Catherine Porter. Cambridge: Harvard University Press, 1993.

Margulis, Lynn and Lynn Sagan. *What is sex?* New York: Simon & Schuster, 1997.

Olkowski, Dorothea. *The Universal (In the Realm of the Sensible).* Edinburgh and New York: Edinburgh and Columbia University Presses, 2007.

Shapin, Steven and Simon Schaffer. *Leviathan and the Air-Pump: Hobbes, Boyle and the Experimental Life.* Princeton: Princeton University Press, 1985.

Schneider, Eric D. and Dorion Sagan. *Into the Cool. Energy Flow, Thermodynamics and Life.* Chicago: University of Chicago Press, 2005.

Stengers, Isabelle and Ilya Prigogine. *Order Out of Chaos. Man's New Dialogue with Nature.* New York: Bantam Books, 1984.

Tasic, Vladamir. *Mathematics and the Roots of Postmodern Thought.* Oxford: Oxford University Press, 2001.

Weber, Max. *The Protestant Ethic and the Spirit of Capitalism.* Trans. Stephen Kalberg. London: Blackwell, 2002.

9
Katrina
John Protevi

Hurricane Katrina was an elemental and a social event. To understand it, you first have to understand the land, the air, the sun, the river and the sea; you have to understand earth, wind, fire and water; you have to understand geomorphology, meteorology, biology, economics, politics, history. You have to understand how they have come together in the past to form, with the peoples of America, Europe and Africa, the historical patterns of life of Louisiana and New Orleans, the bodies politic of the region, bodies you need to study with political physiology. You have to understand what those bodies could do and what they could withstand, and how they intersected the event of the storm.

The land and the river

We could start by talking about plate tectonics, or, for that matter, stellar nucleogenesis, for at the limit everything is connected, and to tell the story of Katrina would be to tell the story of all of the earth, all of the cosmos. But everything is connected, not for a God's eye view, but just past the limit of actualization, just beyond the limit when things slow down enough for them to take form. Not even a God could see everything past that limit, because past that limit there is nothing, nothing fixed, though there are elements, as well as relations and singularities; there are multiplicities. Past the limit everything is connected as sets of related changes, rates of change smoothly changing in relation to other rates of change, up until the singular points where those relations change drastically rather than smoothly. Everything is connected, but nothing is fixed, there is no thing there, nothing anyone, not even God, could see or know. So let us begin, as always, *in medias res*, and talk about the land of Louisiana. To do that, we have to talk about the river. The Mississippi drains a vast swath of the North American land mass. Almost all of the water that falls between the Appalachians and the Rockies, from the Allegheny river in upstate New York to the Missouri River in Montana, almost all of it that does not evaporate or

165

stay in the soil or in a lake, almost all of it drains down the Mississippi, the "father of waters."

Like all rivers that flow into the sea, the Mississippi snakes about at its head, creating a delta as it floods its banks and lays down sediment picked up upstream by those drops of water, which, when they flow fast enough, pick up bits of earth to carry with it. How big those bits of earth are depends on how fast the water flows, and how fast the water flows depends on how much is behind it, pushing it, and on how big a channel it flows through. The bigger the mass of water and the narrower the channel, the faster the flow and the more earth it carries, and the greater the chance the river will overflow the banks and drop that sediment as it slows down and trickles over the face of the earth. As the water flows, it will eddy and swirl, depending on the configuration of the banks and bed, for sometimes a singular configuration will trigger a different pattern of flow. The relation of turbulence and smooth flow will vary, then, not by calculable laws, as would velocity in a perfectly smooth channel, but according to singular points in the configuration of bank and bed. This process went on for some time, as these singularities of the configuration determined which actual pattern of the river's flow would emerge from the differential relations (velocity and turbulence) of the elements (water and earth). At another level, they also determined the historical pattern of that emergence of actual flow, the rhythm of the river's flow and flood.

Slowly, however, after European settlements progressed, flood control via levee construction began. The Europeans built military outposts and then cities on the edge of the river, to the north on the bluffs above the river (Baton Rouge, Natchez, Memphis), and to the south in the swamps along the edge of the river, between the river and the inland sea they called, strangely, a "lake" (New Orleans). The river, when it floods repeatedly, as it does as part of its natural cycle, leaves a ridge of sediment, highest on the banks and sloping gently away, down toward the swamps, the heaviest sediments dropping out first, building up the natural levee, which tends to subside as it dries out. It was on that natural levee that Bienville founded New Orleans in 1718. The floods still came, of course, since the ridge was a *natural* levee. On top of that levee the Europeans used their slaves to build more levees, all up and down the river. What happens when you build levees upstream of a point? You change the actualization structure of the water flow, the values that incarnate the multiplicity, the set of linked rates of change and its singular points. There are many ways a flow can occur, given its differentials and singularities, the way its elements intersect the configurations of its channel. In this case, you squeeze the same amount of water into a narrower channel (for now you contain the same relation of mass and velocity that would have flooded the previously lower banks), so the water downstream rises still higher. And so you build higher levees. As this positive feedback loop continues, you break the river's old rhythms, you change its relation to the earth around it, you de-territorialize it.

However, you cannot stay ahead of the river all the time. Researchers have found that river systems tend toward a state of self-organized criticality, producing a power law distribution regarding riverbank failures and flooding. In other words, one of the guiding principles of human ecology (analogous to treating forest fires) should be that with levee construction you can stop many little floods, but when the river does flood finally—and the probability of its flooding at some point increases the larger the time scale—then those floods will be big ones (Fonstad and Marcus 2003). The most famous and destructive of those big ones happened in 1927, and, to save New Orleans, the powers that be, out of panic, dynamited the levee southeast of the city, flooding Saint Bernard Parish and endangering if not killing many poor people. (In fact, levee failures upstream, which came the very next day, meant that the city was never really in danger. We will discuss failures of judgment in panic situations later.) In response to 1927, the relation of the federal government to the states changed, for another factor was then in place: the mass media. Radio, newspapers, telegraph, photos transmitted the verbal and visual images of the flood's effects—which were even more severe in Mississippi's Delta region than in Louisiana—to the rest of the nation, including the sight of thousands of African–Americans stranded on levees for days with no food or water and later herded into relief camps. The trauma of the viewers and listeners in other parts (for you can be traumatized by seeing and hearing images of pain caused to others—here we would have to talk about mirror neurons and natural sympathy, which we will do later in discussing the use and misuse of Hobbes in discussing Katrina) created a demand to control the river even more. So the Corps of Engineers, along with state and local boards, built more and better and bigger levees, enough, they said, to withstand a Category 3 hurricane, determining that level of protection adequate after a cost–benefit analysis (Barry 1998).

The effects of the Corps projects on the river and the land were profound. First, there is the problem of "subsidence" as the sediment that would ordinarily build up the land is stopped by upstream dams trapping sediment and levees all along the river preventing flooding. Exacerbated by New Orleans's canal and pump system, untreated subsidence means the city is sinking ever faster, as pumping the city dry means drier, more compacted soil, creating a positive feedback loop making the city even more vulnerable. Furthermore, many ship channels and canals were cut into the wetlands. In particular, a deep channel (the notorious Mississippi River Gulf Outlet) was created close to New Orleans so that ocean-going ships would have a short, direct path from the Gulf to New Orleans, where they could intersect another flow carried by the river's flow: the flow of goods from the middle of the country (Bourne 2000). Coupled with the denial of flooding, this shipping channel (plus hundreds of smaller canals and channels created by the oil industry) changed the ratio of salt and fresh water in the bayous, killing many species of plants and animals. The cypress trees were logged, another insult to the wetlands, for sediment had previously collected around the roots. Another positive feedback loop was set up as coastal erosion accelerated, and another coastline was actualized, one much closer to New Orleans. The other famous positive feedback loop with the river, the "dead zone" in the Gulf of Mexico from fertilizer runoff, did not play a role in Katrina's impact, at least to my knowledge, but we can fit it into our story, when we turn to discussing another element, the sun.

The sun

The sun figures in our story in many ways, in evaporation, wind production, and bio-energy. Bio-energy is solar energy, mediated by carbon. Photosynthesis forms carbohydrates out of water and carbon dioxide, releasing oxygen and trapping solar energy in the chemical bonds of the complex molecules. Animal metabolism burns this fuel, combining oxygen with the carbohydrates and releasing the energy, plus water and carbon dioxide. The process is a little more complicated than that, as nitrogen mediated by microbes in and around plant roots plays a role (and producing concentrated nitrogen fertilizers takes a lot of petroleum or carbon-stored solar energy, and they run off down the Mississippi, producing algae explosions that suck the oxygen out of the Gulf and killing the local sea fauna, creating the "dead zone"), but it is basically pretty simple: solar energy becomes bio-energy through the mediation of carbon.

The most efficient form of bioavailable carbon-mediated solar energy capture is sugar cane. Cane production under European supervision was shifted from the Atlantic islands to the Caribbean throughout the sixteenth to the eighteenth centuries, using slave labor, at first the native Caribs, then the Irish (let us not fall for the old "indentured servant" line: when Cromwell, in the midst of the Wars of the Three Kingdoms, intensified the shipping of thousands of Irish to Barbados they were slaves in all but name), then the Africans. As sugar consumption in Europe worked its way down the class stratifications, displacing other forms of bio-energy, some of which came mixed with protein and vitamins, it played a key role in a positive feedback loop described by Marx in the primitive accumulation chapter of *Capital*, for a good percentage of modern state revenue came from consumption taxes, which, together with national debt, funded the military branches of the colonial enterprise. (Marx's analysis centers on Britain, of course, so we would have to work out, *mutatis mutandis*, the effect on France and hence on the all-important Franco–British relation.) The fact that increased sugar consumption played a role in decreased height for British Army recruits from 1780 to 1850 was offset by various differentials: increased mechanization and internal complexity of the armed forces for one thing (they call the Queen's Guards the Beefeaters for a reason).

Here we see another multiplicity: the differential elements are carbon-mediated solar energy and human muscle power, the relations are hypertrophy|atrophy|dystrophy, and the singularities are genetic potentials scattered in the population.

To explain how Africans came to Louisiana, we also have to talk about how they came to the Caribbean. To see that, we need to talk about yet another multiplicity in which solar energy is the key; the heat exchange system of the planet. The elements here are sun and water and air, the relations are heating and cooling, and the singularities are the thresholds for Ice Ages, grand and small, as well as smaller but meteorologically interesting events, such as the Atlantic Multidecadal Oscillation or AMO (Goldberg et al. 2001). The AMO competes with global warming (Webster et al. 2005) as an account for the increased frequency of high-intensity Atlantic hurricanes in the post-1990 era. In this exchange system, we find ocean currents, including the Gulf Stream (whose heat-carrying capacity and its climatological effects help explain how Northern Europe can carry the density of population it has had over the past few millennia by affecting the types of carbon-mediated solar energy it can cultivate in producing bio-energy via agriculture), and the North Equatorial Current, which, along with the Trade Winds, helped Europeans travel from the coast of Africa to the Americas. (A shorter ocean and wind current loop system just off the coast of Africa might explain why it seems African sailors rarely went too far off shore, for it was enough for their navigational purposes up and down the western coast of Africa. Of course it should be admitted that historical research into the exploits of early African sailors has not been pursued with nearly the same enthusiasm as for early European sailors, the Vikings in particular.)

So, meshing the multiplicities of the global heat exchange system and the bio-energy system, we see the actualization we call the Atlantic slave trade. The differential elements here are human muscle power and production processes, or, more precisely put, the amount of force over against the precision of direction of that force, or as we put it, skill. The Atlantic slave trade was internally complex. Just to take the example of the eighteenth-century trade to French Louisiana, which was virtually complete by 1731, we must take into account the multiplicity that links the European appreciation of real and imagined differences between Wolof and Bambara slaves, as well as Portuguese competition for slaves in Africa, and competition from buyers in the French West Indies. The resumption of slave trade to Louisiana under Spanish rule is even more complex, as the trade was no longer almost exclusively from Senegal, but included slaves from Central Africa (Hall 1992). In some cases, the Atlantic slave trade tapped into the well established trans-Saharan slave trade, and took the skilled bodies of captive African peasants (we have to overcome our ridiculous urban, that is verbal, prejudices—our Platonic prejudices—and recognize the great embodied skills of peasants), and, after transporting them across the ocean, deskilled or proletarianized them, and set them to work in industrial agricultural practice on Caribbean plantations (industry can occur in plantations as well as factories, as demonstrated by Mintz 1995). I say "in some cases," because not all Africans were subjected to deskilling: the wonderful vernacular Creole architecture of Louisiana, for instance, is directly traceable to the work of skilled African slave architects who used Senegambian practices to build homes from native cypress—beginning the deforestation whose effects on coastal erosion we noted above. Among other transplanted economic skills, we should also note that the origin of rice cultivation in the Americas was due to skilled African agriculturalists (Carney 2002). In fact, proletarianization as part of industrial production on large plantations was the (highly profitable) exception to the rule, as many Africans worked at a wide range of skilled tasks (Fogel and Engerman 1974). (It should go without saying that all this talk of skills and deskilling refers to the strictly political–economic concept of proletarianization, and has no bearing on the magnificent creativity or awe-inspiring resistance and resilience of African–American culture, during and after the period of slavery. This resistance, often armed and violent, began, of course, with revolts in the slave depots of Africa, continued on the ships of the slave trade, and persisted on the American mainland.)

As this system developed, the French colony of Saint Domingue became one of the most profitable, if not the single most profitable, agricultural complexes in the world, growing sugar and coffee. (We could tell another story about how coffee and sugar go together, as well as how tea and sugar go together, and how that difference between French and English tastes ties into the fact that the English beat the French out of India, and then later how the importance of India allowed the English to get out of the slave business in the nineteenth century: why bother with slaves in the Atlantic when the future of the Empire lies in Asia? Before we even mention the Opium Wars, we can see a complex system that is quite literally "political physiology.") But at some point in the late eighteenth century the Saint Domingue system passed a threshold: too fast a rise in the importation rate of "fresh" or "unseasoned" Africans, plus a feverish step-up in production, plus the singularities called Boukman and Toussaint, led to a revolt.

As the years of war went on, quite of a few of the *gens libres de couleur* fled Saint Domingue for Louisiana, now a Spanish colony, but with a heavy French heritage, bringing with them their African slaves, and thereby creating one of the factors accounting for the difference between light-skinned and dark-skinned African–Americans in New Orleans and Louisiana. You cannot underestimate the importance of this difference in understanding the social relations of contemporary Louisiana; the ignorance of this difference contributes to the buffoonery of calling the light-skinned Ray Nagin "a black mayor." (This is not to say that Nagin belongs to traditional high Creole society, just that his light skin allows him to play the traditional

Creole role of mediation between the dark "Africans" and the whites. "Mediation" is a polite word: the free people of color of New Orleans reprised their role in Saint Domingue and became the hunters of escaped African slaves. The way the NOPD took up that disciplinary function is another story that needs to be told.)

Commercialized sugar production in Louisiana did not begin until 1795, as under French rule it was suppressed by the French metropolitan government to avoid competition with Saint Domingue, and under Spanish rule the cultivation was never very extensive. Louisiana itself was never a very prosperous or well-managed colony under French rule. Shaken to its roots by the Natchez Rebellion of 1729, Louisiana languished in the middle of the eighteenth century, only beginning to revive in the 1770s under Spanish rule. It is only after the Haitian revolution had begun that sugar production began in Louisiana in 1795, sparked by an influx of refugee planters, and using the skills and labor power of "seasoned" or "creolized" slaves from Saint Domingue. After the transfer of the vast Louisiana Territory to the United States in 1803, sugar production boomed (Rodrigue 2001). The motivation for Napoleon's sale of Louisiana is generally attributed to his realization that Saint Domingue was lost to him (despite Jefferson's offer to help supply Leclerc's expeditionary force in its attempt to reinstall slavery there), and thus that Louisiana's putative role as food supply for the much more profitable Caribbean island was mooted (the question of Napoleon's designs for a North American extension of his empire, with Louisiana as its base, is much more difficult to answer).

However, at the same time that Jefferson bought Louisiana, he signed on with the British in attempting to suppress the Atlantic slave trade. Where would the slaves for Louisiana's sugar plantations—and the slaves for the cotton plantations of the Deep South, now made possible by the cotton gin—come from? Why, from Jefferson's home state of Virginia, among other sources for the internal slave trade. Here we find yet another multiplicity, in which physiology and psychology intersect work and climate, thus determining the reproduction rates for African slaves: negative in most parts of the Americas, but positive in the Chesapeake region. Sugar production in the Caribbean meant overwork (dawn to dusk during planting, around the clock during harvest and processing of the cane, and yearlong, as tropical climate allowed multiple growing seasons per year); such overwork, along with the additional factors of bad nutrition, heat, disease, and torture, explain the life expectancy of seven years in the Caribbean after arrival (Blackburn 1997)—meaning a slave owner's punitive sale of slaves to the Caribbean had a clear disciplinary intent, as clear as if he had killed the slave outright. This low life expectancy necessitated a constant importation of "fresh" slaves to make up for the low birth rates and high infant mortality rates—the children died from disease, and from the infanticide slaves practiced sometimes to spare their children. The reproduction rate of the slave population was positive in the Chesapeake region, because of the multiplicity governing tobacco production, linking the energy expenditure of the workers (linked to the singularities of the tobacco plant: its size, the angles of its stems and leaves, and so on) and the relatively moderate climate, which reduced growing seasons, allowed the winter as a period of recuperation, and eliminated the threat of tropical disease. So the ancestors of the current African–American population of Louisiana came from the Caribbean, directly from Africa (mostly from Senegal), and from the northern states.

One of the things the Caribbean arrivals brought with them was revolution, the hope of it among the slaves and the fear of it among the whites. After the 1729 Natchez Rebellion, the most famous episode in Louisiana colonial history is the 1795 Pointe Coupee slave "revolt." Hall (1992) tells the fascinating story of how this Jacobin-inspired multiracial class revolt occurred at the most radical point of the French Revolution, when the National Assembly had recognized the *fait accompli* of the Saint Domingue revolt by abolishing colonial slavery, and when Republican troops fought all the royal powers of Europe, including the Spanish who owned Louisiana at the time. A conspiracy that included many revolutionary poor whites, it aimed at the propertied interests, rather than at "whites," but it subsequently became mythologized as a race war of black against white. Unleashing a wave of racial oppression, the Pointe Coupee conspiracy became the bogey that put the fear of a racially motivated slave revolt directly into the bodies of white Louisianans. As they had been scared as children with tales of the hoped-for savage reprisals for slavery exacted by rebellious black savages, with murder, looting, and rape prominent among them, a panic threshold was established, triggered at the thought or sight of crowds of blacks without sufficient armed guards around them. This bit of political physiology will play a role as well in the Katrina aftermath, exacerbated no doubt by the securitarian phobias of post-9/11 America.

The wind and the sea

We have talked about the Trade Winds, and about the meteorological multiplicity formed by the global heat exchange system. The Northeast Trade Winds, blowing northeast to southwest off the coast of Africa, tend to converge at certain points, triggering singularities and forming turbulences associated with the instability and displacement of these winds in the movement north and south of the Intertropical Convergence Zone or ITCZ. With the proper ocean temperatures (above 80 degrees) and wind speeds, we can get "tropical waves," or groups of thunderstorms. At other singular points of wind speed and water temperatures, a cyclonic heat engine will be actualized, spinning counterclockwise in the Northern Hemisphere, and a hurricane will be formed. Ocean water evaporates and rises out of a "chimney," releasing energy as it condenses aloft, powering winds and forming bands of thunderstorms. In effect, part of the ocean rises into the air and falls back as rain, while part of the ocean is pushed along by the storm's winds, the famous "storm surge." While most hurricanes form off the West African coast, Katrina formed to the east of the Bahamas. Crossing Florida, it hit upon the Gulf "loop current," a deep hot water current that flows to the west through the Yucatan channel, then north through the Gulf until it exits to the east through the Florida channel to join the Gulf Stream. In its current configuration, the loop current brought Katrina not only a vector aiming it at New Orleans, but also huge amounts of energy, for it conserved the energy contribution of the evaporation rate of the 90 degree surface water, as the deep water churned to the surface by the passage of the hurricane was not as cold as it would have been for a hurricane not following the vector of the loop current. As it happens, a late singularity in metrological conditions caused Katrina to swerve a bit and pass to the east of New Orleans, thus sparing the city the worst winds (in the northeast quadrant for Northern Hemisphere hurricanes, which spin counterclockwise due to the Coriolis effect), but devastating Slidell and the Mississippi coast.

So Katrina hit the eroded Louisiana coast, its still strong winds pushing its storm surge into Lake Ponchatrain and destroying some of the floodwalls and levees along New Orleans's canals. Since a hurricane loses 3–8 inches of storm surge for every mile of barrier islands and coastal wetlands it crosses, the Louisiana coastline of 100 years ago would have weakened Katrina enough that the current lake and canal levees of New Orleans would have held. But the eroded coastline let the storm surge through, and the faulty levees collapsed, rather than being "overtopped." The city was flooded, as predicted. At which point, we have to discuss the "man-made disaster."

Hobbes and the people of New Orleans

The government reaction, at local, state and national levels, needs to be seen in historical context. Louisiana was a slave state, and it is now ranked the forty-ninth poorest state in the Union on many measures. (This is not to imply that slavery's economic and political impact was limited to the South. For just a sample of the political importance, beyond the infamous "three-fifths clause" of the Constitution, consider that ten of the pre-Civil War Presidents of the United States were slave-holders, as well as two post-war who had been slave-holders—a ratio that means that to date, one-quarter of the Presidents in US history were slave-holders. The role of slavery in the global and national economies outside the slave zones, especially as it impacted the capital formation of the "Industrial Revolution," was also profound (Bailey 1998). Thirty-five percent of its population is African–American. This figure is higher in cities, which, following the American pattern, surround mostly black cities with mostly white suburbs. The pattern of white flight, sparked by post-WWII suburbanization, is of course another multiplicity, as racial, income, and wealth population differentials cross automobile ownership rates. This was famously demonstrated in New Orleans, a city with 67% African–American population, and a 23% African poverty rate (national average is 12.7%), a large percentage of whom did not have cars. While 80% of the city evacuated, those who stayed behind—or who were left behind—tended to be black. But not all: the French Quarter, 95% white and located on the natural levee where Bienville started the city, and Uptown, along the same levee westward along the river, had many whites who stayed behind. While the startling images of the Superdome and the Convention Center dominated media coverage, one of the most important untold stories unfolds along the geographical and social differential in which money stays on high ground in New Orleans. In Uptown, the private security companies ("Blackwater USA"), with their M-16s and their retired Special Ops forces, their helicopters and their guard dogs, created an enclave of the protected who avoided media attention (Scahill 2005), except for the ever-vigilant *Wall Street Journal*, which reported the plans of those civic leaders to effect "demographic" changes in the city to be rebuilt (Cooper 2005).

Let us talk now about the famous sites, the Superdome and the Convention Center. The name of Hobbes sprang from the keyboards of the commentators as they heard the breathlessly reported rumors of savagery (murder, looting and rapes) and the repeated 20 second loops of "looters" (you must recognize them by now: the woman holding up the Pampers to shield her face from the cameras, the teenage boy skipping through the puddles with his shopping cart). A "state of nature," they wrote, a "war of all against all," they assured us (Ash 2005; Lowry 2005; Will 2005). But what were the contents of those rumors? A "revolt" at the Orleans Parish Prison. Children gang-raped and thrown, throats slit, into the freezer at the Convention Center. "Snipers" shooting at rescue helicopters. A "lockdown" of downtown after a "riot" in Baton Rouge following the arrival of refugees. All these rumors were unfounded and their similarity to rumors in panics about slave revolt cannot be ignored (Dwyer and Drew 2005).

We will have to await the results of patient historical work to know what "really happened" in New Orleans. What we can say now with good confidence in comparing eyewitness testimony, taking into account the widely documented production of rumor and exaggeration in crisis situations, is this: there were at most a few score gang members at the two sites engaging in predatory behavior. But we have to balance these stories against the small number of bodies found at the two sites. According to the *New York Times* report cited above, "state officials have said that ten people died at the Superdome and 24 died around the convention center—four inside and 20 nearby. While autopsies have not been completed, so far only one person appears to have died from gunshot wounds at each facility." While rumors of rapes abounded, police active in the Superdome found little to back them up. Again according to the *New York Times*,

> During six days when the Superdome was used as a shelter, the head of the New Orleans Police Department's sex crimes unit, Lt David Benelli, said he and his officers lived inside the dome and ran down every rumor of rape or atrocity. In the end, they made two arrests for attempted sexual assault, and concluded that the other attacks had not happened.

(It is true that the NOPD is, by general acclaim, probably the worst police department in the country: underpaid, undertrained, overly corrupt, homicidal and fratricidal, and they would be motivated to cover up violence at the Convention Center and Superdome so they could claim that guarding the French Quarter antique shops instead of the people in the shelters did not really hurt anyone. Still, to propose a conspiratorial cover-up here is more than far-fetched, since many government agencies, not just the NOPD, would have to have been involved. What were far-fetched in the first place were the rumors of mass rape and carnage.)

So at the most, we would have stories of a few groups of 10 to 15 young men who hunted other gang members, robbed people, raped women. What can we say about this predation? The first thing to say is that it does not indicate a "reversion" to a "state of nature," for gang members prey on people in these ways in every city of the country and most cities of human history. For gang predation is not some "natural" state into which societies "fall," but a social process that is part and parcel of civilization as we have known it. All the way back to Achilles and the Myrmidons, young men have formed roving predatory gangs that prey on urbanized populations. The relation of such gang-formation processes to urban population density forms another multiplicity to be explored; in New Orleans, gang-formation tends to follow the pattern of housing projects. The few reported murders during the Katrina aftermath can most plausibly be related to either spontaneous fights or gang members running into each other off their favored territories. Encounters during such episodes of deterritorialization would presumably carry with them high probabilities of violence. However, let us not forget this eyewitness testimony, for it shows that not all gangs were predatory. Denise Moore related the following on the *This American Life* radio program:

> [Interviewer: Tell me about the men roaming with guns.]
>
> They were securing the area. Criminals, these guys were criminals. They were. Y'know? But somehow these guys got together and figured out who had guns and decided they were going to make sure that no women were getting raped. Because we did hear about the women getting raped in the Superdome. That nobody was hurting babies. They were the ones getting juice for the babies.... They were the ones fanning the old people. Because

that's what moved the guys, the gangster guys, the most, the plight of the old people.

[Concerning the looting of the Rite-Aid at St Charles and Napoleon.]

They were taking juice for the babies, water, beer for the older people [chuckles], food. Raincoats so they could all be seen by each other and stuff.... I thought it was pretty cool and very well organized.

[Interviewer: Like Robin Hood?]

Exactly like Robin Hood. And that's why I got so mad, because they're calling *these* guys animals? *These* guys? That's what got me mad. Because I know what they did. You're calling *these* people animals? Y'know? C'mon. I saw what they did, and I was really touched by it and I liked the way they were organized about it, and that they were thoughtful about it. Because they had family they couldn't find too. Y'know. And that they would put themselves out like that on other people's behalf. I never had a high opinion of thugs myself, but I tell you one thing, I'll never look at them the same way again. (Moore 2005)

But while the predatory gangs, as opposed to the protective gangs, now had an especially concentrated population on which to prey, and a police force weakened by desertion and dispersed by the mayor to restore "law and order" (that is, guard the antique shops and restaurants of the French Quarter, the big hotels and Uptown residences being sufficiently guarded by their private security forces), we cannot forget the massive solidarity shown by the people of New Orleans. Why was it that the name of Hobbes came flying off the keyboards of our pundits, and not Rousseau or even Locke? Why the focus on the predation, which occurs everywhere, though admittedly with less intensity because of lower prey population density? What about the solidarity on display?

Hobbes is a brilliant philosopher, a great philosopher. His ruthless materialism far outstrips the limited minds of the current crop of Hobbes-mongers; their incomprehension of Hobbes's relentless focus on material power means that they will produce howlers, like Lowry's decrying of "massive lawlessness" in New Orleans. This is absurd for someone decrying a lack of awareness of Hobbesian philosophy. For Hobbes is crystal clear that, while the sovereign has no obligation to the people, the people in turn have no obligation to a failed sovereign. The sovereign's actions are constantly judged by the people, and, when the sovereign's power fails, then civil law lapses and the laws of nature are the only ones in operation. We can read this in several of the most prominent and important parts of *Leviathan*. For instance, in Chapter 17, "Of the Causes, Generation, and Definition of a Common-Wealth," we find the following: "if there be no Power erected, or not great enough for our security; every man will, and may lawfully [by the 'laws of nature' of course] rely on his own strength and art, for caution against all other men." In Chapter 21, "Of the Liberty of Subjects," we read: "The Obligation of Subjects to the Sovereign, is understood to last as long, and no longer, than the power lasteth, by which he is able to protect them. For the right men have by Nature to protect themselves, when no one else can protect them, can by no Covenant be relinquished." Finally, in Chapter 27, "Of Crimes, Excuses, and Extenuations," the following rings out as clear as a bell: "That when the Sovereign Power ceaseth, Crime also ceaseth; for where there is no such Power, there is no protection to be had from the Law; and therefore every one may protect himself by his own power." Thus Lowry is so far from understanding Hobbes that what he will see as the "massive lawlessness" of New Orleans, the taking of goods when sovereign power has failed, is completely and utterly obedient to the law of nature.

Despite Hobbes's brilliance, the anthropological worth of his state of nature thought experiment is next to nothing, for the atomization he predicts in crisis situation is belied by the massive evidence of spontaneous group formation: with family kernels of course, but also by neighborhood, and also, notably, simply by civic and human affiliation. ("My people" is how Jabbar Gibson described the group of neighbors and strangers he gathered on his commandeered bus.) Hobbes will acknowledge the possibility of temporary alliances in the state of nature, as in this passage from *Leviathan* Chapter 13, "On the Natural Condition of Mankind," where he writes: "if one plant, sow, build, or possesses a convenient Seat, others may probably be expected to come prepared with forces united." So while people can form temporary alliances in the state of nature, they won't be able to form durable political units, if these fall short of the absolute sovereign. Time scales are the key here: the longer the state of nature goes on (and Hobbes is as equally concerned with competition for honor in the state of nature as he is with competition for material goods) the more the atomizing forces take hold.

The stories of New Orleans we tell do not always have to focus on the every man for himself fantasy of the Hobbes-mongers, nor on the panicky rumors with their echoes of the fear of slave revolt, but should also be the stories of the thousands and thousands and thousands of the brave and loving people of New Orleans who refused to leave their old, their sick, their young, their helpless, and who walked miles through the floods to safety, pushing wheelchairs and floating the sick on "looted" air mattresses. Yes, we saw images of helpless poor people waiting to be rescued at the Superdome and the Convention Center, but we should never forget that they rescued themselves prior to that, through heroic solidarity, through what we should not be afraid to call "love" (in the sense of *philia*, for Aristotle the emotional concretion of the political nature of humanity).

How can we give a rigorous differential materialist reading of this *philia*, this solidarity? After all, we would not want to be mere Rousseau-mongers in response to the Hobbes-mongers. The question is that of the emergence of human groups. Emergence is the (diachronic) construction of functional structures in complex systems that achieve a (synchronic) focus of systematic behavior as they constrain the behavior of individual components. Theories of social emergence compete with methodological individualism, which denies that social phenomena are anything but the aggregation of individual behaviors. It is important to note that methodological individualism is far more than a "theory": it is the guiding principle behind the active construction of the atomizing practices whose results are described as natural by Hobbes and his followers (Schwartz et al. 1979).

To piece together the multiplicity behind social group emergence, we could begin by tracking the development of infantile face-recognition and emotional sensitivity, which inscribe brain patterns as they develop in feedback loops with caregivers (Hendriks-Jansen 1996). Here we see that sociality is inscribed in our very bodies, as the actualization of the set of linked rates of change and their singularities lying between the reciprocally determined ideal elements of "infant" and "caregiver." (Since such "caregivers" are enmeshed in a historically mediated web of social relations, we should not fall for any sort of "familialism" in thinking these terms.) Such early bonding is of course a repetition of bondings that stretch back throughout human and primate history (De Waal 1996). As would be expected with complex systems in populations, not all bondings "take," of course, either from caregiver absence or neglect, or from difficulties on the side of the infant. Later-appearing trauma can also interrupt or destroy previously established bondings (Niehoff 1999). Nonetheless, many bondings do take, and the neurological basis of these bondings, as with all brain activity, is found first in resonant cell assemblies, which form out of a chaotic firing background in a modular but decentralized network (Thompson and Varela 2001). In these patterns of brain activity we could isolate what have been called "mirror neurons," whose link to what is experienced and described as empathy is a fascinating research frontier (Gallese 2001). The bondings formed through mirror neuron activity would be reinforced at another level of activity by corporeal entrainment, which also plays a key role in producing group solidarity (McNeill 1995).

Of course, not all group formation is of the sort we are looking for in explaining the solidarity in New Orleans. We have to distinguish between the passive affects of subjected groups and the active affects of group-subjects. Subjected groups are swept into a homogeneous mass whose unity is imposed by a transcendent signifier, like a flag. Being taken up out of yourself to join a larger unit can be a hugely powerful emotional experience. We can even call it "erotic" if we remember that this notion of *eros* is wider than that of sexual union. The symbol of a subjected group is a trigger that evokes that feeling of transport into a larger whole. The rage felt when the signifier is disrespected is directly related to the joy in erotic transport into the group, and that joy is inversely related to the pain felt in being subjected to atomizing practices: the sort of everyday isolation and its concomitant feeling of powerlessness that is well attested to in America. Imagine the power of the emotions we call "patriotism," then: the larger and more powerful the political unit you belong to, and the weaker and more isolated you feel on your own, the stronger the emotional surge, the more sacred the symbols. We can then say that the keepers of those symbols have a vested interest in increasing your pain in isolation in order to increase the power they get from controlling the keys to your joy in union. So an empire of isolated and powerless citizens would be a powerful and dangerous beast indeed!

On the other hand, we have to think the joy of the active group-subject, the immanently self-organized spontaneous group formation we saw in New Orleans and elsewhere. For that solidarity was not just demonstrated among the people of New Orleans, but among many of the people of Louisiana. Alongside the brave men and women of the Coast Guard and the Louisiana Department of Wildlife and Fisheries, let us not forget the hundreds of volunteer rescuers, who came down to New Orleans in their trucks and their boats, pulled somehow by that solidarity to rescue strangers. These rescuers, though able to work the first few days on their own, were eventually refused entry to the area by FEMA, which gave us the worst of all possible governmental responses: not only did they not do it themselves, they refused to get out of the way and let the volunteers do the work. For two reasons: one, the securitarian|racist panic over thousands of blacks together without enough police, and two, because—and here we are at the limit of paranoia, but indulge me—they ideologically want government to fail.

Here we see some of the political consequences of the neoliberal denial of the very truth of solidarity (and hence government as expression of solidarity) and consequent production of atomized behavior (and hence government as transcendent source of order against anarchy). For you can also do the inverse, strain solidarity and increase atomization, via scarcity—in some situations, for, after all, you can also increase solidarity via scarcity. Scarcity is an intensifier of underlying processes, a catalyst. And scarcity is produced, let's not forget, according to a multiplicity whose elements are rich and poor people, "good" and "bad" neighborhoods. Scarcity is produced so that poverty is actualized along the social–geographical differential relation of access to goods, a differential enforced by the police at the singularities of entry points to privileged neighborhoods and bridges to cities across the river. During the post-Katrina weakening of the police presence, we then saw four types of "looting" actualized from this multiplicity: (1) getting necessities of life: food, water, medicine, diapers; (2) taking of non-necessities for future use or resale (one of the things our Hobbes-mongers underplay is that along with the tolerated looting of type 1, type #2 cannot be condemned, for there is no crime in the state of nature, only private judgment as to what is necessary to secure the future); (3) revenge against the rich (you've taunted me with your fancy goods my whole life, so I'll wreck them so you won't have them either); (4) nihilistic rage (you've left me to die, so fuck you, I'm burning it all down).

So the political lesson is not that we need order from above to prevent the anarchy that is supposedly close by, but that the solidarity that holds almost all of us together, the civic and human bonds that led all those thousands to stick together, and led those hundreds of volunteers to head to New Orleans, needs only support from a government which—instead of being systematically dismantled and artificially rendered inadequate so that it can be, in the now horribly ironic words of Grover Norquist, "drowned in a bathtub"—needs to be recalled to its proper function as the organized expression of that solidarity.

Notes

Many thanks to my research assistant, Ryanson P. Ku, and to comments from James Williams, Mark Bonta, Robin Durie, Paul Patton, and Jeff Nealon.

This article was previously published: "Katrina." *Symposium: Canadian Journal of Continental Philosophy / Revue canadienne de philosophie continentale* 10:1 (Spring 2006) 363–381, and is reprinted here with kind permission.

Works cited

Ash, Timothy Garton. "Just Below the Surface." *Guardian Weekly* (September 9, 2005).

Bailey, Ronald. " 'Those Valuable People, the Africans:' The Economic Impact of the Slave(ry) Trade on Textile Industrialization in New England" in *The Meaning of Slavery in the North*. Eds D. Roediger and M. H. Blatt. NY: Garland, 1998.

Barry, John. *Rising Tide: The Great Mississippi Flood of 1927 and How It Changed America*. NY: Simon and Schuster, 1998.

Blackburn, Robin. *The Making of New World Slavery*. London: Verso, 1997.

Bourne, Joel. "Louisiana's Vanishing Wetlands: Going, Going..." *Science* 289 (September 15, 2000): 1860–1863.

Carney, Judith. *Black Rice: The African Origins of Rice Cultivation in the Americas*. Cambridge, MA: Harvard University Press, 2002.

Cooper, Christopher. "Old Line Families Escape the Worst; Plot the Future; Mr. O'Dwyer, at His Mansion, Enjoys Highball With Ice; Meeting With the Mayor." *The Wall Street Journal* (September 8, 2005).

De Waal, Frans. *Good Natured: The Origins of Right and Wrong in Humans and Other Animals*. Cambridge, MA: Harvard University Press, 1996.

Dwyer, Jim and Christopher Drew. "Fear Exceeded Crime's Reality in New Orleans." *New York Times* (September 29, 2005).

Fogel, William and Stanley Engerman. *Time on the Cross: The Economics of American Negro Slavery*. NY: Norton, 1974.

Fonstad, Mark and W. Andrew Marcus. "Self-Organized Criticality in Riverbank Systems." *Annals of the Association of American Geographers* 93:2 (2003): 281–296.

Gallese, Vittorio. "The 'Shared Manifold' Hypothesis: From Mirror Neurons to Empathy." *Journal of Consciousness Studies* 8:5–7 (2001): 33–50.

Goldberg, S. B., C. W. Landsea, A. M. Mestas-Nunez, and W. M. Gray. "The Recent Increase in Atlantic Hurricane Activity: Causes and Implications." *Science* 293 (July 20, 2001): 474–479.

Hall, Gwendolyn Midlo. *Africans in Colonial Louisiana*. Baton Rouge: LSU Press, 1992.

Hendriks-Jansen, Horst. *Catching Ourselves in the Act: Situated Activity, Interactive Emergence, Evolution, and Human Thought*. Cambridge, MA: MIT Press, 1996.

Lowry, Rich. "A National Disgrace." *National Review Online* (September 2, 2005).

McNeill, William. *Keeping Together in Time: Dance and Drill in Human History*. Cambridge, MA: Harvard University Press, 1995.

Mintz, Sidney. *Sweetness and Power: The Place of Sugar in Modern History*. NY: Penguin, 1995.

Moore, Denise. "After the Flood." *This American Life* (September 9, 2005). www.thislife.org.

Niehoff, Debra. *The Biology of Violence*. NY: Free Press, 1999.

Rodrigue, John. *Reconstruction in the Cane Fields*. Baton Rouge: LSU Press, 2001.

Scahill, Jeremy. "Blackwater Down." *The Nation* (October 10, 2005).

Schwartz, Barry, Richard Schuldenfrei and Hugh Lacey. "Operant Psychology as Factory Psychology." *Behaviorism* 6 (1979): 229–254.

Thompson, Evan and Francisco J. Varela. "Radical Embodiment: Neuronal Dynamics and Consciousness." *Trends in Cognitive Science* 5 (2001): 418–425.

Webster, P. J., G. J. Holland, J. A. Curry, and H. R. Chang. "Changes in Tropical Cyclone Number, Duration, and Intensity in a Warming Environment." *Science* 309 (September 16, 2005): 1844–1846.

Will, George. "Leviathan in Louisiana". *Newsweek* (September 12, 2005).

10
Technoecologies of Sensation

Luciana Parisi

With the ingression of a digital architecture in cybernetic culture, media have ceased to be instruments of communication and have become part of an atmospheric grid of connections where distinct milieus adapt together as microclimates in complex weather systems. Whether you are at the airport, the shopping centre or the underground, a mediatic environment unfolds through innumerable resonances through audio-visual, video-telephonic mobile connections ready to envelop a technoculture addicted to constant feeling. This environment may not simply be explained in terms of information overload where too many messages result in a paralysis of communication and sensibility. Rather, it may be that the symptoms of sensory overload in digital atmospheres offer some clues to the transformation of sensing modalities in cybercapitalist culture. Indeed, a rewiring of modes of feeling seems to be at the core of a new regime of cybernetic power no longer operating through perfectly integrated circuits of communication, but through a new interlocking of distinct milieus of information sensing. Such interlocking does not simply indicate a new calculation of sensory probabilities of communication, but rather seems to imply a radical reorganization of the velocities of information sensing resulting into an anticipation of feeling (that is, the feeling of feeling) or the sensation of preemption.[1]

This article argues that changes in technical machines are inseparable from changes in the material, cognitive, and affective capacities of a body to feel. It suggests that current modifications in cybernetic and bioinformatic machines of communication are leading to the formation of a technoecology of information sensing, implying a new level of relatedness between organic and inorganic milieus of transmission. In particular, the article will focus on the bionic tendencies of new media technologies, which involve not simply an extension of sensory perception but a mutation in sensations all together. Here sensation extends beyond sensory perception to expose a *nonsensuous* mode of feeling irreducible to the split between the mental and the physical, the rational and the sensible.[2]

Machinic involution

As Deleuze and Guattari remind us, the question of technology needs to be addressed in an anti-Aristotelian fashion. The technical machine needs to be rethought in terms of a more vague yet more real mechanosphere where machines are not subsets of technics. For Deleuze and Guattari, machines come first (*A Thousand Plateaus* 71). Against the Aristotelian tradition whereby *techne* teleologically creates that which does not exist in nature, Deleuze and Guattari pose us a Spinozian question: what can a machine do? Here it is no longer a question of looking for the creative action of humans on nature by attributing to *techne* the invention of artefacts. Rather, it is a matter of prioritizing a *machinism* in nature, not the optimal functioning of machines, but the breaking down, the anomalies and the escape route of social, aesthetic, technical, cultural machines.

In contrast to structures or systems, constituted by elements meeting at a steady point, a machine cannot be divided into parts without changing its nature. Its components primarily enter additional compositions instigating continual irregularities, unexpected disjunctions at singular points. Deleuze and Guattari use Maturana and Varela's notion of autopoiesis to think of machines as animated processes of self-organization enjoying certain plasticity in their intra-action with the outside. Yet their concept of machines goes further (*Anti-Oedipus* 42). What remains unique to Deleuze and Guattari's notion of machines is an amodal ontology that rejects at once mechanicism and vitalism: the reduction of machines to their constitutive parts or to the formed substance of life, *bios*. To think machinically is to engage with technical machines in terms of semi-concatenations of partial objects running through strata. It entails bringing *abstract matter* back into the analysis of information technologies and the networked body.

So what counts as a machine?

We have a machine each time there is an "ensemble of the interrelations of its components, independent of the components themselves" (Guattari "Machinic Heterogeneities" 41). Thus, a technical object is nothing outside the technological ensemble to which it belongs and from which it can mutate. Far from being a matrix of combinatorial codes, a machine is always already traversed by internal, external and associated milieus, regions of supra-action ready to give way to new protomachines: machines in potential, futurity-machines. There is a whole ecology of machines traversing substantial scales: mental, natural, social, technical dimensions ceaselessly code drift, side-communicate across space and time. Whilst scrambling genealogical evolution, the routes from the simple to the complex based on hereditary filiation, machinic assemblages also add surplus values of code to connection, a viral hyperlink between micro and macro levels of organization resulting in involution or becoming (*A Thousand Plateaus* 238).

Each machinic level is then overpopulated by compossible machines: an open calculus of infinite series ruled by tendencies of convergence and divergences between a thousand worlds (Deleuze, *The Fold* 92).

For Guattari, the series of machinic ensembles—energetic, semiotic, algorithmic, diagrammatic, social, neuronal, desiring—linking transversally material, cognitive, affective, and social machines can be in direct contact with technical-experimental configurations giving rise to new proto-ontological, protoethical and protoesthetic transactions. From this standpoint, this article suggests that a new alliance between biological, neural and affective machines is directly connected to new technical configurations of information sensing coinciding with a new ontology of power. In particular, such an alliance is entangled with the technical applications of bionic technologies experimenting with visceral and neural sensations by assembling together neural, bacterial and silicon milieus of information sensing. This new level of *machinic involution* cannot but be felt by a body registering a new threshold of relatedness between organic and inorganic matter at the core of the mediatic atmospheres of an amodal power operating by the preemption of feeling.[3]

Sensory feedback

Deleuze and Guattari's notion of machinic involution helps us to rethink new media technologies neither in terms of form (technical medium) nor content (the code, the signifier), but as *technoecologies of sensation* intersecting energetic, cognitive, affective capacities of feeling.

Recently, it has been argued that new media are to be viewed as databanks of disorganized information, which is at each time framed by a centre of indeterminate perception—that is, my living body.[4] Whilst the conception of media as databanks may be useful to an understanding of media as more than mere objects of communication, this argument, however, seems to overemphasize the transparency of subjective sensory perception—albeit indeterminate—by overlooking the levels of variation of the centre. Indeed, such a centre is less to be conceived as a constituted subject than a distributed *superject* laying out the nexus of variable percepts emerging from the folds of matter itself.[5] Here each milieu of information can apprehend itself whilst apperceiving its external variations.[6] It would, then, be impossible to maintain that new media—information databanks—are enframed by my biological—lived—body since the latter can only subsist in its intricacies with sub and super layers of percepts in matter that are not governed by an actualized point.

Media technologies, and new media, have most often been discussed in terms of cybernetic feedbacks of information where a continuous adaptation between the inside and outside of the system, rather than the fixed positions of the subject and object, is privileged. When Norbert Wiener argued that technical machines—as well as living systems—receive information through their kinesthetic organs, he also highlighted that modern automatic machines no longer function according to the logic of the clock, where future action is predetermined by past behaviors (21–22). Cybernetic automata "possess sense organs; that is, receptors for messages coming from the outside" (23) through which they adapt and adjust to contingent rather than expected conditions (23). The sensory calculation of past events enables the future tendency towards energy dissipation to become a probability. Hence uncertainties are turned in negentropic order, where past sensory experiences constitute the base of calculations for the future.

Cybernetic media

Thus, the cybernetic principles of media communication point out that sensory feedbacks regulate message patterns in analog and digital communication. As Kittler argues, all media, whether analog or digital, are information systems since they store, record and reproduce data for our sense perception (Gramophone, Film, Typewriter 34). Decoupling information from digitality, Kittler suggests that modern mass media—telegraphy, telephony, radio, and television—already count as information systems. Unlike printing and writing, modern media are information systems since they "make use of physical processes which are faster than human perception and are only susceptible of formulation in the code of modern mathematics" (Kittler, "What's New about New Media?"). Yet modern media are to be distinguished from digital media, since the latter not only handle the transmission and storage of data but also control, through more sophisticated mathematical algorithms, the processing of commands.

For Manovich, digital or new media also constitute, unlike analog media, an information database, which, like the computer, helps to discriminate, control and exploit the smallest variabilities, timetables and orientations (1999). As Kittler reminds us, this digital processing of commands involves the speeding-up of information rate, which erases all distinction between media and renders human perception obsolete (*Gramophone* 31). Unlike McLuhan, Kittler highlights that media are not simply extensions of the human senses. Media have a historicity of their own based on strategic feedbacks amongst themselves irreducible to sensory perception (Kittler 32). Whilst McLuhan affirms that each electronic medium coincides with bodily senses, for Kittler the digitalization of information has given way to a sort of "embryonic sack supplied through channels that serve the purpose of screening out the real background: noise, the night and the cold of an unlivable outside" (32).

The mathematical catalyzer of digitalization, Kittler argues, not only excludes noise from binary calculation, but also renders physical sensing superfluous. Yet, Kittler's explanation of the progressive evolution of media towards digital nonhuman rates of probabilities seems to miss the direct relation that technical machines have with energetic, mental and affective machines. In short, Kittler overlooks the machinic intricacies of the physical and nonphysical and opts for the binary opposition between the sensible and the rational, which explains the new media age as a disincarnated circuit of information bits. Yet, a machinic view of new media implies not the ultimate evolution towards a digital matrix of perfect neuro-communication, but the viral intersection of digital computation with its own differential analog calculus: the double transduction from analog waves to digital particles and back to the analog does not simply render the physical sensorium redundant but rather exposes the extrasensory (sensuous and nonsensuous) capacities of a body to feel. Here, the digital dubbing of analog frequencies entails an amplification of the physical and the nonphysical at the threshold between micro and macro perceptions, whereby feeling cannot be disentangled from thinking (Murphie 199).

Bioinformatic integration

Yet, it has often been argued that the digitalization of physical systems—the binary codification of analog organisms—has been crucial for the application of cybernetics to biology, developing into the technoscience of bioinformatics. Donna Haraway famously stated that integrated binary circuits neutralize the difference between humans, animals and machines (1991). Her view on the Integrated Circuit of Command and Control may be read together with Deleuze and Guattari's notion of societies of control, where the relation between humans and machines "is based on internal, mutual communication" (*A Thousand Plateaus* 458). Yet Haraway, contrary to Deleuze and Guattari's antipathy for the model of communication, does not challenge the metaphysics of the digital cyborg as the only possible field of corporeal politics.

At the core of the cyborg, but also central to Kittler's conception of media, is a notion of information transmission derived from Shannon's theory of communication (*Mathematical Theory of Communication*). Here information patterns are individuated quantities integrating together all qualitative differences, where redundant patterns of information enable the message to pass through all channels of transmission without change.[7]

In bioinformatics, Shannon's theory is used to classify genetic information into digital bits, to preserve genetic data transferred across media channels. Thacker, for example, argues that bioinformatics tends to isolate wet information from the sterile silicon bed so as to prevent mutation during the transmission through different channels (DNA in a plasmid and DNA in silicon) (Thacker 53). Yet, he suggests that this bioinformatic view of the relationship of genetic and silicon data is not based on a reduction of the genetic code to a string of 0s and 1s, but to patterns of relationships across different material substrates: a cross-platform preservation of specific patterns according to a bio-logic that privileges the interaction between components on a new nonbiological level (Thacker 54). If bioinformatic machines integrate biological and digital material substrates, what remains to be challenged, it will be pointed out here, is the ontological condition of such binary combinatorics.

The question of information transmission, from analog to digital media, from genetic codes to digital databanks, is often problematically posed in terms of reproducibility—reproducing the same across different milieus—and in terms of discrete codes opposed to continual waves, numbers opposed to qualities, technical to biological, information to sensation, culture to nature, mind to body. There are questions that encompass these problems at once: what does the movement of transmission entail? To what extent can information be disentangled from sensation?

Recently, Brian Massumi argued that whilst the digital defines a set of actualities ready to be recombined, it is far from giving us the virtual (137–138). The virtual is of another nature from the digital. It is the abstract immanent phase space of each actuality and never exhausts itself in one realization or another. The virtual is a full body of vibratory activities remaining in potential, enfolding bodies of intensities, where particles annex to waves and waves split into particles. The analog, as Massumi suggests, is yet more inclined to part companionship with the field of the virtual. The analog expresses continual variations of the direction of the wave according to the pressures of its environment of action.

However, the bioinformatics transduction of the biological into the digital realm of data points to a direct contact between a-signifying codes—algorithms—and a-semiotic encodings—DNA—a machinic involution. The preservation of biological into digital data seems not simply to engender a nonbiological platform of interaction but, more importantly, an ontology of a-biological inevitabilities: an amodal activity inciting the biological and digital into a new machinic arrangement. This is when a transversal connection—a contagious surplus value of code—takes over the form and substance of content and expression of analog and digital media to unleash its proper force of invention raising from a changing relatedness between organic and inorganic layers in matter. Digital media do not use biological media as sources for a new bio-logic. There is no dialectical quarrel between these modes of transmission, only a viral transduction able to spin out microvariations during the transfer.

It is then possible to rethink the analog–digital, biological—technical relation outside the ontology of the One. If there is a superiority of the analog, it will privilege neither phenomena of subjective perception nor those of objective transportation of the bio-logic to a digital level. Rather, it will affirm the tendency of binary systems to code drift, of numbered numbers to enter the fuzziness of numbering numbers, of atoms and molecules to unleash their own micropercepts, where information transmission coincides with rarified areas of vague sensations, with the velocities of felt-thought. Here units or codes are neither logical nor organic, neither based upon pieces nor formed or prefigured by those units in the course of a logical development or of organic evolution (Deleuze, *The Fold* xiv). Units, codes are propagating ciphers, ensembles of contagious numbers turning the mathematics of probabilities into the infinitesimal potentials of feeling-thoughts. This is an incorporeal pack of information sensing housed by a body however small, however inorganic. At all levels of matter there is immaterial corporeality ready to entertain various degrees of togetherness beneath the unity of codes and organisms. The particle and the wave are together enveloped in the infinitesimal speeds of matter defining nanozones of feeling, adding new levels of physical and nonphysical perception, affection, and mentalities to the body.

Bionic sensorium

If we look at recent bionic technologies, such as neuromorphic chips, cochlear electronic implants, synthetic and engineered retinas, electronic tongues and extended limbs (see Geary), it remains difficult to hold onto a notion of information that preserves its instructions whilst passing through different milieus of transmission. Unlike the cyborg, bionic technologies highlight not the dematerialization of the body in information patterns, or the rematerialization of the biological in extrabiological context, but the biomathematical relatedness of distinct milieus of information, a nexus of felt relationality between inorganic and organic rates of sensing. Such biomathematical correlation has a long history and was first explored in the nineteenth century by D'Arcy Thompson, who explored the mathematico-geometric continual variations between biological and mechanical forms.[8]

Such relatedness is synthesized by the biochip or BioMEMS (biological MicroElectroMechanical Systems) which, during the 1990s, replaced electronic transistors with small strands of bacterial DNA entrapped in silicon wafer that could be directly connected to the brain.[9]

Contrary to the bioinformatics of integrated codes, biochips such as *in vivo* blood pressure sensors (with wireless telemetry), DNA chips, *in vivo* drugs probes, cannot function without inciting an information trade between biomolecules, MEMS devices, and data signals. Here information is not transmitted between the environment, body and machines, but an entire ecology of information sensing is at play in the movement of trans-mission between channels. What is at stake here is the extrapolation of numbers from the *unnatural* conjunction between milieus of information sensing. This is not a recombination of probabilities, but, as Deleuze suggests, the packing together of extensions or intensive quantities (*The Fold*).

The extension of feeling

With bionics—a term coined by Clines and Klives in 1958—information sensing has entered the realm of corporeal prosthetics, where bodily parts are not only extended but are themselves engineered to become semi-biological, semi-mechanical and semi-electronic devices. Although biotechnology may be supplanted by nanodesign, combining genetic engineering with robotics, bionic technologies seem crucial for extracting sensing potentials below and above frequencies of habitual sensory perception. Indeed, bionic technologies seem to directly ingress sensuous and nonsensuous perception, or, as Whitehead remarks, perception of immediate presentation (sensory perception of the here and now) and perception of causal efficacy (thought perception of the there and then) (Process and Reality 180–181). Bionic technologies seem to directly connect with the causal field of sensation, accounting not simply for sensory–motor perception, but, more importantly, for the causal intricacies of the physical and the nonphysical, whereby thought itself is felt.

Bionic technologies thus count neither as mechanical extension nor as digital dematerialization of the physical sensorium. Rather, the bionic sensorium is above all implicated in the machinic extension of the nonphysical and physical capacities of feeling, entailing the rearrangement of sensation at the shortest span of time. Thus, bionic technologies are not simply the sensory enhancement of cybernetic systems—media, humans, animals—but more importantly are in the process of constituting a veritable technoecology of sensation—a machinic intricacy of organic and inorganic milieu of information sensing preceding sensory perception, resulting in an inarticulate sensation, that is, an unframed feeling. This is not the emotional or the sensational, but, as Deleuze argues, a synthetic sensation rejecting all figurations and representations of sensing and directly acting "on the nervous system, the levels through which it passes, the domains it traverses" (*Logic of Sensation* 39).

Technoecologies of sensation install themselves in the machinic field of code-drifting communication ready to engender surplus values of sensing at all scales of transmission. Here distinct information milieus combining at certain speeds add molecular zones of sensing to perception. For example, whilst until now artificial retinas used in the vision systems of robots or smart missiles have been based on silicon, researchers are developing a biochip that would use a protein found in bacteria as a digital storage medium.[10] The protein, called bacteriorhodopsin, is photosensitive: it changes properties when exposed to laser beams of differing wavelengths. This protobacterial artificial retina, able to react in only a few microseconds to changes in light intensity, adds microdurations to optical perception.

Whilst bionic technologies push neurosensorial perception towards microecologies of sensing, a new physiological resculpturing of technical machines is also coming into place. James Geary points out that, as biochips enable human bodies to enhance their senses, so computers are increasingly able to see, smell, taste and touch (3). Indeed, the new tendency towards biomechanics and nanotechnological engineering seems to emphasize the sensitivities of mechanical and digital devices able to fuse more directly with the highly complex layers of biophysical responsiveness.

Moving beyond the "yes|no" logic of conventional computers, Rosalind Picard argues for the need to include emotional responses in the designing of computers able to interact intelligently with humans.[11] At the MIT "Affective Computing Lab," she has been involved in the designing of sensing devices (gloves, masks, sensor-laden jewelry and clothing) and wearable computers to give digital machines a new sense of physicality compared with the traditional box on a desk.

Affective computers are outfitted with bionic senses: videocameras watch gestures and facial expressions, speech-recognition devices monitor voice intonation, and a network of biosensors—unobtrusive and lightweight computers that are embedded in everything from clothing to jewelry—keep track of physiological signals such as pulse, respiration and skin conductivity.

Picard's notion of the affective computer derives from the neurobiology of affect discussed by Antonio Damasio. Damasio argues that, thanks to the interplay of the brain's frontal lobe and limbic systems, our ability to reason depends in part on our ability to feel emotion (58). To explain emotion, Damasio, and indirectly Picard's *Affective Computing*, draw on the Spinozist notion of affect (*Ethics* II). In particular, Damasio uses Spinoza's concept of the mind as a "feeling brain," a brain that registers neural maps of somatic affective states (36).

By analyzing neurological and chemical pathways, Damasio states that emotion and reason link neurologically (55). In strong resonance with William James, Damasio claims that emotions do not cause bodily symptoms but are caused by the symptoms: we do not cry because we are sad; we are sad because we cry (105). The emotional behavior comes first; conscious feelings are its later by-product. Damasio shows that, far from being intangible experiences as feelings are commonly thought, joy or sadness generates patterns of brain activity recognizably associated with each feeling. Despite engaging with the feeling of the mind, and thus linking sensation to thinking, it may still be difficult to suggest that these neurobiological reformulations of cognitivism provide a way to engage with the affective dimensions of bionic machines.

Picard's and Damasio's use of the Spinozist notion of affect seem not to fully follow its most crucial implication: that affect is neither in the feeling subject nor in the objective neural patterns recording emotions. Affect is above all a direct feeling of the virtual: the sensation of invisible forces acting on a body; the abstract dimensions of sensation falling out of step from emotional responses and neural mapping. What comes first here is not the neural representation of the states of bodily feeling, but the direct inarticulate sensation of change: the arrest or snapshots of perpetual motion, the residual rhythm traversing the sensing–thinking regions of a body.

Affective transmission thus involves not a linear correspondence between sensory perception and mental states, qualified emotions and neural patterns, the sensible and the mental, but only an action at a distance connecting infinitesimal degrees of variations between distinct layers of information sensing: proprioception, exteroception and interoception (Massumi, *Parables* 58–62), including the rhythms of cells, molecules, atoms and elementary particles. This is not a sensation of different orders, but these are different orders of the same sensation: a polyrhythmical feeling of the resonances between distinct milieus of information sensing. Here sensation at a particular domain, order, world, enters in contact with virtual forces that exceed every domain and traverse them all.

Symbiosensation

Affective bionics, then, may entail that biochip sensors connect neural networks—the patterns of nerve cells that conduct chemical and electrical traffic inside our bodies—with the senses in a new way.

Proprioception or kinesthetic sensibility—the feeling of movement in the muscles and ligaments of a body—is entangled with the tactile sensibility of the skin (exteroceptive sensations of the five senses) and the visceral sensibility of the guts (the interoceptive sensing gathering information from the senses immediately before getting to the brain) (58–62). Kinesthetic sensibility feels the movement of the body as if in strict resonance with the velocities of information sensing captured by the skin and the guts. Information travels through rates of sensibilities between proprioception, exteroception and interoception. The body registers the in-betweenness of these rates, which connects the action of anonymous forces upon a body with new sensations in the thinking-flesh.

As bionic technologies intervene in milieus of information sensing, reconnecting through biochips, implants, neural networks to the sensing of movement or the sense of touch or hearing, what will a body be able to feel, which domains of sensation will it enter?

Bionic technologies are described as a sort of neuroprosthetics where neurons are aided by bacterial and silicon transmitters to feel at faster rates, that is, feel more, feel before, to optimize all levels of sensing. Yet bionic transmission seems to do more than that. Biochip technologies enter the relation between kinesthetic, synesthetic and visceral sensibilities through a symbiotic assemblage of silicon, neural and bacterial information sensing, giving rise to a sort of bionic sensorium ready to feel the infinitesimal proximities of organic and inorganic matter.

From this standpoint, it is possible to suggest that bionic technologies do not just extend sensory–motor perception but confront abstract feeling or nonsensuous sensation spinning out of the feedback circuits of information sensing linking the velocities of bacterial communication with those of neural firing and silicon light-sensitive transfer. For bionic technologies to tackle the sensation of movement, sensory perception, and visceral sensibility it is not sufficient to reduce the transfer of information sensing to digital codes able to translate incompatible milieus into an extra-biological platform of smooth communication.

To grasp sensations, the unframed feeling–thought traversing a body prior to sensory–motor response, it is necessary that bionic machines enter in contact with the abstract yet concrete dimension of *"microperceptions* occupying the rungs of interlocking strata before they move to the molar level..." (Massumi, *Shock to Thought* xxx).

The problems that bionic technologies encounter are the infinitesimal percepts and affects proper to virtual matter out of which proprioceptive, extraceptive and introceptive sensations sprawl. The problem of how to make a body affected—how the infra-action between distinct milieus of information sensing generates certain sensations which determine sensory response at a particular moment—remains central to the tendency of biodigital technoecology to infinitesimally calculate the nonsensuous feelings of a body.

The bionic assemblage of bacterial, neural and silicon velocities of sensing implies a new degree of variation in feeling directly experienced by a body, ready to perceive the smallest transitions in sensing. Call such a feeling symbiosensation: the felt experience of a nonsensuous relatedness between organic and inorganic matter, adding on a new gradient of feeling in the thinking-flesh.

It would be misleading, however, to suppose that symbiosensation is caused by bionic technologies, since the affective relatedness between organic and inorganic matter precedes and exceeds the bifurcation between the natural and the artificial. Such relatedness indeed defines the felt experience of continual transition in the mechanosphere of sensations.

Brian Massumi has discussed the way microlayers of perception—lying beneath the thresholds of consciousness—are felt experiences occurring at the shortest span of time, the incipient momentum of half a second too short to be consciously registered yet long enough to be felt (*Shock to Thought* 29). Here a body is a transducer: it captures the shortest degrees of change between distinct information milieus as sensations in the flesh. The body enters warps of time by feeling before sensory recognition, experiencing the anomaly of the déjà sensed occurring at the intensive overloaded interval between what has been and what is yet to be felt, between past and future. In William Gibson's *Pattern Recognition*, Cayce Pollard, a professional "cool hunter," uses her allergy to trademarks to spot new trends, and advise advertising agencies and marketers how best to commodify their products. Her allergy equips her body with nonsensuous sensibility, a subperception that is neither sensory nor mental, neither emotional nor cognitive, but affective in so far as it captures the transition, the feeling of change between the past and the future in the present. Whenever exposed to trademarks, her body transduces the nonsensuous feeling of information variations in the brandscape into chemical reactions for the logos in her flesh, resulting in allergic responses, such as fainting spells and sneezing fits.

It is mesoperception, according to Massumi, to define the synthetic sensation of distinct velocities of sensing—prioprioceptive, exteroceptive, introceptive—entering in contact with the nonhuman forces that traverse them all (*Parables* 62). If mesoperception is the sensation of the amodal connection between distinct domains feeling the outside, then symbiosensation is the machinogenesis of a novel relatedness between organic and inorganic milieus of information sensing: a concrescence, to borrow from Whitehead, the growing of "novel togetherness" of actual occasions (*Process and Reality* 21–22), the felt nexus of distinct societies or worlds across scales and milieus, adding new dimensions of nonphysical feeling to the body. Symbiosensation is not direct perception, but prehension. Whitehead uses such a notion to reject representative perception or consciousness. Prehensions are feelings—at once conceptual and physical, nonsensuous and sensuous—experienced by a body entering slabs of duration by repeating the feeling of the past and anticipating the feeling of the future (*Adventures of Ideas* 192–193).

Prehensive feeling or symbiosensation indeed implies a readiness to perceive, to anticipate the incipience of new kinesthetic, synesthetic, visceral sensations in the flesh. Symbiosensation marks the capacities of a body to protosense its molecular mutations, entering nonlinear durations. But when exactly do we have symbiosensation and how does it relate to the biodigital ecology of information sensing?

Let us recapitulate.

In the first place, symbiosensation accounts for how a body feels an expanded environment of bodies as if it were part of its spinal cord, its own kinesthetic movement. For example, the expanded proprioception proper to the autonomic functions of a body acting autonomously from visual perception: driving a car without looking at your feet, swimming without observing your arms, dancing without watching your steps. Machinic assemblages between organic and nonorganic milieus of information sensing engender an extended proprioceptive sensation whereby movement or spatiotemporal orientations have become ecological. The kinesthetic interdependence between milieus of information sensing entails environments of preinteraction, the feeling of movement before movement, the anticipation of the sensation of orientation triggered by the rhythms of autonomic responses. Far from defining a prosthetic extension of sensory–motor perception, a machinic, ecological, conception of the sensation of movement, or proprioception, rather points to a nonsensuous perception of movement whereby an ecologically expanded body (an already bionic body) is ready to feel motion before the sensory perception of actual movement.

In the second place, we have symbiosensation when a body feels the haptic contactedness in matter or synesthetic contiguity amongst distinct senses. Sensory perception—or exteroception—defines the registering of external stimuli through sight, hearing, taste, touch, and smell. Yet the five senses, before becoming determinate by their specific channels of sensing, share a common field of forces impinging upon the skin first. This synesthetic condition is manifested, for example, by the overlapping of distinct senses. For example, the smelling of sounds, the shades of touch, the hearing of light, the taste of color and so on. What remains common to such conditions is that senses share a capacity of being affected or impinged upon by external forces—light, frequencies, pressures, and temperatures—running at certain speeds on a synesthetic skin. This entails a prior symbiotic connection between sensory organs: a link of an intensive nature trading the velocities of data through the surface of the skin, before reaching specific channels–organs of perception.

This level of synesthetic symbiosensation entails less a mere physical sensing than an action at a distance between unarticulated sensations entertaining a continual skin relation that suspends the actualization of specific sensory qualities. Such synesthetic symbiosis indicates the diffused condition of a continual overlapping of information sensing in new media ecologies.[12]

For example, take the synesthetic symbiosis of new media gathering together a multicapacity of sensing in one small gadget. Take the self-evident example of the mobile phone, a proto-machine par excellence, constantly adding new capacities for sensing by overlapping and thus suspending the sensory perception of hearing and seeing. Such condition of suspension, however, is not only derived from software able to smooth distinct milieus of information into the integrated circuit of perfect communication. Indeed, and more importantly, the continuous interruption of sensory perception also seems to derive from the soft design of these no longer media objects but rather mediatic *objectiles*: objects imbued with potentialities of coexten-sion with all kinds of media, a membranic coexistence with compossible objects. These objectiles expose the limbo of the synesthetic condition of feeling enveloped by the coextensive skin of sensing objects. On such a membranic surface, the software potentialities for thinking are paralleled to the hardware potentialities for sensing, each time demanding a feeling of thought and a thought of feeling: a synesthetic experience prior to sensory perception. Rather than a multimedial addition of actual sensory functions, such experience is entangled with the coextensive envelope of media objects, their folding together into the symbiotic architecture of feeling at the core of new media.

Last but not least, symbiosensation is here used to define a direct nonsensory (nonsensuous) perception of enmeshing durations in a nexus of past–present–future. In particular, visceral sensibility immediately registers excitation before being fully processed not only by the brain, but also by exteroceptive organs. Yet this internal sensibility is not of a constituted inside (for example, a self). Instead, such in-depth sensibility is proper to the involutionary foldings of matter: a proto-feeling preceding and exceeding the organization of information sensing into neurosensorial channels. What is peculiar to visceral sensibility is the catching of the passing of time. Hence it does not concern the feeling of the present, what lies now before the body, but what passes through a body. Whitehead argues that the sense perception of the contemporary world is always accompanied by the perception of the *withness* of the body. "It is this withness that makes the body the starting point for our knowledge of the circumambient world. We find here our direct knowledge of causal efficacy" (*Process and Reality* 81). Such a feeling of causality or causal relations is a vague yet concrete extrasensory (intuitive) experience of worlds pressing against skins.

According to Massumi, visceral sensation subtracts quality from sensory excitation (*Parables* 33). It envelops intensity, the polyrhythms of nonsensuous perception. Thus, it acts in the interval between stimuli and response, the quantum leap before and after excitation, the low-frequency vibrations running beneath the flesh. This extrasensory experience of nonlinear, intensive durations each time extends the realm of the lived body outside itself on the autoaffective plane of matter: the time-stuff of spatial abstraction (33).

What visceral sensibilities grasp is a schizo-continuum in duration, *aion* time.[13] This entails the feeling of what happens that clashes with what has happened and what is about to happen; a sensibility towards incorporeal time, which Massumi defines in relation to the fear of whatever or whoever dominating the global political sphere in control societies (*Politics of Everyday Fear* 6–9). Whilst fear is linked to the incumbent threat of the unknown, it is also defined as an impersonal sensation of anticipation calling forth future tendencies into present possibilities. Similarly, going back to the example of the mobile phone, the sensation of anticipation is here derived from the constant state of low-level awareness of being potentially contacted, a ceaseless awaiting for the arrival of the yet-to-be contact.[14]

Symbiosensation entails a visceral prehension of the speeds of time: the nonsensuous feeling of distinct temporal vectors leading to a feeling of anticipation or anterior future, where the past and the future ceaselessly rewind in the present. For example, the bionic sensorium enmeshing together media and bodies, bacterial, neural and silicon information sensing, involves the nonsensuous feeling of a new synthesis of time, the way sensory perception indeed depends on the new level of causal relation between organic and nonorganic velocities of matter, where molecular bodies not only affect human experience but are in turn affected by it. Here a visceral sensation exposes the short-term intuition between what comes before and what comes after, a new level of extrasensory feeling of the vague yet concrete velocities of transmission added on by bionic machines.

Thus, whilst it can be argued that symbiosensation already points to the bionic quality of media ecologies here argued to imply that all sensorium depends on extrasensory feeling or sensory-perception or non-sensuous sensibility, it can be argued that bionic technologies rather mark a new degree of conjunction between the organic and the inorganic, deploying a technoecology of sensation extending at all levels of matter. Here not only is linear evolution from the biological to the technical turned into a symbio-genesis of bacterial, silicon and neural information sensing, but also the entire biostratum is exposed to abiotic sensibilities in matter, the nonsensu-ous prehension of nonlived matter up to and including subatomic particles, which entails a radical modification of feeling. This is the sense in which media technologies can be conceived with Deleuze and Guattari's notion of machinic involution: at the thresholds of prehension between the most minute and the most vast, the most ancient and the most advanced, the most vague and the most concrete, the growing together of all kinds of entities involves an ontomutation of matter resonating across organic and nonorganic feeling bodies.

Although the bionic sensorium has not yet been actualized, the virtual action of symbiosensation is already an object of preemptive control. What is at stake here is the anticipation of nonsensuous variations entailing the way sensation becomes biodigitally linked to bacterial and silicon sensibilities in the biochip.

How biochips will impact on gut feelings, the variable ratio between the senses, and proprioception, the feeling of an expanded movement, remains an open question. However, such a question may start to suggest that the expansion of a bionic architecture is inseparable from a technoecology of preemptive sensation, a readiness that an intricate nexus of—organic and nonorganic—bodies have, to perceive spatiotemporal activities before their sensory actualization. Yet there is more to this. Such technoecology involves a new ontogenetic phase transition of feeling arranged by the machinic involution of distinct milieus of information sensing. Here the intensifica-tion of symbiotically enmeshed capacities of sensing across scales and vectors of evolution results in a preemptive technoecology of visceral control, governed not by an informational confusion of sensibility, but by an intuitive anterior decision taken by thinking-dancing nanoparticles, where the anticipation of sensations defines the causal relations of sensations yet to come.

Deleuze and Guattari already envisaged the spatiotemporal sophistication of technical machines able to directly connect with virtual worlds of feeling-thoughts running beneath, above, and across the world of actual sensory perception. What remains to be added, however, is how machinic ecologies of sensation may allow for the symbiotic invention of microsocialities twisting the lines of preemptive sensation into a constructive protoethics of feeling.

Notes

1. On preemptive power, see Brian Massumi, 1993, 20–22.
2. On prehensions, see Alfred N. Whitehead, 1978, 121–126. See also Deleuze, 2003, 41–42.
3. Amodal power implies a virtual combination of sovereign, disciplinary and control orders of power slipping into present recurrences through the preemptive anticipation of futurity. See Massumi, 2005.
4. See Hansen's theory of new media arguing for a phenomenological relation between information and materiality. Hansen, 2004, 6–9.
5. As Deleuze suggests, this Whiteheadian notion of the superject indicates that every point of view is a point of view on variation. See (*The Fold* 19–20).
6. Contrary to Descartes, who argues that perception cannot occur without a physical body (that is, it can only be sensory), Leibniz, for example, highlights that perception exists in thought itself as the microperception of an incorporeal materiality.
7. Cyborg technologies define how information units are preserved in their transmission from biological to digital domains.
8. The study of topological deformation was central to the first biomathematician D'Arcy Thompson, 1961.
9. Developed within the field of biotechnology, the biochip is at the core of genomics, computational biology, proteomics, neuroprosthetics, opticbionics, nanobiogenetics. These miniaturized laboratories of information sensing are able to synthesize simultaneous biochemical reactions in living and nonliving systems, using tiny strands of bacterial DNA to latch onto and quickly enter into communication with thousands of genes at a time.
10. On the bionic use of bacterial photoreceptors see Palkar, Uma S., 2005, 65–71(7); Koyama, K., Yamaguchi, N., and Miyasaka, T., 1994, 762–765; Koyama, K. and Miyasaka, T., 1993, 6371.
11. On the history of emotional computing as agent-based interaction, see Sloman, A., and Scheutz, M., 2002, 169–176; Sloman, A. and Croucher, M., 1981; Christoph Bartneck and Michio Okada, 2001; Picard, 1997.
12. The concept of media ecologies has been developed by Matt Fuller, 2005.
13. For Deleuze, "Aion is the past-future, which in an infinite subdivision of the abstract moment endlessly decomposes itself in both directions at once and forever sidesteps the present." 1992, 77.
14. On the mobile phone as media altering temporal perception, see Belinda Barnet, 2005.

Works cited

Barnet, Belinda. "Infomobility and Technics: some travel notes." *ctheory* (October 27, 2005). http://www.ctheory.net/articles.aspx?id=492. Last accessed September 14, 2008.

Bartneck, C. and M. Okada. "eMuu—An Emotional Robot." (2001) http://www.bartneck.de/work/bartneck_robofesta.pdf. Last accessed July 7, 2007.

——. "Robotic User Interfaces" http://www.bartneck.de/publications/2001/roboticUserInterfaces/bartneckHC2001.pdf. Last accessed July 7, 2007.

Deleuze, Gilles. *Francis Bacon. The Logic of Sensation*. London and New York: Continuum, 2003.

——. *The Fold.* Minneapolis: University of Minnesota Press, 1993.

Deleuze, Gilles and Félix Guattari. *Anti-Oedipus, Capitalism and Schizophrenia.* Preface by Michel Foucault. Trans. Robert Hurley, Mark Seem, and Helen R. Lane. London: The Athlone Press, 1983.

——. *A Thousand Plateaus, Capitalism and Schizophrenia.* Trans. B. Massumi. London: The Athlone Press, 1987.

Damasio, Antonio R. *Looking for Spinoza.* New York: Harcourt Brace, 2003.

Fuller, Matthew. *Media Ecologies, Materialist Energies in Art and Technoculture.* Massachusetts: MIT Press, 2005.

Geary, Jay. *The Body Electric: An Anatomy of the New Bionic Senses.* New Brunswick, NJ: Rutgers, 2002.

Gibson, William. *Pattern Recognition.* New York: G.P. Putnam's Sons, 2003.

Guattari, Félix. "Machinic Heterogeneities." *Reading Digital Culture.* Ed. D. Trend. Massachussetts and Oxford: Blackwell, 2001, 38–51.

Hansen, Mark. *New Philosophy for New Media.* Cambridge, MA: MIT Press, 2004.

Haraway, Donna. "A Cyborg Manifesto: Science, Technology, and Socialist-Feminism in the Late Twentieth Century." *Simians, Cyborgs, and Women: The Reinvention of Nature.* London: Free Association Books, 1991, 149–181.

Koyama, K., N. Yamaguchi and T. Miyasaka. "Antibody-mediated bacteriorhodopsin orientation for molecular device architectures." *Science* 265: 5173 (1994): 762–765.

Koyama, K. and T. Miyasaka. "Image sensing and processing by a bacteriorhodopsin-based artificial photoreceptor." *Applied Optics* 32: 31 (1993): 6371. http://www.opticsinfobase.org/abstract.cfm?URI=ao-32-31-6371. Last accessed July 5, 2007.

Kittler, Friedrich. "Gramophone, Film, Typewriter." *Literature, Media, Information Systems: Essays.* Ed. John Johnston. Amsterdam: GB Arts International, 1997, 28–49.

——. "What is new about new media?" *Mutations.* Ed. Koolhass et al. Bordeaux: Actar, 2000.

Manovich, Lev. *The Language of New Media.* Cambridge: MIT, 2001.

Massumi, Brian. *Parables for the Virtual, Movement, Affect, Sensation.* Durham: Duke University Press, 2002.

——. "Everywhere You Want to Be: Introduction to Fear" in *The Politics of Everyday Fear.* Ed. B. Massumi. Minnesota: University of Minnesota Press, 1993, 3–37.

McLuhan, Marshall. *Understanding Media, the Extensions of Man.* London: Routledge, 1964.

Murphie, Andrew. "Putting the Virtual back into VR." *A Shock to Thought.* Ed. B. Massumi. London & NY: Routledge, 2002, 188–214.

Palkar, Uma S. "Natural Photoreceptors: Structure, Function and Applications." *Frontiers in Natural Product Chemistry* 1: 1 (2005): 65–71.

Picard, Rosalind. *Affective Computing.* Cambridge: MIT Press, 1997.

Sloman, A. and M. Scheutz. "A framework for comparing agent architectures." *Proceedings of UK Workshop on Computational Intelligence,* 2002, 169–176.

Sloman, A. and M. Croucher. 'YOU DON'T NEED A SOFT SKIN TO HAVE A WARM HEART. Towards a computational analysis of motives and emotions' (1981), http://www.cs.bham.ac.uk/research/cogaff/sloman-croucher-warm-heart.pdf. Last accessed July 5, 2007.

Shannon, C. E. and W. Weaver. *The Mathematical Theory of Communication.* Urbana, Illinois: The University of Illinois Press, 1949, 1–28.

Spinoza, Baruch. *Ethics, Treatise on the Emendation of the Intellect and Selected Letters.* Trans. and ed. S. Shirley and S. Feldman. Cambridge, Indianapolis: Hackett Publishing Company, 1992.

Thacker, Eugene. *Biomedia.* Minnesota: University of Minnesota Press, 2004.

Thompson, D'Arcy. *On Growth and Form.* Cambridge: Cambridge University Press, 1961.

Whitehead, Alfred N. *The Concept of Nature.* Amherst New York: Prometheus Books, 2004.

——. *Process and Reality; An Essay in Cosmology.* Corrected Edition. Ed. David Ray Griffin and Donald W. Sherburne. New York: The Free Press, 1978.

——. *Adventures of Ideas.* New York: The Free Press, 1993.

Wiener, Norbert. *The Human Use of Human Beings: Cybernetics and Society.* London: Free Association Books, 1989.

11
Eco-Aesthetics: Beyond Structure in the Work of Robert Smithson, Gilles Deleuze and Félix Guattari

Stephen Zepke

There is a break that traverses both the art of Robert Smithson and the philosophy of Gilles Deleuze and Félix Guattari. It is the break between Robert Smithson's Site–Nonsite works (1966–1969) and his Earthworks (1970–1973), it separates the structure of "Nature's" signs from Nature as a process of production, and divides conceptual art's immaterial ideas from the vitality of the earth. Gilles Deleuze's work of the same time also manifests this break. The transition from *Difference and Repetition* (1968) to *Anti-Oedipus* (1972) takes us from a conceptual mapping of structures to a material machinery of production.[1] Deleuze was to later say that "*Anti-Oedipus* marks a break," a break whose name is Félix Guattari (*Negotiations* 144). For Deleuze, Guattari had "gone further than I had," and their collaboration allowed Deleuze to find what he felt was missing in his own work, an engagement with real political processes (*Negotiations* 13). Similarly, Smithson's break marks his abandoning of the institution in favor of an art of direct intervention, the Earthworks confronting one of the most pressing political concerns of his (and our) time, the destruction of the earth. Beyond the break then, both Deleuze|Guattari and Smithson explore an eco-aesthetics, one that affirms the artistic construction and expression of Nature itself.

Despite this break between the Site–Nonsite and Earthworks, they emerge in Smithson's thinking at the same time. In both his designs of outdoor sculptures for the Dallas–Fort Worth Airport, produced in 1966–1967, and the essay "A Tour of the Monuments of Passaic, New Jersey" (1967) Smithson begins to think about sculptures unconfined by the gallery. For the airport project, Smithson contributed designs for sculptures—abstract forms of natural materials such as earth or sand—to be located between the runways, and to be viewed from the air. These works were not realized, and his plans for large-scale Earthworks remained on maps and in diagrams. The next few years saw Smithson exploring this relation between a "Site" and its representation within a discursive "Nonsite" that was understood as both a linguistic structure and an institutional frame. This approach was consistent with Deleuze's structuralist understanding of the dialectic, inasmuch as the Site–Nonsite work dramatizes the relation between a virtual and intense dialectical inequality (Site–Nonsite), and its actual expression in signs (Deleuze, *Difference* 222). Smithson's work therefore considers its own structure, offering aesthetic solutions to the dialectical "problem" of the Site–Nonsite.[2] In this respect the work is typical of conceptual art, inasmuch as the work investigates its own conceptual conditions of possibility, or, in Deleuze's terms, its own "sufficient reason" (*Difference* 222).

Smithson's Site–Nonsite works present material fragments of the earth along with maps and photographs of the sites from which they were taken. This produces what Smithson calls a "range of convergence" between the opposed terms, through which we can discover "the rules of this system of signs." It is worth quoting Smithson at length here:

> The range of convergence between Site and Nonsite consists of a course of hazards, a double path made up of signs, photographs, and maps that belong to both sides of the dialectic at once. Both sides are present and absent at the same time. The land or ground from the Site is placed *in* the art (Nonsite) rather than the art placed *on* the ground. The Nonsite is a container within another container—the room. The plot or yard outside is yet another container within another container. Two-dimensional and three-dimensional things trade places with each other in the range of convergence. Large scale becomes small. Small scale becomes large. A point on the map expands to the size of a land mass. A land mass contracts into a point. Is the Site a reflection of the Nonsite (mirror), or is it the other way around? (Smithson 153)

Following his earlier sculptural work with reflection and displacement Smithson locates the immaterial dialectic of Site–Nonsite within the paradoxical indeterminacy of its signs. In this sense the work is not so much concerned with the appearance of the Site within the Nonsite, as with the differential relations which do not appear—Site and Nonsite not as actual places, but as a difference without positive terms—but act as the conditions of appearance as such. The "convergence" of the Site and Nonsite is therefore better understood as an actualized indeterminability expressing the intense structural difference that remains, according to one perceptive commentator, the "blind spot" of the work, a "set of hidden assumptions" that produces its "enantiomorphism" (Reynolds vxii).[3] "Convergence" indicates a reversibility of the relations determining the work's signs that actualizes its differential virtual structure, but only by canceling it inside its linguistic and institutional terms.

"*This is 1967*," Deleuze writes in his article "How Do We Recognize Structuralism?," and it is much more than the date he shares with Smithson.

When Deleuze writes that structure is not a sensible form, nor an imaginary figure, but a "combinatory formula supporting formal elements" (*Difference* 173), he could be describing Smithson's Site–Nonsite works. For Deleuze, a work is "structural" when it attempts to express its own virtualities, (*Difference* 186) its own "transcendental topology" (*Difference* 174). This is how Smithson's Site–Nonsite works function; they actualize a structural difference that cannot itself appear within their system of signs, but nevertheless acts as the work's genetic condition.[4] It is precisely this structure that determines in advance the work's relation to the Site, and to Nature, a structure that will be dissolved in the "rotary" of the *Spiral Jetty*, where matter and experience will "converge" on a single plane (of immanence).

The problem that emerges from Deleuze's structuralism is that, although actual signs find their genesis in virtual differentials this intense difference is "cancelled" in actuality.[5] As a result, changes in the actual world must be (and can only be) accounted for on the level of virtual structure. Smithson is close to Deleuze in this respect when he explains how the dialectical intensity of Site and Nonsite are connected to others (inside|outside, two|three dimensions, points|masses, etc.), forming a network of virtual differentials (what Deleuze calls "the differentiation of the Idea" (*Difference* 221)) that are actualized in signs. The problem remains, however, that although this differentiation structures the Site–Nonsite works, marking Smithson's introduction of structuralist concerns into conceptual art's fascination with "language," this differentiation cannot actually take place. The Site–Nonsite work can only actualize its conditioning difference, and remains caught within a conceptual paradox; its immaterial conditions, understood as the "passive genesis" of its dialectical structure, can only be actualized by its disappearance in signs. This is the limit of discursive systems; they signify the differential forces of "Nature," but in doing so contain "Nature" within their representational signs. Smithson's acceptance of this fact marks the end of his interest in Conceptual Art's project, as it does for Deleuze's interest in structuralism, as both choose to engage directly with the compositional—and entirely material—forces of the Earth. Smithson and Deleuze, together, abandon the structural Site once they see that its construction is not entirely immanent with what it expresses.

Let us return to the second emergence of Smithson's interest in the earth, the essay "A Tour of the Monuments of Passaic, New Jersey" (1967). This essay introduces the dialectic in a similar way to the Site–Nonsite works, but in an entirely different register. Where the Site–Nonsite works maintain a detached scientific aesthetic of maps and samples, the Passaic essay is mostly ironic. It presents a trip Smithson makes to his old hometown in the form of a tourist guide to the city's "monuments," quasi-artworks composed of mostly industrial objects (pipes, pumps, an outfall, a bridge, a sand-pit). Here Smithson sees things in the world as art works, but his whimsical found objects remain tied to a process of documentation that returns us to the Site–Nonsite, even though this process has now entered things: "Noon-day sunshine cinema-ized the site, turning the bridge and the river into an over-exposed *picture*…. The sun became a monstrous light-bulb that projected a detached series of 'stills' through my Instamatic into my eye" (Smithson 70). Smithson is attempting to relocate the Site–Nonsite outside of its institutional frame and into the every-day, and this introduces the element of time, an element that will be central to Smithson's Earthworks. Time appears in the duration of the tour (one day), and in Smithson's constant description of banal details—what he is reading (Brian Aldiss's *Earthworks*), street signs and the label on his film package—that place us directly in the here and now. But this immediacy emerges only in opposition to a much greater temporal horizon, an eternity that will resonate through all of Smithson's work to come: entropy. "The last monument," he writes, "was a sandbox or a model desert. Under the dead light of the Passaic afternoon the desert became a map of infinite disintegration and forgetful-ness" (Smithson 74). Here, and *at once*, the familiar dialectic (the Site and Nonsite of the sandbox and the desert) reaches ground zero, *and* something emerges that seems to overflow it: Nature-as-entropy. "All that existed were millions of grains of sand, a vast deposit of bones and stones pulverized into dust." And when Smithson tells us that "every grain of sand was a dead meta-phor that equalled timelessness" (Smithson 74), he seems to suggest that entropy is a natural force entirely beyond the structuralist dialectic. So although the Passaic essay brings together the Site–Nonsite work and what we could call proto-Earthworks, they appear in fact either side of a break called "entropy." It is precisely this appearance of entropy that the Site–Nonsite works failed to produce, and where they do address the presence of entropy at the Site (for example, *Nonsite "Line of Wreckage," Bayonne, New Jersey* 1968, or *Nonsite (Oberhausen, Germany)* 1968) this remains a documented reflection, rather than an embodiment. Once entropy has pulverized both Site and Nonsite into dust, once it has overwhelmed every spatial and temporal boundary, the virtual and genetic dialectical structure has nowhere to stand.

"This is 1967." While Smithson was seeing the dusty force of entropy appear in the sandbox, Deleuze was seeing something similar, but unlike Smithson he was not amused. In *Difference and Repetition* Deleuze explicitly rejects entropy, claiming it only explained what happens in the actual world, without account-ing for "many things of another order." In fact, Deleuze writes, "[r]eality is not the result of the laws which govern it" (*Difference* 227), but instead actualizes a virtual ontogenesis. As a result: "The principle of degradation [i.e., entropy] obviously does not account either for the creation of the most simple system or the evolution of systems" (*Difference* 255). Although entropy posits intensive differences (unequal temperatures) as the cause of changes in physical sys-tems, these changes serve to equalize and ultimately negate difference. This makes the undifferentiated the ontological ground, and establishes entropy as "a transcendental physical illusion" (*Difference* 223, see also 229). Deleuze places Nietzsche's eternal return against this "eschatological" illusion of entropy, because it is the eternal return "which opens up the possibility of

difference having its own *concept*" (*Difference* 40–41, italics added).[6] Deleuze deploys the eternal return against entropy as a concept of difference in itself, and as the philosophical antidote to our contemporary version of scientific nihilism; the entropic fate of the universe. The eternal return is opposed to entropy because it produces differences rather than their equalization. Eternal return, writes Deleuze, is no longer "merely a theoretical representation: it carries out a practical selection among differences according to their capacity to produce—that is, to return or to pass the test of the eternal return" (*Difference* 41). Deleuze's heroic (precisely, the "structuralist *hero*" ("How Do We" 191)) attempt to *think* the genetic principle of being as becoming culminates in permanent revolution, but this revolution remains philosophical inasmuch as its production finds its genetic principle in a *concept* that remains almost impossible to think, let alone actualize (*Difference* 87).[7]

Deleuze's use of the eternal return will, strangely enough, find its complement in Smithson's use of entropy. For both, what appears in the most "extreme" form of time is a creative movement in which both subject and world lose their coherence. But Smithson will find something in entropy that Deleuze will not in the concept of the eternal return, and that will be a materialism that will break with his previously *conceptual* work and thrust him into the productive forces of Nature. Both sides of this break appear in the Passaic essay. On one, Smithson clearly associates entropy with the equalization or banalization of the Site–Nonsite dialectic he finds in the monument of a Passaic parking lot:

> Everything about the site remained wrapped in blandness and littered with shiny cars—one after another they extended into a sunny nebulosity. The indifferent backs of the cars flashed and reflected the stale afternoon sun. I took a few listless, entropic snapshots of that lustrous monument. (Smithson 73)

Here, Smithson figures the documentation process as a moment in a general malaise, an entropic decline expressing the undifferentiated banality of a universal suburb. Passaic replaces Rome as the "eternal city" (Smithson 74). But this pessimism is closely followed by another appearance of entropy, this time in the desert of the sandbox. Now Smithson shifts rhetorical gears, and we move from the suburban cliché of Passaic into the epic spaces of an entropic sublime where the "dissolution of entire continents, the drying up of oceans" (Smithson 74) appear in the "open grave" of the sandbox. But as a metaphor for entropy the sandbox operates only through the agency of another element, the creativity of children. These children are the mechanism by which the black and white sand in the sandbox becomes an indifferent grey, and they are what makes the universal process of Nature—entropy—optimistically appear as "cheerful play" (Smithson 74). Smithson seems to anticipate Deleuze's later suggestion that "Art says what children say" (*Essays* 65). Children, Deleuze and Smithson argue, are like (land) artists; they are inseparable from what they produce, and what they produce is inseparable from the world. The Earthwork will express these at once playful and entropic forces in what Deleuze tellingly calls "maps of intensity" that construct and express the world's "becomings" (*Essays* 64). These maps|becomings are both Site *and* Nonsite, at once forces and images, "both extensive and intensive" (*Essays* 65–66) and art, Deleuze says, "renders their mutual presence perceptible" (*Essays* 67).[8] In his Passaic essay Smithson already senses what Deleuze will need Guattari to point out to him, that the dialectical relations of intense differences and their cancellation in extension, of virtual and actual, of structure and sign, must be moved beyond if we are to experience the productive processes that are continually making and unmaking the world. What Smithson still had to do, however, was to find a form by which entropy could be turned creative.

In 1968 Smithson wrote the essay "A Sedimentation of the Mind: Earth Projects" where he describes entropy as an ontogenetic force immanent to earth and mind, and composing the horizon of Nature itself.

> One's mind and the earth are in a constant state of erosion, mental rivers wear away abstract banks, brain waves undermine cliffs of thought, ideas decompose into stones of unknowing, and conceptual crystallizations break apart into deposits of gritty reason. Vast moving faculties occur in this geological miasma, and they move in the most physical way.... The entire body is pulled into the cerebral sediment, where particles and fragments make themselves known as solid consciousness. A bleached and fractured world surrounds the artist. To organize this mess of corrosion into patterns, grids and subdivisions is an esthetic process that has scarcely been touched. (Smithson 100)

Smithson now sees the work of art emerging from the collapse of the conceptual definition of the Nonsite into the materiality of the Site. This "slump" describes a Nature that encompasses the Mind, and organizes the entropic process into "patterns" and "subdivisions." Smithson's emphasis of the Site over the Nonsite coincides with his association of the Site with a creative process of entropy. "At the low levels of consciousness," Smithson writes, "the artist experiences undifferentiated or unbounded methods of procedure that *break* with the focussed limits of rational technique. Here tools are undifferentiated from the material they operate on, or they seem to sink back into their primordial condition" (Smithson 102, italics added). This passage reflects the influence of Anton Ehrenzweig's book *The Hidden Order of Art* (1967) from which Smithson takes both the term "de-differentiation" and its association with entropy as part of the productive process.[9] Ehrenzweig associated entropy with a "death drive," but unlike Freud he placed it in a creative rhythm with Eros that produced "an optimal threshold for further increases in differentiation," and acted as "the hidden order of art" (Ehrenzweig 219). De-differentiation is, Smithson continues, an experience of "a suspension of *boundaries*" (Smithson 103) between the "self and the non-self" (Smithson quoting Ehrenzweig 103) that allows a new figure of the artist to emerge. This figure bears a striking resemblance to the "schizophrenic" of Deleuze and Guattari's *Anti-Oedipus*, who "sought to remain at that unbearable point where the mind touches matter and lives its every intensity, consumes it" (*Anti-Oedipus* 20). This experience is the one that Smithson describes as a "plunge into a world of noncontainment" (Smithson 102), a "sensation" (Smithson 103) of being "physically engulfed" (Smithson 104). This schizo-artist composes matter, as Smithson puts it, in order "to give evidence of this experience…of the original unbounded state" (Smithson 104). This means, as Deleuze and Guattari put it: "What the schizophrenic experiences, both as an individual and as a member of the human species, is not at all any one specific aspect of nature, but nature as a process of production" (*Anti-Oedipus* 3). This aesthetic evidence, this production, no longer actualizes a transcendental topology (the Site–Nonsite dialectic) in a (linguistic) sign, because it extends into a de-differentiated and asignifying materiality where, Smithson writes, "The names of minerals and the minerals themselves do not differ from each other because at the bottom of both the material and the print is the beginning of an abysmal number of fissures. Words and rocks contain a language that follows a syntax of splits and ruptures" (Smithson 107). Words and rocks, in other words, are becoming de-differentiated according to an entropic logic by which "mind and matter get endlessly confounded" (Smithson 107). The artist composes this material heap into new expressions—Smithson's description is precise and resolutely dialectical; composition is "a three-dimensional *perspective* that has broken away from the whole, while containing the lack of its own containment" (Smithson 111). Smithson no longer thinks of his Site-Nonsite work as conceptual *in* nature, but as a vital–material expression *of* nature. This expression does not negate entropic processes but rather "reclaims" them, attempting to turn their production artistic.

In 1969 Smithson will describe the genetic moment of the Site–Nonsite work as a "low level scanning" (Smithson 189) of the Site giving a "vision" of de-differentiated Nature. This "primary process" will become from this point on the condition of all Smithson's creative activity, an entropic de-differentiation of the ego that enables a creative and asubjective aesthetic differentiation. Although Ehrenzweig sees these "two opposing structural principles" as being in "conflict" (Ehrenzweig 19), Smithson is more interested in their rhythmical complementarity, their reciprocal determination. Consequently Smithson rereads his Site–Nonsite works as "a key to where the site is and then you can operate within that sector." The Nonsite no longer "contains" the Site; instead the Nonsite is generated from the de-differentiated experience of the Site, and serves as its "designation" (Smithson 189). Smithson is attempting to subsume Site and Nonsite in a de-differentiated experience of Nature, but this retroactive reading nevertheless remains dialectical, retaining the Site–Nonsite structure as a limit beyond which he must go to discover the artistic mechanism capable of constructing Nature's expression.

In July 1968, Smithson, Nancy Holt and Michael Heizer made a trip to the Mojave Desert and Death Valley in Nevada. It was a trip Holt was later to say produced a "new paradigm" in Smithson's work, one in which the Site–Nonsite dialectic was dissolved. Smithson's move into the desert "swallows up boundaries" and allows a new aesthetic to emerge, one in which the artist "makes contact with matter" (Smithson 103), a de-differentiation that begins what Smithson calls a "rhythm," "that swings between 'oceanic' fragmentation and strong determinants" (Smithson 103).

Entropy emerges here as the de-differentiation necessary for the production of an expressive materiality, and this means, at last, the emergence of Nature itself. Nature is no longer contained in the Site *or* the Nonsite; entropy achieves a de-differentiation that unleashes Nature from its structure. As a result, the Earthworks explore a Nature that cannot be distinguished from what it produces; its expressions are inhuman sensations that construct its plane of immanence. In this sense—as entropy made visible—these works are a sensation of our own disappearance, and Smithson will explore this catastrophe in terms of the two major streams of political activism from the 1960s, ecstasy (*Spiral Jetty*) and social movements (*Partially Buried Woodshed*). Deleuze, *post* Guattari, will also explore this Nature, and its two modes of political transformation, although Deleuze and Guattari will take their Nature from Spinoza. In his earlier discussions of Spinoza Deleuze had favored "God" (*Expressionism in Philosophy*) or "Substance" (*Difference and Repitition*) as terms for Spinoza's plane of immanence. In *Anti-Oedipus*, however, "Nature" becomes a central term describing the ontogenic process of production. This shift in relation to Spinoza can be understood in terms of Deleuze's break with structure, and his eventual suspicion of a virtual genesis in excess of its actualization. This was in fact Deleuze's criticism of Spinoza, that being (Substance) and becoming (the modes) are not absolutely immanent, a problem he "corrects" with Nietzsche's concept of the eternal return.[10] But this *concept* remains insufficient, and Deleuze and Guattari correct this correction with the Body without Organs of Antonin Artaud (*Anti-Oedipus* 327 and *Thousand Plateaus* 154).[11] Nature is the de-differentiated BwO producing "partial objects," objects that express and construct their body,[12] and are not the result of a "de-differentiated" (the term is Deleuze and Guattari's) organism simply being "stuck back together" in a Hegelian sublation forming a greater whole (*Anti-Oedipus* 326). These terms can be used to understand the break instituted by Smithson's Earthworks, which express a de-differentiated Nature (the BwO) as it constructs the artwork (as partial object). Here, the artwork and Nature

> are at bottom one and the same thing, one and the same multiplicity....
> Partial objects are the direct power of the body without organs, and the

body without organs, the raw material of the partial objects. The body without organs is the matter that always fills space to given degrees of intensity, and the partial objects are these degrees, these intensive parts that produce the real in space starting from matter as intensity=0. (*Anti-Oedipus* 326–327)

The question, as it is so often, is how? Deleuze's answer, at least as far as art is concerned, is found in his book on *Francis Bacon*. There he argues that sensations (as art's "partial objects") are experienced as an intensive *descent* of matter relative to its de-differentiated BwO, intensity=0. As a result: "The fall is what is most alive in the sensation, that through which the sensation is experienced as living. [...] The fall is precisely the active rhythm" (*Bacon* 82). This, Deleuze argues, is the "irrational logic, or this logic of sensation, that constitutes painting" (*Bacon* 83). And not just painting, as "[m]ost artists...seem to have encountered the same response; the difference in intensity is experienced as a fall" (*Bacon* 81). Indeed, this is precisely Smithson's understanding of entropy, it is the "fall," the "rhythm" between a de-differentiated matter (BwO) as intensity=0 and the intensities, the sensational partial objects, that animate it and make it live. In entropy, defined in this Deleuzian sense, Smithson dissolves his dialectical structures in Nature, in a lake of immanence.

Smithson begins his 1972 essay on the *Spiral Jetty* by pointing out that it was his "Mono Lake Site-Nonsite" of 1968 that led to the *Spiral Jetty*, and he attaches to this observation a long footnote concerning the various dialectical relations relevant to the work (Smithson 143). A little later, however, he dramatically claims the *Spiral Jetty* gave birth to itself in a process that rejected his earlier use of the dialectic altogether:

This site was a rotary that enclosed itself in an immense roundness. From that gyrating space emerged the possibility of the Spiral Jetty. No ideas, no concepts, no systems, no structures, no abstractions could hold themselves together in the actuality of that evidence. My dialectics of site and nonsite whirled into an indeterminate state, where solid and liquid lost themselves in each other....No sense wondering about classifications and categories, there were none. (Smithson 146)

A whirl, a fall, a de-differentiation. Sensation consumes our vision. Smithson the schizo: "My eyes became combustion chambers," indiscernible from Nature: the "flaming chromosphere" (Smithson 148). In *Spiral Jetty* the Site appears in its de-differentiated state, intensity=0, an entropic landscape composed of "melting solids," "dead ends," "incoherent structures," and "abandoned hopes" (Smithson 145–146). But this de-differentiation simultaneously gives rise to something new, to "protoplasmic solutions, the essential matter between the formed and the unformed" (Smithson 149). Between the formed and the unformed, inhuman forces begin to compose their material plane: "As I looked at the site, it reverberated out to the horizon only to suggest an immobile cyclone while flickering light made the entire landscape appear to quake. A dormant earthquake spread into the fluttering stillness" (Smithson 146). At this point the BwO appears to itself, and—psychedelic—according to its own organ, "a floating eye adrift in an antediluvian ocean" (Smithson 148). The dialectic is reborn as a rhythmic and material pulsation in which subjective and objective coordinates are "dissolved into a unicellular beginning" (Smithson 149) From this "beginning" a "scale of centres" differentiates itself: a dislocation point, a wooden stake in the mud, the axis of the helicopter propeller, and James Joyce's ear channel (Smithson 150). "Spinning off" from these centers is a "scale of edges," various forces of becoming: "slipping out of myself," sunstroke, dizziness, ripples, vaporization, "an arrangement of variables spilling into surds" (150). *Spiral Jetty* is a cartography of the plane of Nature, an hallucinatory vision emerging from entropic de-differentiation, a fall that is a constant rearrangement, alive with sensation.

Smithson's Site–Nonsite dialectic has been revalued: "The rationality of a grid on a map sinks into what it is supposed to define" (Smithson 147). The body of the *Spiral Jetty* de-differentiates the Site and Nonsite, grinding our discursive structures to dust. This now includes the process of documentation, the film Smithson makes of *Spiral Jetty* being, as he says, simply more "masses of impenetrable material" (Smithson 150). But crawling from this wreckage are sensations—individuations—whose trajectories are composed by the forces of Nature. *It is 1972* and Smithson states it simply: "I am for an art that takes into account the direct effect of the elements as they exist from day to day apart from representation" (Smithson 155).[13] These nonrepresentational expressions of the elements, these Earthworks, require our subjective de-differentiation in order to be seen. But Smithson's *trip* into the desert has also unearthed political concerns. It forces him to consider the relationship of technology and Nature, but in a strange "environmentalism" this causes him to attempt a revaluation of the "machine."

Art's materialization of Nature's forces in a sensation works through machines. Not simply the dump trucks and bulldozers of *Spiral Jetty*, because "everything," Deleuze and Guattari claim, "is a machine" (*Anti-Oedipus* 2). Indeed,

[t]here is no such thing as either man or nature now, only a process that produces the one within the other and couples the machines together. Producing machines, desiring machines everywhere, schizophrenic machines, all of species of life: the self and the non-self, outside and inside, no longer have any meaning whatsoever. (*Anti-Oedipus* 2)

"Art as abstract machine" (*Thousand Plateaus* 547); these aesthetic machines exist when, as in Spinoza, the world appears in living diagrams, and we consider ourselves "as if it were a question of lines, planes and bodies" (*Ethics* III, Pref.)

Smithson's Earthworks are an expression and construction of Nature, each within "one and the same process." This is nature in its Spinozian sense, active, affirmative, and perfect. Here, Deleuze and Guattari tell us, "everything is production: *production of production*" (*Anti-Oedipus* 4). This direct reference to Marx's theory of machines encourages us to consider how Smithson's Earthworks operate in the wider field of social production. For Marx the emergence of machines as the *means* of capitalist production marks the beginning of biopolitics, because when living labor is usurped by machinery capitalism succeeds in making itself, rather than human needs, the *end* of "immaterial" as well as material production. Following Marx's grounding of value in labor, Deleuze and Guattari nevertheless argue that "technical machines are not an economic category, and always refer back to a socius or a social machine that is quite distinct from these machines, and that conditions reproduction" (*Anti-Oedipus* 32). Departing from Marx, however, Deleuze and Guattari claim that capitalism operates on the level of social production primarily through a "primal psychic repression" or "social repression" (*Anti-Oedipus* 32) they call "Oedipus." The production of subjectivity is normalized|neuroticized through the Oedipus complex, which reterritorializes desire in the human individual. Deleuze and Guattari suggest that Oedipus is capitalism's condition because it overcodes the de-differentiated and schizophrenic "production of the real" (*Anti-Oedipus* 32). Although this emphasis upon production invokes the sublime ecstasies of a "molecular revolution,"[14] it can also be turned on capital as a weapon of immanent critique. The de-differentiated schizo-sensations of Smithson's Earthworks explore capitalism's own tendencies towards de-differentiation (most particularly environmental degradation), while producing sensations that feed these schizo tendencies back into the social realm. This reterritorialization of art onto capitalism's power of deterritorialization is a succinct definition of Smithson's final development of entropy as a mechanism for social production—his *environmentalism*.

For Smithson any political intervention had to begin from the indiscernibility of human activity and Nature, inasmuch as man and Nature are, as Deleuze and Guattari put it, "one and the same essential reality, the producer-product" (*Anti-Oedipus* 5). Smithson's dismissive public statements about political activism must be understood in this context, and rather than rejecting politics he advocates "a geopolitics of primordial return" (Smithson 150–151). This politics of de-differentiation returns the question of social production to that of a productive Nature, and in this sense Smithson's last works attempt to turn the machines of capitalism into aesthetic mechanisms that would give back to the earth its "primary process" of production.

Although I have emphasized Smithson's move into the desert as the turning point in his work, the break it marks also enabled him to return to the landscape he had previously abandoned, the industrialized world he explored in Passaic. After *Spiral Jetty* Smithson's major Earthworks projects, with the exception of *Amarillo Ramp* (1973), were constructed on old industrial sites. *Spiral Hill* and *Broken Circle* (1971) occupy an abandoned sand quarry in Emmen, Holland, and exemplify Smithson's thinking about reclamation. "With my work in the quarry," Smithson said, "I somehow reorganised a disrupted situation and brought it back to some kind of shape" (Smithson 253). Aesthetic reclamation is thus a way of "collaborating with entropy" (Smithson 256), a way of using the environmental catastrophe to compose new life. This is life "revalued" beyond its natural or organic definition, as a form of vitality that is able to survive entropy by turning the chaos it unleashes creative. Entropy and negentropy, we could say, have been de-differentiated.

Smithson's exploration of an eco-aesthetic "industry" ignored distinctions between man and nature, constructing machines in which "tools are undifferentiated from the material they operate on" (Smithson 102). Deleuze and Guattari give us a sense of the implications of this: "Industry is then no longer considered from the extrinsic point of view of utility, but rather from the point of view of its fundamental identity with nature as production of man and by man" (*Anti-Oedipus* 4). Here Smithson's *industrial* art—qua production without utility—turns capitalism's catastrophes into living sensations-signs. In relation to his reclamation Earthworks in Holland, Smithson argued: "A dialectic between land reclamation and mining usage must be established. The artist and the miner must become conscious of themselves as natural agents" (Smithson 376). This is a concise statement of Smithson's final version of the dialectic, the material dialectic of Nature moving between an entropic de-differentiation and the production of a new future. The exploitation of the earth and its reclamation in an aesthetic experience are here part of the same process, part of the "industry" of eco-aesthetics. This is to posit aesthetics as the industry of Nature itself, as the machinery of the eternal return. The political function of art therefore lies in the immanence of sensation and Nature, an immanence that includes and attacks both capitalism and us. The concept of eternal return—its conceptual "art"—has been materialized, and, in Smithson as much as Deleuze and Guattari, the future is reclaimed as the horizon of art.

Smithson's interest in exhausted mining Sites does not attempt to return them to a "natural" Nature, but turns entropic industry productive in an aesthetic sense. This "reclamation" project would therefore utilize the entropy (de-differentiation) produced by Capital, in order to create machines that avoid "alienation" (Oedipus), and are therefore free to play on the edge of time. In Smithson's late work art has become a way of creating the future, a productive repetition of the destroyed Site that redirects its force of de-differentiation, via art, via the machinery of the *aesthetic paradigm*, into social production.

Smithson gives a more detailed description of this production process in a 1970 response to a questionnaire about "The Artist and Politics." Smithson suggests that recent student riots marked the emergence of "a primal contingency—not a rite but an accident." In this sense the students "are a 'life force' as opposed to the police 'death force'" (Smithson 135). This life force was the subject of a work Smithson had produced earlier in the year at Kent State University called *Partially Buried Woodshed*. The work involved piling 20 truckloads of earth onto a shed on the University's grounds, until its central beam cracked. Less than five months later the National Guard shot and killed four Kent State students protesting the Vietnam War. In the context of Smithson's comments the work affirms entropic events—both natural and political—as being catastrophic in a revolutionary sense. Smithson's unquenchable faith in the future affirms beauty in every disaster; an echo of Nietzsche: no creation without destruction.

Partially Buried Woodshed avoids the hallucinatory trip of *Spiral Jetty*, and instead aims its entropic force against the institution. Smithson dramatizes the moment of collapse, the inevitable, necessary, fall. The students' "life force" will be released in a revolutionary catastrophe, and here Smithson's work expresses a seismic impact. Smithson's revolutionary diagram is "geo-political"; it dramatizes the structure's fall into the earth. He compares *Partially Buried Woodshed* to an earthquake that cleared a part of Anchorage that then became a park, and to a volcanic eruption in the Vestmann Islands that created a "buried house system" (Smithson 305). Both are examples of productive and creative processes emerging from the chaos of an entropic catastrophe. This, Smithson explains, is "an interesting way of dealing with the unexpected, and incorporating it into the community" (305). In Smithson's dramatization of entropy in art-machines, however, the catastrophe is not employed as part of a spontaneist political movement, but as the first *intervention* of a constructivist practice.

Deleuze and Guattari suggest something similar, claiming chaos is not something we hopelessly resist, but something we must "plunge into," for "we defeat it only at this price" (*What Is Philosophy?* 202). Although chaos exists as the "infinite speeds that blend into the colorless and silent nothingness they traverse" (*What Is Philosophy?* 201), art can "defeat" it by going through a "catastrophe" to leave "the trace of this passage...from chaos to composition" (*What Is Philosophy?* 203). This revolutionary plunge into de-differentiation produces a social and subjective *break*, a break with clichés and opinions—the reign of the same—which attempt to "protect us" from chaos by imposing upon us "a conformity with the past" (*What Is Philosophy?* 202). This, and not chaos, is "the misfortune of the people" (*What Is Philosophy?* 206). Smithson's Earthworks proclaims his own kind of environmentalism, but it is not one dedicated to reclaiming energy that is imagined as dissipating, or to conserving what remains. "It seems," Smithson said instead, "that one would have to recognize this entropic condition rather than try and reverse it" (Smithson 307). The Earthworks embrace catastrophe as an act of reclamation, they turn catastrophe expressive, and produce a sensation. "Art struggles with chaos," Deleuze and Guattari argue, "but it does so in order to render it sensory," to give us a new and "enchanted landscape" (*What Is Philosophy?* 205).

In this sense Smithson has a political vision. "How we *see* things and places," Smithson wrote in relation to a reclamation project, "is not a secondary concern, but primary" (Smithson 380). It is precisely in composing this "vision" in sensation, Smithson argues, that "art can enter the social and educational process at the same time" (Smithson 379). From the catastrophe and its explosion of chaotic de-differentiation comes the composition of a new social, aesthetic and political body organized through sensation. In this sense, and as Smithson laconically remarks, "wreckage is often more interesting than structure" (Smithson 257). Deleuze and Guattari put this rejection of structure precisely, summing up in a sentence one of the trajectories of this essay: "The monument [i.e., art work] does not actualize the virtual event but incorporates or embodies it: it gives it a body, a life, a universe" (*What Is Philosophy?* 177).[15]

We are still to unfold this body, this life, this universe. Beyond the break introduced by Smithson and by Deleuze and Guattari we must understand for ourselves the ways in which we can embody revolution. This is nothing but the definition of the "contemporary," the way in which art is transformed into eco-aesthetics, and the way in which eco-aesthetics can open up for us a new future.

Notes

I would like to thank Ralph Paine for his significant contribution to this essay. A collaborator by any other name.

1. This shift has already been explored by Éric Alliez (2003) and Alberto Toscano (2006), and has been described by Alliez as a "Break, *breakthrough* without which materialism remains an *Idea* [...]; without which the conceptual operations can't be *made as physical ones*" ("The BwO" 19–20). What I hope to do here is to extend this break towards its aesthetic and ecological horizons. Simon O'Sullivan (2006) and Felicity Coleman (2006) have both traced the similarities between Smithson's practice and Deleuze's philosophy. Both however, offer readings that focus on the consistencies in Smithson's practice rather than its break.

2. For Deleuze, "Each dialectical problem is duplicated by a symbolic field in which it is expressed" (*Difference* 179).

3. Although Reynolds' insistence on Smithson's use of enantiomorphic structures is sharp at this point, her attempt to read all of Smithson's work in this way fails precisely at the point of the Earthworks emergence, which she only briefly discusses. Deleuze's description of "enantiomorphic bodies" (bodies formed through a mirror reflection that is not entirely symmetrical) as "structuralist" is more useful for our purposes. Smithson's Site–Nonsite works actualize an *"internal difference"* that acts as their "transcendental principle" but which cannot, in itself, appear (*Difference* 231). As we shall see, it is precisely this enantiomorphism that the Earthworks abandon.

4. Rosalind Krauss, in her seminal essay "Sculpture in the Expanded Field" (1979) saw Smithson's work as organized "through a universe of terms felt to be in opposition within a cultural situation." Krauss argues that Smithson's work is structured by the "landscape—architecture" opposition, but she does so only in order to define a "post-modern" practice that traverses all of Smithson's work (Krauss 289).

5. Deleuze writes: "For difference, to be explicated is to be cancelled or to dispel the inequality which constitutes it....Difference of intensity is cancelled or tends to be cancelled in this system, but it creates this system by explicating itself" (*Difference* 228).

6. Deleuze had already explored Nietzsche's critique of thermodynamics in *Nietzsche and Philosophy*, where he writes: "Science is part of the *nihilism* of modern thought. The attempt to deny differences is a part of the more general enterprise of denying life, depreciating existence and promising it a death ('heat' or otherwise) where the universe sinks into the undifferentiated" (*Nietzsche and Philosophy* 45). This "terminal state of becoming" is an ascetic ideal opposed to Nietzsche's world as will to power, where "The law of the conversion of energy demands *eternal recurrence*" (*Will to Power* 1063). Rather than implying an eventual loss of creative force within a system whose energy remains constant (entropy as the tendency towards equilibrium), Nietzsche argues that the first law of thermodynamics (the constant level of energy) requires the eternal return of becoming to itself.

7. See (Toscano 174) on this point.

8. "De-differentiation" is, Smithson claims, "Anton Ehrenzweig's word for entropy" (Smithson 110).

9. "De-differentiation" is, Smithson claims, "Anton Ehrenzweig's word for entropy" (110).

10. "Substance must itself be said *of* the modes and only of the modes. Such a condition can be satisfied only at the price of a more general categorical reversal according to which being is said of becoming, identity of that which is different.... Repetition in the eternal return, therefore, consists in conceiving the same on the basis of difference...it carries out a practical selection among differences according to their capacity to produce" (*Difference* 40–41).

11. In *A Thousand Plateaus* Deleuze and Guattari offer their definitive definition of a Spinozian BwO, and it is one that proclaims the absolute immanence of intensity and matter: "A continuum of all substances in intensity and of all intensities in substance. The uninterrupted continuum of the BwO. BwO, immanence, immanent limit" (154).

12. Éric Alliez has pointed out the necessity of understanding Deleuze and Guattari's grasp of Spinoza's Nature as the indiscernibility of expression and construction, of intensity and extension. He has also extended this indiscernibility far into the realm of art. See Alliez, 2004, chapter 3 and Alliez and Bonne, 2007.

13. Art, Deleuze writes, materializes other "elementary forces like pressure, inertia, weight, attraction, gravitation, germination" (*Bacon* 57). Deleuze also traces Smithson's rejection of representation in favor of expression to Spinoza. "The opposition of expression and signs," Deleuze writes, "is one of the fundamental principles of Spinozism" (*Expressionism* 181–182).

14. The title of Chapter 21 of Timothy Leary's *The Politics of Ecstacy*. London: Paladin, 1970.

15. It is important to note that Deleuze does not abandon the virtual/actual distinction along with its Structuralist interpretation, but they do become much closer, being, for example, "totally reversible" in the crystal-image of *Cinema 2*, which does not cancel the virtual in the actual, but is "actual and virtual at the same time" (*Cinema 2* 69).

Works cited

Alliez, Éric. "The BwO Condition or; The Politics of Sensation," in *Biographen des organlosen Körpers*. pp. 11–29. Eds Éric Alliez and Elisabeth von Samsonow. Vienna: Turia+Kant, 2003.

——. *The Signature of the World. What is Deleuze and Guattari's Philosophy?* Trans. Eliot Albert and Alberto Toscano. New York and London: Continuum, 2004.

Alliez, Éric and Jean-Claude Bonne. "Matisse and Dewey and Deleuze." *Pli*, 18 (2007), pp. 1–19.

Coleman, Felicity. "Affective Entropy: Art as Differential Form." *Angelaki, Journal of the Theoretical Humanities*, 11(1) (April 2006), pp. 169–178.

Deleuze, Gilles. *Nietzsche and Philosophy*. Trans. Hugh Tomlinson. New York: Columbia University Press, 1982.

——. *Cinema 2: The Time-Image*. Trans. Hugh Tomlinson and Robert Galeta. Minneapolis: University of Minnesota Press, 1989.

——. *Expressionism in Philosophy: Spinoza*. Trans. Martin Joughin. New York: Zone Books, 1992.

——. *Difference and Repetition*. Trans. Paul Patton. New York: Columbia University Press, 1994.

——. "What Children Say," in *Essays Critical and Clinical*. pp. 61–67. Trans. Daniel W. Smith and Michael A. Greco. Minneapolis: University of Minnesota Press, 1997.

——. *Francis Bacon: The Logic of Sensation*. Trans. Daniel W. Smith. London: Continuum, 2003.

——. "How Do We Recognize Structuralism?" in *Desert Islands and Other Texts 1953–1974*. pp. 170–192. Trans. Michael Taormina. New York: Semiotext(e), 2004.

Deleuze, Gilles and Guattari. Félix. *Anti-Oedipus, Capitalism and Schizophrenia*. Trans. Robert Hurley, Mark Seem, and Helen R. Lane. Minneapolis: University of Minnesota Press, 1983.

——. *What Is Philosophy?* Trans. Hugh Tomlinson and Graham Burchell. New York: Columbia University Press, 1994.

Ehrenzweig, Anton. *The Hidden Order of Art*. Berkeley: University of California Press, 1995.

Krauss, Rosalind. *The Originality of the Avant-Garde and Other Modernist Myths*. Cambridge, MA: MIT Press, 1986.

Nietzsche, Friedrich. *Will to Power*. Trans. Walter Kaufmann and Richard Hollingdale. New York: Random House, 1967.

O'Sullivan, Simon. *Art Encounters Deleuze and Guattari, Thought Beyond Representation*. Basingstoke and New York: Palgrave Macmillan, 2006.

Reynolds, Anne. *Robert Smithson, Learning from New Jersey and Elsewhere*. Cambridge, MA: MIT Press, 2003.

Smithson, Robert. *Robert Smithson: The Collected Writings*. Ed. Jack Flam. Berkeley: University of California Press, 1996.

Spinoza, Benedict de. *The Ethics*. Trans. Edwin Curley. Princeton: Princeton University Press, 1994.

Toscano, Alberto. *The Theatre of Production: Philosophy and Individuation between Kant and Deleuze*. Basingstoke and New York: Palgrave MacMillan, 2006.

12
The "Weather of Music": Sounding Nature in the Twentieth and Twenty-First Centuries

Bernd Herzogenrath

> Climate is what you expect. Weather is what you get.
> (Robert Heinlein. *Time Enough for Love*.)

What is the "weather of music?" In seminal works of Classical music which refer to the seasons (Schumann's *Symphony No 1* ("Frühling"|"Spring"), Gershwin's "Summertime," or Vivialdi's *The Four Seasons*), and|or to the weather, such as Beethoven's *Symphony No 6* ("Pastorale"|"The Pastoral Symphony"), with its fourth movement "Thunderstorm," composers were primarily concerned with an *acoustic|musical translation* of *subjective sense perceptions*, that is, with a *representation* of nature and natural forces. Sometimes, the representation even threatened to *erase* nature itself—Gustav Mahler, when his friend, the conductor Bruno Walter, visited the composer in Steinbach at Attersee in the mountainous region in Upper Austria and was impressed by the spectacular vista of the Höllengebirge, is reported to have commented: "You won't have to watch it anymore—I have already composed it away..." (Walter 30)—as in Mark Tansey's *Still Life*.

This essay, however, is interested in the question of whether there is another connection between nature, weather and music *beyond representation*, if weather phenomena *themselves* can be music, and if music *itself* can be "meteorological." I argue that whereas the composers of the eighteenth and nineteenth centuries were mainly interested in the *representation* of the subjective effects of weather phenomena, the avant-garde more and more focuses on the *reproduction* of the processes and dynamics of the weather as a system "on the edge of chaos." I will correlate this with the additional claim that a particular American modernist tradition in music from Charles Ives via John Cage to John Luther Adams not only *starts* with the writings of Henry David Thoreau—Thoreau already provides an aesthetics of music, the

Figure 1

radicalism of which is only followed up on today and culminates in John Luther Adams's search for an "ecology of music."[1]

Let me first point out the particularity of Thoreau's musical aesthetics and "musical ecology." In 1851, Thoreau notes an auditory experience in his journals that reveals his particular sensibility to his sonic environment— "Yesterday and to-day the stronger winds of autumn have begun to blow, and the telegraph harp has sounded loudly...the tone varying with the tension of different parts of the wire. The sound proceeds from near the posts, where the vibration is apparently more rapid" (*Journal* III: 11). This was far from being an isolated case; Thoreau focuses on the "sound of nature"— and in particular the "sound of the weather"—in various other entries in his journals: "Nature makes no noise. The howling storm, the rustling leaf, the pattering rain are no disturbance, there is an essential and unexplored harmony in them" (I: 12). Thoreau is exploring the audible world like a sound-archaeologist, carefully distinguishing "sound" from "music":[2] "now I see the beauty and full meaning of that word 'sound.' Nature always possesses a certain sonorousness, as in the hum of insects, the booming of ice...which indicates her sound state" (I: 226–227). What Thoreau is pointing at is the fact that nature *itself* produces what one might call "ambient sound." Thoreau's sensitivity for environmental sounds heralds an avant-garde aesthetics in music that begins with the work of Charles Ives. That Ives, and Cage and Adams as well, were effectively influenced by Thoreau is beyond question. However, I am interested more in which *particular* inspirations these composers draw from Thoreau's aesthetics, and how they made this inspiration fruitful for their own ecology of music. Thus, let me begin with Ives' reading of Thoreau and Ives' "Weather of Music" as *representation*.

Charles Ives—The "Weather of Music" as *Representation*

Far from the madding crowds, metropolises and music centers of the world, awry to every "trend" in the classical music of the declining nineteenth century, Ives was composing his music in Danbury, Connecticut, a music that was intimately related to New England Transcendentalism, a literary–philosophical "movement" that can be understood both as a secular brand of American Puritanism, and as "American Romanticism," since it drew its inspiration from that of which America had in abundance—Nature (with a capital N). American Transcendentalism is inextricably intertwined with the names Ralph Waldo Emerson and Thoreau. While Emerson's metaphysical and idealistic (in the sense of a Hegelian Idealism) brand of Transcendentalism made him the philosophical spokesman of the movement, his disciple Thoreau followed a much more materialist and "physical" philosophy, without, however, completely casting off the Emersonian Metaphysics. This seemingly small difference in the initial conditions will have important effects, since Thoreau's ambivalence in this matter will result in the contrasting readings of his work by Ives and Cage respectively.

Ives is a dyed-in-the-wool Transcendentalist, and Thoreau is promoted to the private patron saint of his own conception of music. In an essay on Thoreau, Ives emphasizes that "if there shall be a program for our music, let it follow [Thoreau's] thought on an autumn day of Indian summer at Walden..." (*Essays* 67). In these 1920 *Essays Before a Sonata*, which Ives conceived as a literary counterpart to his *Piano Sonata No 2* ("Concord"-Sonata), Ives summarizes his understanding of the central idea of Emersonian Transcendentalism, which is also the guiding tenet of his own work:

> Is it not this courageous universalism that gives conviction to [Emerson's] prophecy, and that makes his symphonies of revelation begin and end with nothing but the strength and beauty of innate goodness in man, in

Nature and in God—the greatest and most inspiring theme of Concord Transcendental philosophy...? (*Essays* 35)

If Ives' phrase "symphonies of revelation" obviously refers to both Emerson's visionary power and the "musicality" of his oratorical prose, Ives also finds these very qualities in Thoreau's writings:

> Thoreau was a great musician, not because he played the flute but because he did not have to go to Boston to hear "The Symphony." The rhythm of his prose...would determine his value as a composer. He was divinely conscious of the enthusiasm of Nature, the emotion of her rhythms, and the harmony of her solitude. (*Essays* 51)

The reason for Ives' reference to both Emerson and Thoreau can be found in the observation that Ives reads Thoreau's "materialist" sound-aesthetics on the foil of Emerson's Idealism, according to which nature is the expression (and effect) of reason—"the whole of nature is a metaphor of the human mind" ("Nature" 24). As has been pointed out, in Emerson's work, the "subject's triumph over nature" takes center stage (see Schulz 117). Defining intuition and imagination as primary sources of a creative comprehension of Truth, for Emerson, the creative subject attains a Divine status—"Whoever creates is God" (*Journals* V: 341).

To Emerson mind, not matter, is of prime importance—matter is only a manifestation of the mind. Thoreau, in contrast, stresses the material and sensual aspects of nature—"We need pray for no higher heaven than the pure senses can furnish, a *purely* sensuous life...Is not Nature...that of which she is commonly taken to be a symbol merely?" (*A Week* 307). Thoreau does not *read* nature, does not interpret nature according to a spiritual principle external to it—such a principle, because of nature's manifoldness, is *immanent* to it. For Thoreau, nature and its "music" are not only "God's voice, the divine breath audible" (*Journal* I: 154), but also—and maybe even first and foremost—"the sound of circulation in nature's veins" (I: 251). It is in this stress on nature as sensuous experience and materiality that Thoreau "deviates" from Emerson. Thoreau focuses on (the music of) nature as a material, physical process, not as an Emersonian emblem of reason—"The very globe *continually transcends* and translates itself....The whole tree itself is but one leaf, and rivers are still vaster leaves whose pulp is intervening earth" (*Walden* 306–307). "Transcendentalism" is understood by Thoreau as completely "physical"—the natural, dynamic process of metamorphosis, of continuous change...transcendence becomes immanence.

Deleuze rarely mentions Thoreau in his writings. Yet, in a prominent passage where he actually *does*, he refers precisely to Thoreau's

> affirmation of a world *in process*, an *archipelago*. Not even a puzzle, whose pieces when fitted together would constitute a whole, but rather a wall of

loose, uncemented stones, where every element has a value in itself but also in relation to others; isolated and floating relations, islands and straits, immobile points and sinuous lines. (*Critical and Clinical* 86)

For Emerson, in contrast, nature is the manifestation of the spirit, of reason, and the "music of nature" is spirit|reason expressing itself, is thus pure transcendence, pure metaphysics.[3]

Even if Ives is following Thoreau in his music aesthetics and makes the sonority of the world his main principle, he is mainly interested in the sonority of the *human* world, which he does not *reproduce*, but *represent*, and which he generates from various quotations and samples taken from European Classical music, American popular tunes and liturgical music, as well as from the "compositorial transformation" of sounds of everyday human life. Ives does not only compose "nature," but complex cityscapes|soundscapes, impressions of man's urban "second nature"—thus, in *Over The Pavements*, he layers the irregular movements, speeds, and rhythms of people, carts, and horse-carriages into a polyrhythmic, albeit to an extremely high degree controlled, ensemble:

In the early morning, the sounds of people going to and fro, all different steps, and sometimes all the same...I was struck with how many different and changing kinds of beats, time, rhythms, etc. went on together—but quite naturally, or at least not unnaturally when you get used to it. (*Memos* 62)

In a similar vein, his *Holidays Symphony*, according to Ives, paints "pictures in music of common events in the lives of common people" (*Memos* 97–98), and his hymns "*represent* the sternness and strength and austerity of the Puritan character" (*Memos* 39, emphasis added). With Ives transferring the Emersonian Transcendentalism's "correspondence" of spirit and nature to the realm of music,[4] his "weather of music" always coagulates into a *representation*, that is, a sonic *picture* of the weather—for example, in the *Holidays Symphony* into an acoustic "picture of the dismal, bleak, cold weather of a February night near Fairfield" (*Memos* 96), with the weather itself in turn "reflecting the sternness of the Puritan's fibre" (*Memos* 96n1). If, according to Emerson, "[l]anguage clothes nature as the air clothes the earth" (*Journals* V: 246), and if the use of "compositorial languages," such as tonality or a-tonality, according to Ives, depend "a good deal—as clothes depend on the thermometer—on what one is trying to do" (*Essays* 117), on which representational effect one is aiming at, then, quite obviously, bad weather (or weather at all) does not really seem to exist for Ives—only inappropriate outdoor-gear.

Thus, even if Ives explicitly refers to Thoreau, his relation to Thoreau, with all of Ives' interest in experimental soundscapes, is a one-sided and single-minded affair at most. For Ives, listener and composer are aural equivalents to Emerson's almighty and visual *me*—as with Emerson, music for Ives is not only "purely a symbol of a mental concept" ("Correspondence" 115),[5] but the almost mystical revelation of the Emersonian "Over-Soul," with the composer's role matching the one of Emerson's Poet:

> For Poetry was all written before time was, and whenever we are so finely organized that we can penetrate into that region where the air is music, we hear those primal warblings...The men of more delicate ear write down these cadences more faithfully, and these transcripts though imperfect, become the songs of the nations. ("The Poet" 449)

Like Emerson, Ives stresses the need of the representation and translation of those "primal warblings," since he, notwithstanding his acceptance of "sounds," always emphasizes the need of a "subjective corrective" to bring out|about the sounds' "ethereal quality" ("Music and Its Future" 192). For Thoreau, however, the "music of nature" needs and requires no translation— "This earth was the most glorious musical instrument, and I was audience to its strains" (*Journal* II: 307). Thoreau lets the sounds rest and dwell in their semantic indeterminacy, focusing on "the language which all things and events speak without metaphor" (*Walden* 11) instead, a "language" of the real that, beyond the seemingly stable and fixed realm of representation, is an open, dynamic system that is "permeated by unformed, unstable matters, by flows in all directions, by free intensities" (Deleuze|Guattari *Thousand Plateaus* 40), a real that, because of its machinic set-up, in fact "is the abolition of all metaphor" (69).

Ives, in his privileging of the "idea" over the "senses," does not follow Thoreau in his deviation from (and maybe even re-conceptualization of) Emerson's idealistic Transcendentalism, and is thus closer to Emerson than to Thoreau.[6] However, when Ives takes over from Thoreau the development of a conception of art that is indifferent to the source of its materials, and also to more traditional aspects of "form," he begins a "turn" in music that proceeded to become an aesthetic dictum with avant-gardists such as John Cage.

John Cage—The "Weather of Music" as *Mapping*

"[I]s not all music program music? Is not...music...representative in its essence?" (*Essays* 4). Cage would have definitely answered Ives' rhetorical question with a firm "No!" Cage deviates from Ives in that he precisely puts Thoreau's shift of emphasis towards nature's materiality center stage in his own aesthetics. Cage came across Thoreau's Journals for the first time in 1967, and has since made Thoreau not only the addressee of numerous compositions, but his "[retroactive] muse": "Reading Thoreau's *Journal*, I discover any idea I've ever had worth its salt" ("Diary" 18). One of the challenging ideas that

Cage saw already "prefigured" in Thoreau is the nondualistic conception of the world that counters Emerson's doctrine of the "metaphoricity of nature" and the "partitioning of the world" into *Me* and *not-me* with a fundamental coexistence of both spheres—according to Thoreau, "[a]ll beauty, all music, all delight springs from apparent dualism but real unity" (*Journal* I: 340). Beauty (and music) for Thoreau and Cage are explicitly *not* Hegel's "idea made real in the sensuous" (284)—and Emerson and Ives would certainly have embraced the Hegelian concept. Nature, for Thoreau and Cage, is not a function of the idea, perceptions are not *interpretations* of the world, but *part* of that world. This notion completely contradicts both the Idealism inherent to Emerson's Transcendentalism and its claim that the subject imposes its power on matter. If Emerson claims that "the poet conforms things to his thoughts...and impresses his being thereon" ("Nature" 34), that the creative subject in-forms matter in the first place, then for Thoreau, in contrast, "[t]he earth I tread on...is not a dead inert mass. It is a body, has a spirit" (*Journal* II: 165). With regard to the "telegraph harp" being "played" by the weather, the resulting music of which he claimed to be "the most glorious music I ever heard" (*Journal* III: 219), Thoreau states: "the finest uses of things are accidental. Mr. Morse did not invent this music" (*Journal* III: 220).[7] Cage finds in Thoreau thus both the focus on the materiality of nature, which Ives still had "subjectified" into human and symbolic music, and the accidental, which Ives always had attempted to control. It is exactly these parameters that Cage turns into the center of his compositions. Against the traditional composer's attempt at control, Cage envisions "a composing of sounds within a universe predicated upon the sounds themselves rather than the mind which can envisage their coming into being" (*Silence* 27–28). For Deleuze and Guattari, this focus on "pure sounds" and random processes

> is a question of freeing times...a nonpulsed time for a floating music....It is undoubtedly John Cage who first and most perfectly deployed this fixed sound plane, which affirms a process against all structure and genesis, a floating time against pulsed time or tempo, experimentation against any kind of interpretation, and in which silence as sonorous rest also marks the absolute state of movement. (*Thousand Plateaus* 267)

In his absolute reduction of "subjective control" and his valorization of "sound" Cage combines two other maxims of Thoreau—"the music is not in the tune; it is in the sound" (*Journal* IV: 144), and "[t]he peculiarity of a work of genius is the absence of the speaker from his speech. He is but the medium" (III: 236). The radical difference in Ives's and Cage's reference to Thoreau can be illustrated by recourse to a passage from *Walden*:

> Sometimes, on Sundays, I heard the bells, the Lincoln, Acton, Bedford, or Concord bell, when the wind was favorable...At a sufficient distance over

the woods this sound acquires a certain vibratory hum, as if the pine needles in the horizon were the strings of a harp which it swept. (*Walden* 123)

Ives repeatedly quotes this passage and always emphasizes the "spiritualizing effect" of the sound described by Thoreau—the symbolic meaning of the distant church bells, a "transcendental tune" (*Essays* 69), a mere echo of a more divine "sphere music." Cage, in contrast, combines Thoreau's "auditory observation" with his remark of the accidental Aeolian music of the telegraph harp. This merging of sound and indeterminacy becomes his "Music for Carillon," a composition for chimes, for which Cage "translates" nature "without metaphor," by transferring the natural patterns of the wood's grain into musical notation. By drawing stave-lines onto the wood, Cage lets the musician "read" the knotholes and grain patterns as notes—in a similar way, in "Music for Piano," Cage uses the material irregularities of a sheet of paper to determine the position of notes.

Thoreau's preference for sounds and for the accidental makes him a progenitor of a decidedly avant-garde musical practice in Cage's eyes, a practice which does away with the individual as locus and agency of control. This also means that this aesthetics is not dealing with the *representation* of affectations and sensations anymore, but with the *reproduction* of the dynamics of natural processes. Thus Cage does not only state the importance of sounds (and silence) for him as a composer, but claims much more fundamentally: "the function of the artist is to imitate nature in her manner of operation" (*Silence* 194). Here, I argue, music becomes "meteorological," since nature operates according to extremely complex dynamics, probabilities and improbabilities... just like the weather! What is fascinating about the weather is thus not just the power of its atmospheric special effects, the combined LucasArts™ of thunder and lightning, but most and foremost the fact that the weather is a highly complex, dynamic, open, and thus in the long run unpredictable and uncontrollable system of forces and intensities. For Cage's aesthetics and compositional practice, this means that they reveal a line of flight, a vector "away from ideas of order towards no ideas of order" (*Silence* 20), with the stress being on *ideas* of order, that is, a *mental* order as against a "natural order" with its own "manners of operation" (and self-organization). This introduction of indeterminacy and chance into the compositional process molecularizes it, frees it from the molar regime of representation and makes it form rhizomatic connections with the virtualities of the environment. Cage's swerve in his compositional "plan" from "ideas of order" towards "chance" can be read parallel to Deleuze's distinction between the (molar) plan of organization and the (molecular) "plan of composition" (*Practical Philosophy* 128), in which

[t]here is no longer a form, but only relations of velocity between infinitesimal particles of an unformed material. There is no longer a

subject, but only individuating affective states of an anonymous force. Here the plan is concerned only with motions and rests, with dynamic affective charges. (*Practical Philosophy* 128)

Against the Emersonian stress on the representation (and control) of nature by the individual, Thoreau and Cage emphasize *perception* as a practice (of both art and life). According to Chris Shultis, Cage and Ives posit "the two poles of self (...the coexisting and controlling) in American experimental music, connecting contemporary concerns to a nineteenth-century past" (xviii), two poles already prefigured in Thoreau and Emerson. The compositional complexity of Ives's work is due to an *intertextual* interweaving of "samples of culture" that are ultimately (re)inscribed in a higher (transcendent) unity. The modernist Ives folds cultural quotations (and cultural quotations of nature, nature *as* a cultural quotation) into each other and thus implicitly refers to a "supplementary dimension" underlying these quotes—and, as Deleuze and Guattari remark with regard to the cut-ups of William S. Burroughs, "[i]n this supplementary dimension, unity continues its spiritual labor" (*Thousand Plateaus* 6). Like the work of Joyce, Ives's work "affirm[s] a properly angelic and superior unity" (6)—"this is to say the fascicular system does not really break with dualism, with the complementary between a subject and an object, a natural reality, and a spiritual reality" (6). With regard to that "higher unity" in Ives, David Nicholls has poignantly referred to it as an "organized chaos" (67). For Cage, it is a "purposeless play," an *against organization|control* that is (at) the "heart" of life and art:

This play...is an affirmation of life—not an attempt to bring order out of chaos nor to suggest improvements in creation, but simply a way to wake up to the very life we're living, which is so excellent once one gets one's mind and desires out the way and lets it act of its own accord. (*Silence* 12)

It is that attitude in Thoreau to "the very life we're living" that makes him such an inspiration for Cage—"Thoreau only wanted one thing: to see and hear the world around him...he lets things speak and write as they are" (*For the Birds* 233–234). Instead of "painting" symbolic sonic pictures, Cage's compositions rather construct *maps* and *charts*, that is, *topographies* of natural processes. It is thus more than a coincidence that the motif|motive of the "map" is so important in Cage's work—see on the one hand his many compositions based on atlases or celestial charts, such as *Atlas Eclipticalis*, or the *Etudes Australes* and the *Etudes Boreales*, and on the other hand Cage's various "graphic notations"...maps, not tracings, in that they are "entirely oriented toward an experimentation in contact with the real" (*Thousand Plateaus* 12).

Cage's *mapping* of the "weather of music" finds its maybe most direct and "literal" reflection|precipitation in his *Lecture on the Weather*. In this composition, passages from Thoreau's journals (determined by chance operations), read simultaneously by various speakers in tempi of their own choice, and field recordings of wind, rain, and thunder condense into a commentary on the political climate of the United States in the mid-1970s. In a performance of *Lecture on the Weather* at the *Cage*-Fest in Strathmore, Maryland, on May 5, 1989,

> doors were open to the outside where a storm began to be audible and visible...this had the interesting effect of eradicating the distinction between "inside" and "outside"—the meteorological display over Strathmore Hall was continuous with what was going on in the room where Cage's more gentle storm included the weather of predetermined and coincidental conjunctions of sound and voice variables. (Retallack 248)

the performance in fact "is not *about* weather; it *is* weather" (Perloff 25).

Cage's "weather of music" thus can be understood as an assemblage of sonic intensities and natural processes—the compositions become "meteorological systems" themselves. These systems, however, are, as Cage himself admits and regrets, still "framed"—even silence has the precise temporal coordinates of 4:33. And this is one of the decisive differences between Cage and the sound installations of John Luther Adams that aim at reproducing the "weather of music" as a dynamic *ecosystem*.

John Luther Adams—The "Weather of Music" as *Ecosystem*

As Gigliola Nocera has emphasized, the living and working conditions of Ives and Cage were comparable to Thoreau's isolation at Walden Pond—Ives was working far off the "art centers" in Danbury, Connecticut, and Cage in Stony Point, in New York State (see Nocera 356). Isolation is an even bigger issue with the composer John Luther Adams,[8] who lives and works in Fairbanks, Alaska, approx. 125 miles south of the Arctic. Adams's work is highly influenced by his environment, this arctic, "hyperborean zone, far from the temperate regions" (Deleuze *Critical and Clinical* 82), far from equilibrium.

From his early works onwards he has always pointed out that he wants his music to be understood as an interaction with nature—as a site-specific "contact" with the environment that he calls "sonic geography" ("Resonance of Place" 8).

Adams's sonic geography comprises a cycle called *songbirdsongs* (1974–1980), consisting of various imitations of Alaskan birds reminiscent of Olivier Messiaen's *Catalogues d'oiseaux*. Although Adams in the compositional process and the transcription brings birdsong on a "human scale" in terms of tempo, modulation, pitch, etc., he conceptualizes the different melodies—or "refrains"—as a "toolkit," so that during the performance ever-new aggregations of phrases and motifs come into existence, an open system, indetermined in combination, length, intonation, tempi, etc. *Earth and the Great Weather* (1990–1993), an evening-long piece—or opera—consisting of field recordings of wind, melting glaciers, thunder in combination with ritual drummings and chants of the Alaskan indigenous people, was "conceived as a journey through the physical, cultural and spiritual landscapes of the Arctic" ("Sonic Geography"), a traversing of smooth "Eskimo space" (Deleuze|Guattari *Thousand Plateaus* 494).[9]

In a further step, Adams combined his "sonic geography" with the concept of what he calls "sonic geometry" ("Strange and Sacred Noise" 143). Adams is more and more interested in the "noisier" sounds of nature and refers to findings of Chaos Theory and Fractal Geometry in order to find sonic equivalents for nature's *modus operandi*—*Strange and Sacred Noise* (1991–1997) is an example of this approach.[10]

To date, the culmination of Adams's sonic geography|geometry has been his recent project *The Place Where You Go To Listen*, the title of which refers to an Innuit legend according to which the shamans hear the wisdom of the world in (and get their knowledge from) the whisper of the wind and the murmur of the waves, being sensitive to what Deleuze, with reference to Leibniz, calls "little perceptions" (*Difference and Repetition* 213).[11]

Adams aims at the realization of a "musical ecosystem,...A work of art...that is directly connected to the real world in which we live and resonates sympathetically with that world and with the forces of nature" (Mayer "Northern Exposure")—Adams does not only *imitate* nature in its manner of operation, as Cage still does; he taps into nature's dynamic processes *themselves* for the generation of sound and light. Adams developed this project in close collaboration with geologists and physicists—as Adams stated in an interview, "[a]t a certain level, it was like...they were the boys in the band" (Mayer *Living on Earth*).

Adams—like Deleuze—is thus interested in "the relations between the arts, science, and philosophy. There is no order of priority among those disciplines" (*Negotiations* 123) for both Deleuze and Adams. Whereas science involves the creation of functions, of a propositional mapping of the world, and art involves the creation of blocs of sensation (or affects and percepts), philosophy involves the invention of concepts. According to Deleuze|Guattari, philosophy, art, and science are defined by their relation to chaos. Whereas science "relinquishes the infinite in order to gain reference" (*What Is Philosophy?* 197), by creating definitions, functions and propositions, art, on the other hand, "wants to create the finite that restores the infinite" (197)—and it was exactly this creation of *finite* objects that bothered Cage. In contrast, "philosophy wants to save the infinite by giving it consistency" (197).

Yet, since "sciences, arts, and philosophies are equally creative" (5), it might be fruitful, as Deleuze proposes, "to pose the question of echoes and resonances between them" (*Negotiations* 123)—to pose the question of their ecology, which is what Adams's installation does.

As Deleuze specified in one of his seminars, "between a philosophical concept, a *painted* line and a musical sonorous bloc, resonances emerge, very, very strange correspondences that one shouldn't even theorize, I think, and which I would prefer to call 'affective'... these are privileged moments" ("Image Mouvement Image Temps").[12] These moments privilege an affect where thought and sensation merge into a very specific way of "doing thinking" *beyond* representation and categorization, a moment that might be called "contemplation," which also fittingly describes the "audience's" approach to Adams's installation. Already Thoreau had pointed out the anti-Cartesian tenor of contemplation—"Western philosophers have not conceived of the significance of Contemplation" (*A Week* 110).

In Adams's installation, real-time data from meteorological stations all over Alaska and from the five stations of the Alaska Earthquake Information Center are collected, coordinated, and made audible through "pink noise filters." As Curt Szuberla, one of the physicists involved in the project, explains, "[t]he strings and bells and drumheads are plucked, bashed and banged based on the geophysical data streams. And the geophysical data streams... are the fingers and mallets and bells that hit things and make things sound" (Mayer *Living on Earth*). *The Place Where You Go To Listen* is a permanent installation at the *Museum of the North* in Fairbanks, where sound and light are generated in real time through data processing of the day and night rhythms, the rhythm of the seasons, of the moon phases, the weather conditions, and the seismic flows of the magnetic field of the Earth—nature itself, as well as the music it produces, operates according to its own times and speeds (and slownesses). Hours, even days (and more), might pass between perceivable seismic changes or changes in the magnetic field of the Earth. *The Place* is an open system, a machinic aggregation operating according to what Deleuze calls "differences of level, temperature, pressure, tension, potential, *difference of intensity*" (*Difference and Repetition* 222)—just like the weather. Adams's noise-filter-machine is plugged into the sun-machine, and also into the wind-machine, rain-machine etc.; these in turn couple together to form the weather-machine. Digital machines cut into the flows of nature, but within a machine|nature ecology|ontology which is not based on the strict separation of these two spheres, where nature is either a fixed, unchanging essence, or the mere retro-effect of culture and representation, but an ecology|ontology of dynamics and production. Adams's installation thus presents "modes of individuation beyond those of things, persons or subjects: the individuation, say, of a time of day, of a region, a climate" (Deleuze, *Negotiations* 26).

The Place Where You Go To Listen focuses on nature as process and *event*—in an almost Stoic emphasis on *becoming* versus *being*, Adams privileges time-sensitive *dynamics*, not clear-cut *states*. In his study *La théorie des incorporels dans l'ancien stoïcisme*, to which Deleuze refers in *The Logic of Sense*, Emile Bréhier states that, according to Stoic "thought, one should not say, 'the tree is green,' but 'the tree greens'...what is expressed in this proposition is not a property, such as 'a body is hot,' but an event, such as 'a body becomes hot'" (Bréhier 20–21).[13] This *becoming*, writes Deleuze, passes the line "between the sensible and the intelligible, or between the soul and the body" (*Dialogues* 63)—or nature and culture—and places itself "[b]etween things and events" (*Dialogues* 63). By getting rid of the *is* of representational thought, where an object's quality is at least potentially related to a subject that expresses this quality as an attribute, by replacing fixity with process as both the subject's and the world's "manner of operation," these "infinitive-becomings have no subject: they refer only to an 'it' of the event" (*Dialogues* 64). Adams's installation goes further in the direction of the event than Ives and even Cage—although these two composers had also already pondered the conflict between the processuality of nature, and the means of art. Ives asked himself:

> A painter paints a sunset—can he paint the setting sun?...[Is] [t]here...an analogy...between both the state and power of artistic perceptions and the law of perpetual change, that ever-flowing stream, partly biological, partly cosmic, ever going on in ourselves, in nature, in all life? (*Essays* 71)

Ives tried to master this problematic by way of the ever-increasing complexification of his compositorial means. Cage also emphasized that he did not think it correct to say "the world as it *is*"—

> it *is* not, it becomes! It moves, it changes! It doesn't wait for us to change...it is more mobile than you can imagine. You're getting closer to this reality when you say as it "presents itself;" that means that it is not there, existing as an object. The world, the real is not an object. It is a process. (*For the Birds* 80)

However, Ives was still the subject in control of chaos, and Cage, in spite of all indeterminacy, regretted that he was still creating "clear-cut" objects. Adams solves this problem by leaving the executing|processing energy to the processual forces of nature *itself*. Music and environment thus become an ecosystem of a dynamics of acoustic and optic resonances interacting in|with an environment in constant flux. Thus, "music" in this sense becomes for Adams something entirely different from a "means" of human communication about an external world:

> If music grounded in tone is a means of sending messages to the world, then music grounded in noise is a means of receiving messages *from* the world....As we listen carefully to noise, the whole world becomes music. Rather than a vehicle for self-expression, music becomes a mode of awareness. ("Ecology of Music")

Thus, *The Place Where You Go To Listen* leaves the conceptualization of a music *about* nature, of music as a means of the *representation* of nature and landscape, on which Ives, for example, still relied, and creates music as a part of nature, as coextensive with the environment—"Through attentive and sustained listening to the resonances of this place, I hope to make music which belongs here, somewhat like the plants and the birds" (Adams "Resonance of Place" 8). Even more direct than Cage, Adams emphasizes nature's "manner of operation" in not only taking them as a model, but directly "accessing" and relating to the becoming of a site-specific environment and creating works that *are* this relation—a music of place, of a place where you go to listen. Even if Adams does not explicitly refer to Thoreau, his work is indebted to Thoreau's sound aesthetics—even more so, I argue, than Ives's or even Cage's work. As it is for Adams, music for Thoreau is already part of the environment—nature has no need to be translated or represented; nature and the environment already *sound*, already *express themselves*. In *Walden*, Thoreau writes that "making the yellow soil express its summer thought in bean leaves and blossoms rather than in wormwood and piper and millet grass, making the earth say beans instead of grass— this was my daily work" (*Walden* 157). If Thoreau calls this "natural expression" by the name of "saying," he is evoking a correspondence between "expression" and "production" of nature—a correspondence that goes far beyond the level of representation. The "expression" of nature on the side of "production" arrives in the subject as "impression," so that, from the perspective of culture, what we call representation is already rooted in nature—"every word is rooted in the soil, is indeed flowery and verduous" (*Journal* I: 386).

"A history of music would be like…the history of gravitation"—with regard to Adams's physico-musical ecosystem, this sentence perfectly makes sense as a postmodern credo of New Music…however, it is taken from Thoreau's journals (I: 325). Here we have come full circle, to Walden Pond, where in 1851, Thoreau was experiencing an ecology of music which only today is being realized.

Notes

1. And it should be noted that this radicalism makes Thoreau a patron saint not only of music, but also of ecology.
2. See also Thoreau's essay "Walking" and his|its concept of "wildness"—"sound" can be read as "wildness" with regard to "music" (as sound organized by a traditional composer)…the unformed, unintended, untamed in comparison to John Sullivan Dwight's canonization in Thoreau's time of European Classical Music (and in particular the compositions of Beethoven) as *the* paradigm for a future American Music.
3. In the above quotation from *Essays Critical and Clinical*, Deleuze explicitly refers to Thoreau *and* Emerson, and does not discriminate between their respective brands

of Transcendentalism—for reasons that hopefully become clear in this essay, I would see the need for a more precise distinction between the two.

4. For the theory of correspondence, see, for example, Emerson's "Nature."

5. See also: "'the music' as being the character of the idea or spirit, quite apart from its embodiment in sound" (John Kirkpatrick's footnote in *Memos* 242).

6. Betty E. Chmaj calls Ives the "Emerson of American music" (396). On the relation Ives|Emerson, see also (Shultis) and (Mehring).

7. Thoreau goes even further and envisions the coexistence of the "telegraph harp" with the greater cycle of nature—"What must the birds and beasts think where it passes through woods, who heard only the squeaking of trees before! I should think that these strains would get into their music at last. Will not the mocking-bird be heard one day inserting this strain in his medley?" (*Journal* III: 219). There is a loud and clear "Yes!" to Thoreau's question—today's birds have integrated radio jingles and cell phone ring tones in their song...

8. Adams, it has to be noted, is also an environmental activist and founder of Alaska's Green Party. Mitchell Morris thus dubs Adams a "'Green' composer" (131), referring, however, to the notion of ecology as in *Deep* Ecology, whereas I would suggest placing Adams firmly within a Deleuzian Ecology that is based on a nondualist ontology.

9. In fact, the first movement of *Earth and the Great Weather* is already named "The Place Where You Go To Listen"...

10. *Strange and Sacred Noise* is a concert-length cycle of six movements for percussion quartet. Its first and last movements ("...dust into dust..." and "...and dust rising...") are based on the Cantor set and Cantor dust (the two-dimensional version of the Cantor set). These fractals model the behavior of electrical noise, which Adams takes as a diagram for the percussion set to explore "the dynamic form of the Cantor dust, whereby in an infinite process, line segments are divided into two segments by the removal of their middle third" (Feisst, "Music as Place"). See also Feisst, Sabine M. "Klanggeographie—Klanggeometrie. Der US-amerikanische Komponist John Luther Adams." *MusikTexte* 91 (November 2001): 4–14.

11. A direct Leibnizian reference can be found in his *New Essays on Human Understanding*:

    To hear this noise as we do, we must hear the parts which make up this whole, that is the noise of each wave, although each of these little noises makes itself known only when combined confusedly with all the others, and would not be noticed if the wave which made it were by itself... [w]e must have some perception of each of these noises, however faint they may be; otherwise there would be no perception of a hundred thousand waves, since a hundred thousand nothings cannot make something. (55)

    Such a "sonorous ocean," it can be argued, the becoming-perceptible of micro-sounds "underneath the [human] radar," also provides a more materialist version of the Pythagorean idea of "sphere music": contrary to the harmonious universe rotating according to "well-tempered" intervals, it would refer to the multiplicity of sounds of "the world"—nature changes constantly, everything moves, and everything that moves oscillates according to a certain frequency, the total result of which would be "white noise" (the murmur of the universe). Such a concept, I argue, also defines much of today's electronic music (see, for example, (Murphy), in particular 161–162).

12. My translation of: "Alors je dirais que le concept philosophique n'est pas seulement source d'opinion quelconque, il est source de transmission très particulière, ou entre un concept philosophique, une ligne picturale, un bloc

sonore musical, s'établissent des correspondances, des correspondances très très curieuses, que à mon avis il ne faut même pas théoriser, que je préférerais appeler l'affectif en général…. Là c'est des moments privilégiés."
13. My translation of: "On ne doit pas dire, pensaient-ils: 'L'arbre est vert,' mais: 'L'arbre verdoie'…Ce qui s'exprime dans le jugement, ce n'est pas une propriété comme: un corps est chaud, mais une èvénement comme: un corps s'échauffe."

Works cited

Adams, John Luther. "Resonance of Place." *The North American Review* CCLXXIX: 1 (January/February 1994): 8–18.

———. "Sonic Geography of the Arctic. An Interview with Gayle Young." (1998) http://www.johnlutheradams.com/interview/gayleyoung.html

———. "Strange and Sacred Noise." *Yearbook of Soundscape Studies. Vol. 1: "Northern Soundscapes."* Eds R. Murray Schafer and Helmi Järviluoma. Tampere 1998, 143–146.

———. "In Search of an Ecology of Music." (2006) http://www.johnlutheradams.com/writings/ecology.html

———. quoted in Amy Mayer. "Northern Exposure: A museum exhibit converts activity in the Alaskan environment into an ever changing sound show." *Boston Globe* April 16, 2006.

———. quoted in *Living On Earth*. Radio Interview with Amy Mayer, see www.loe.org/shows/segments.htm?programID=06-P13-00016&segmentID=5

Bréhier, Emile. *La théorie des incorporels dans l'ancien stoicisme.* Paris: Librairie Philosophique J. Vrin, 1970.

Cage, John. "Diary." M: Writings '67-'72. Hanover, New Hampshire: Wesleyan University Press, 1969, 3–25.

———. *Silence.* Hanover, New Hampshire: Wesleyan University Press, 1973.

———. *For the Birds. John Cage in Conversation with Daniel Charles.* Boston and London: Marion Boyars, 1981.

Chmaj, Betty E. "The Journey and the Mirror: Emerson and the American Arts." *Prospects* 10 (1985): 353–408.

Deleuze, Gilles. *Spinoza: Practical Philosophy.* Trans. Robert Hurley. San Francisco: City Lights Books, 1988.

———. *Difference and Repetition.* Trans. Paul Patton. New York: Columbia University Press, 1994.

———. *Negotiations.* Trans. Martin Joughin. New York: Columbia University Press, 1995.

———. *Essays Critical and Clinical.* Trans. Daniel W. Smith and Michael A. Greco. Minneapolis: University of Minnesota Press, 1997.

———. "Image Mouvement Image Temps." Cours Vincennes – St Denis: le plan – 02/11/1983. www.webdeleuze.com/php/texte.php?cle=69&groupe=Image%20Mouvement %20 Image%20Temps&langue=1, last accessed April 2, 2008.

Deleuze, Gilles and Felix Guattari. *A Thousand Plateaus: Capitalism and Schizophrenia.* Trans. B. Massumi. Minneapolis: University of Minnesota Press, 1987.

———. *What Is Philosophy?* Trans. H. Tomlinson and G. Burchell. New York: Columbia University Press, 1994.

Deleuze, Gilles and Claire Parnet. *Dialogues.* New York: Columbia University Press, 1987.

Emerson, Ralph Waldo. "Nature." *Ralph Waldo Emerson. Essay and Lectures.* Ed. Joel Porte. New York: Library of America, 1983, 5–49.

———. "The Poet." *Ralph Waldo Emerson. Essay and Lectures.* Ed. Joel Porte. New York: Library of America, 1983, 445–468.

——. *The Journals and Miscellaneous Notebooks of Ralph Waldo Emerson*. 16 Vols. Ed. William H. Gilman et al. Cambridge, Mass.: Harvard University Press, 1960–1983.

Feisst, Sabine M. "Klanggeographie – Klanggeometrie. Der US-amerikanische Komponist John Luther Adams." *MusikTexte* 91 (November 2001): 4–14.

——. "Music as Place, Place as Music. The Sonic Geography of John Luther Adams" (unpublished manuscript).

Hegel, G. W. F. *Aesthetics: Lectures on Fine Art, Vol. I*. Trans T.M. Know. Oxford: Oxford University Press, 1998.

Ives, Charles. *Essays Before a Sonata, the Majority, and Other Writings*. Ed. Howard Boatwright. New York and London: W. W. Norton & Company, 1999.

——. *Memos*. Ed. John Kirkpatrick. London: Calder & Boyars, 1973.

——. "Music and Its Future." *American Composers on American Music. A Symposium*. Ed. Henry Cowell. Palo Alto: Stanford University Press, 1933, 191–198.

——. "Correspondence with Clifton Joseph Furness, July 24, 1923." Ives Collection, Yale University. Quoted in Charles W. Ward "Charles Ives's Concept of Music." *Current Musicology* 18 (1974) 114–119.

Leibniz, G. W. *New Essays on Human Understanding*. Eds and trans. Peter Remnant and Jonathan Bennett. New York: Cambridge University Press, second edition, 1996.

Mehring, Frank. *Sphere Melodies. Die Manifestation transzendentalistischen Gedankenguts in der Musik der Avantgardisten Charles Ives und John Cage*. Stuttgart: Metzler, 2003.

Morris, Mitchell. "Ectopian Sound or The Music of John Luther Adams and Strong Environmentalism." *Crosscurrents and Counterpoints*. Ed. Per F. Broman et al. Göteborg: 1998, 129–141.

Murphy, Timothy S. "What I Hear Is Thinking Too: The Deleuze Tribute Recordings." *Deleuze and Music*. Eds Ian Buchanan and Marcel Swiboda. Edinburgh: Edinburgh University Press, 2004, 159–175.

Nicholls, David. *American Experimental Music, 1890–1940*. Cambridge: Cambridge University Press, 1990.

Nocera, Gigliola. "Henry David Thoreau et le neo-transcendentalisme de John Cage." *Revue d'Estetique* (1987–1988): 351–369.

Perloff, Marjorie. *Radical Artifice. Writing Poetry in the Age of Media*. Chicago and London: University of Chicago Press, 1991.

Retallack, Joan. "Poethics of a Complex Realism" in *John Cage: Composed in America*. Eds Marjorie Perloff and Charles Junckerman. Chicago: University of Chicago Press, 1994.

Schulz, Dieter. Amerikanischer Transzendentalismus: Ralph Waldo Emerson, Henry David Thoreau, Margaret Fuller. Darmstadt: Wissenschaftliche Buchgesellschaft, 1997.

Shultis, Chris. *Silencing the Sounding Self. John Cage and the Experimental Tradition in Twentieth-Century American Poetry and Music*. Boston: Northeastern University Press, 1998.

Szuberla, Curt. Quoted in *Living On Earth*. Radio Interview with Amy Mayer, see www.loe.org/shows/segments.htm?programID=06-P13-00016&segmentID=5

Thoreau, Henry David. *The Journal of Henry David Thoreau*. Eds Bradford Torrey and Francis H. Allen. In 14 volumes (bound as two). New York: Dover Publications, 1962.

——. *The Illustrated Walden*. Ed. J. Lyndon Shanley. Princeton: Princeton University Press, 1973.

——. *A Week on the Concord and Merrimack Rivers*. Harmondsworth: Penguin, 1998.

Walter, Bruno. *Gustav Mahler. Ein Portrait*. Berlin und Frankfurt: Fischer, 1957. (First Edition: 1936.)

13
Deleuze and *Deliverance*: Body, Wildness, Ethics

Mark Halsey

> *Bobby*: It's true Lewis, what you said. There's something in the
> woods and the water that we've lost in the city.
> *Lewis*: We didn't lose it. We sold it. (28 mins)
> > *Deliverance*. A John Boorman Film

Gilles Deleuze once remarked, "There is no original spectator. There is no beginning, there is no end. We always begin in the middle of something. And we only create in the middle by extending lines that already exist in a new direction or branching off from them" (*Two Regimes* 216). Taking Deleuze's lead, my objective in this chapter is to build upon previous conversations concerning depictions of nature or "the environment" in popular culture. In a slightly unorthodox manner, I have chosen to focus on the confluence and effects of three textual machines: the novel *Deliverance* by James Dickey; the film of the said novel directed by John Boorman (screenplay by Dickey); and select aspects of the oeuvres of Gilles Deleuze and Felix Guattari. Deleuze and Guattari (*Thousand Plateaus* 334) liken machines to keys insofar as each permits entry into various rooms or assemblages. Just as there are different kinds of keys there are also different kinds of machines—some permit entry only (machines of axiomization), some permit one to do things to the room or assemblage once within it (machines of relative deterritorialization), and some permit the space entered to be changed beyond all recognition or to become part of other rooms, worlds, problematics, and so forth (machines of absolute deterritorialization). As Guattari writes, "Machines arrange and connect flows," and this is no less true of the coagulations, displacements, and new universes associated with reading a book or viewing a film (46).

Dickey and Bormann have created two distinct yet closely related machines—the book functioning chiefly as a chirographic machine and the film functioning mainly as a visual-sonorous machine. In the former, the reader is permitted into the rooms and spaces of *Deliverance* through the affective dimensions induced within the reader by words on a page. In the film, the viewer is permitted entry through the combination of images, words and sounds emanating from the screen; a literary and a cinematic event both with their own intensities and durations and with their own unique means of bringing about transformations (minor or major becomings) in their audiences. Both separately and together, the book and the film evince passages and themes which bring to life the revolutionary concepts of the body, nature, and law extant in the writings of Spinoza, Nietzsche, and more aptly Deleuze and Guattari.

Applying Deleuze to *Deliverance* requires one to part with the image of thought typically conjured by references to the film or book. There are things more important and of greater nuance which can be gleaned from these works than the molar refrains having to do with "conquering nature," or finding one's "inner self," or of "overcoming the odds" or of dealing with questions of sexuality and\or masculinity (Barnett). These are, to be sure, real and relevant themes but they are by no means the only ones invested in these works. *Deliverance* is not *Apocalypse Now* in disguise. It is not a southern stylized *Heart of Darkness*. The similarity with these cultural texts begins and ends only with the journeying down a river. *Deliverance* has its own alterity and must be remarked as such. With reference to key Deleuzean concepts, I hope to engage with *Deliverance* in terms of what it says about the body, about urbanity and wildness, about notions of law and ethics, and, as would seem entirely relevant to current times, about the "larger" matter of late capitalism and its propensity to overcolonize in questionable and irrevocably damaging ways the mental and environmental realms of the planet.

Before proceeding to the critical sections of this chapter, I need briefly to address three issues. First, I am cognizant of the fact that there are differences (most of a fairly inconsequential kind) between the way the story of *Deliverance* unfolds in its detail in the book as opposed to the film. The dialogue in the film is spare and departs often from that of the book and particular scenes are brought into sharper focus and dwelt on in the film that are mentioned only cursorily in the written text. Having said this, both media reflect and cling to the same series of events and it is this consistency which makes a commentary on both media possible and important. Moreover, the affective dimensions accompanying a reading of the book are distinct from those of viewing the film and these differences are crucial to understanding the relevance and productivity of a Deleuze\Guattarian analysis in this context.

Second, I am also keenly aware of Deleuze's works on cinema and the unique insights generated by such work. Clearly, *Deliverance* is a film constructed around the movement-image insofar as it involves central characters reacting to events in a chronological manner. However, the current chapter is not about the philosophical aspects of film making or of viewing images. It is instead about the socioecological "lessons" which can be made to emerge from a particular film, and from a particular book, when these latter two machines are made to pass through particular examples of Deleuze and Guattari's concepts.

Third, and finally, for the purpose of clarity I need to attempt a summary of the story *Deliverance* in order to orient the reader to the task ahead. The dust jacket of the book reads:

> The setting is the Georgia wilderness, where the state's most remote white-water river awaits. In the thundering froth of the river, in its steep, echoing stone canyons, four men on a canoe trip discover a freedom and exhilaration beyond compare. And then, in a moment of horror, the adventure turns into a struggle for survival as one man becomes a human hunter who is offered his own harrowing deliverance (Dickey).

The back cover of the DVD is more to the point:

> Four ordinary men in two canoes navigate a river they only know as a line on a map, taking on a wilderness they only think they understand (Warner).

Neither of these statements adequately capture the true nature of the book— not because of their brevity but because there is no essence or central truth to be found within the story. However, there is a plot which at its most basic level can be fairly well agreed upon. Four men from the city, a scheduled two-day canoe trip on a remote river, the unexpected encounter with two mountain men, the struggle to survive, and the wrestling with one's conscience about what has been done to each and what each has had to do to others whilst navigating their way down and eventually clear of the river. One commentator has noted that *Deliverance* proceeds according to "the basic structure of all great narrative" (Butterworth 71) in that it is a tale built around the archetype of *descent* and *return*. I fully concur with this reading and my objective now is to explore how this dynamic plays out with regard to the constructions of nature in *Deliverance*. What happens to nature during the processes of descending into the Cahulawassee and returning from it? And how does nature impress upon and offer up alternative renderings of the body, of urbanity, and of law within the text(s) which deliver *Deliverance* to its audience(s)?

Body

The bodies populating *Deliverance* in its early stages are not those presenting at the conclusion of the story. The names of the central characters persist, and they all (with the exception of Drew Ballinger, who is shot and killed on the rapids by a mountain man) return to their previous city-based lives and vocations. But each is psychically and physically transformed. Ed, the central character, is haunted by what he was required to do in order to survive, Bobby is traumatized by the sexual abuse suffered in the forest, and

Lewis grapples with the idea that he is not invincible and that his body broke precisely at the point when it needed to work best. Ed's early recountings of what Lewis used to say to him about the body are important in this context—"It's what you can make it do…and what it'll do for you when you don't even know what's needed. It's that conditioning and reconditioning that's going to save you" (Dickey 29).

This statement asks that we suspend the view that thought (mind) controls and directs the body (substance)—that one's body is merely an appendage for doing the work of the mind. What Lewis points to is the fact that modern life generally removes the necessity for us to go back to the body or to consider what it means to have a body. *The body has become an inconvenience*—something which gets in the way of desire, and which is taken to be subordinate or base due to its primordial (irrational) nature. The modern body has become overcoded by the tropes of lack and deficiency. It is incessantly judged and spoken of against the contours of the perfected body circulating in the mass media, in general chatter, or in the conversations one has with oneself. We have, in other words, neglected to consider the body as a force to be reckoned with—as something we know little about, and which therefore has untold capacities and potentials. Spinoza's remark comes to the fore here. "We do not know," he writes, "what the body can do" (Deleuze, *Spinoza* 17)—which, in this context, is shorthand for saying we do not consider what our bodies can do beyond their ability to provide nourishment to the mind, to provide the encasing for our organs and for our "true" and immutable identity, or to provide the wherewithal for moving (walking, running) from one place to the next.

For Deleuze and Guattari, our bodies have taken on an overwhelmingly molar dimension—they have been organized and stratified by all manner of binary machines—particularly those "of social classes, sexes (men|women), ages (child|adult), races (black|white), sectors (public|private), and subjectivizations (ours|not ours)…" Furthermore, "[these machines] are dichotomizing rather than dualistic; and they can work diachronically. If you are neither *a* nor *b*, then you are *c*; the dualism has been transposed, and no longer concerns simultaneous elements to be chosen, but successive choices; if you are neither black nor white, you are a mulatto; if you are neither a man nor a woman, you are a transvestite" (*On the Line* 77).

Of the four main characters in *Deliverance*, only Lewis considers his body as something other than a molarity—as something which is always already escaping what it is labeled or perceived to be. Bobby and Drew, on the other hand, are entirely content to occupy the rational urban world[1] and let their bodies subsist as an afterthought (arising only to prominence when their corporeal components or the world begins to break down—a dodgy knee, signs of blood in the urine, shortness of breath, a flash flood, an earthquake). Against each of these, Ed's character unfolds in a very different fashion. He slowly reveals himself to be imbued by an ongoing restlessness with who he is, and this restlessness—this molecularity—shows itself most clearly in his interaction with Lewis.

> You've been sitting in a chair that won't move. You've been steady. But when that river is under you, all that is going to change. There's nothing you do as vice-president of Emerson-Gentry that's going to make any difference at all, when the water starts to foam up. Then, it's not going to be what your title says you do, but what you end up doing. You know: *doing*...I just believe...that the whole thing is going to be reduced to the human body, once and for all...I think the machines are going to fail, the political systems are going to fail, and a few men are going to take to the hills and start over. (Dickey 41–42)

The whole of *Deliverance*, I would argue, can be read as a meditation on the molecular transformation of the body of Ed—the tension generated by his simultaneous satisfaction and disgust with who he is or might become. Very early on in the book, Ed, sitting at work, reflects upon and is paralyzed by the mundane nature of his life:

> I sat for maybe twenty seconds, failing to feel my heart beat, though at that moment I wanted to. The feeling of the inconsequence of whatever I would do, of anything I would pick up or think about or turn to see was at that moment being set in the very bone marrow. How does one get through this? I asked myself. By doing something that is at hand to be done was the best answer I could give; that and not saying anything about the feeling to anyone. It was the old mortal, helpless, time-terrified human feeling, just the same. (Dickey 18)

This is something more than existential angst. Instead, this situation goes to the flow of desire (that is, "everything that exists *before* the opposition between subject and object, *before* representation and production") and its capacity to force a crack in the way one ascribes value to or perceives oneself and the world (Guattari 46). Ed is moving somewhere, but the location of that somewhere is yet to be struck upon.[2] All that is given is that the things which bind him to the world and provide comfort and security—work, family, friendships— produce, ironically, the sense of a wasted and humdrum life in need of repair. Deleuze and Guattari use the term molecularity to describe the little movements within one's body which have a transformative effect or which produce an encounter with the "self" one had inadvertently (even intentionally) morphed into. In contrast to the rigid lines of molarity, molecular lines are "much more supple" (*On the Line* 69). As Deleuze writes,

> Many things happen along this second type of line—becomings, microbecomings—that don't have the same rhythm as our "history"...We can

no longer put up with things the way we used to, even as we did yesterday.[3] The distribution of desire within us has changed, our relationships of speed and slowness have been modified; a new kind of anguish, but also a new serenity, have come upon us. (70, 73)

Lewis's presence—his physique, his philosophy, his seriousness, his spirit—is infective and affective. He gets into people's psyches and makes them feel uncomfortable about who they are. Lewis puts in train all manner of incorporeal events upon others' bodies—making each examine their social standing, their professional and personal trajectories, and their proximity to urban and rural speeds. This intermingling with Lewis's body is that which induces a molecular reorientation of Ed (and, to a lesser extent, of Bobby and Drew).

> *Ed*:   Well the system's done all right by me.
> *Lewis*:   Oh yeah, you got a nice job, you got a nice house, nice wife, nice kid.
> *Ed*:   You make that sound rather shitty Lewis.
> *Lewis*:   Why do you go on these trips with me Ed?
> *Ed*:   I like my life Lewis.
> *Lewis*:   Yeah, but why do you go on these trips with me? (26 mins)[4]

It is at Lewis's behest that they (Ed, Bobby and Drew) join *him* for a weekend away in the mountains, not the other way round. It is Lewis's rhetoric—the way he unsettles and dislodges their attachment to comfortable urbane life—which underpins the assembling of the group. This is why the events taking place on and around the Cahulawassee River are as much about the emergence of bodies (or a new-found respect and understanding of the body) as it is about the ruin or suppression of bodies (by gunshot, by burial, by arrow, by drowning). Everything, for Lewis (but also for each of the other characters in the book and film), comes back to and hits against the body. This is perfectly illustrated by Lewis's realization that his private air-raid shelter ultimately functioned not as a place of closure and protection but as an event which brought the world and its vitality into unbearably stark relief. As with Deleuze or Nietzsche, the challenge associated with having a body is to experiment with it—to maximize the number of connections and affective moments one is capable of entering into. This means facing the world head-on, as it were, not retreating from it. As Lewis remarks concerning his shelter,

> I went down there one day and sat for a while. I decided that survival was not in the rivets and the metal, and not in the double-sealed doors and not in the marbles of Chinese checkers. It was in me. It came down to the man, and what he could do. The body is the one thing you can't fake; it's just got to be there. (Dickey 42)

There is no interiority or final point of capture or closure for the body and its senses (or its capacity to produce and enter into streams of sense and non-sense). Even if one were never to emerge from an air-raid shelter the "dead" body would in any case continue as a multiplicity and return to earth its flows of carbon, marrow, water, and so forth. It is the suffocating binaries attached to Cartesian thought which provide us with the belief in discrete and safe interiors impermeable to the outside (of thought, of earth), or of pure or redeemable souls beyond the reach of bodies, or of an enduring rationality expunged of all emotion, or of a calculated action that is no longer at the mercy of risk.[5] Lewis cuts through all this nonsense and fear-mongering by adopting an unconditional zest for and interest in the body. For he realizes he has no choice—since it is, after all, "the one thing you can't fake." One can wish the body away, hack at its appendages, do violence to its sensory elements, even change its shape and size. But one cannot be done with the body. Like the death of God, the body is omnipresent and must be reckoned with in any process of becoming. In order to understand more about the body which descends and returns in *Deliverance* it is necessary to comment on its collisions with urbanity and wildness, and, subsequently, with law and ethics.

Urbanity and wildness

I read and re-read and, more commonly, view and re-view—in short, consume—*Deliverance*, because it literally takes me on a journey. The book affects me in ways different from the film insofar as the latter provides a far more intensive experience than the written word alone (even though such words bleed into all kinds of images). Mostly, though, the film provides the perfect medium through which to depict the infiltration of urban—more particularly, industrial—flows and speeds upon rural (or even remote) portions of the United States. Lewis captures this notion in the way he narrates the Cahulawassee as something hidden, unknown, and largely inaccessible (to outsiders).

> Bad roads in there, but my God Almighty, the little part of the river I've seen would knock your eyes out. The last time I was near there I asked a couple of rangers about it, but none of them knew anything. They said they hadn't been up in there, and the way they said "in there" made it sound like a place that's not easy to get to. Probably it isn't, but that's what makes it good. From what I saw, the river is rough but not too rough just south of Oree. But what's on down from that I don't have any idea. (Dickey 39)

It is the visual and chirographic *absence* of the city, of urbanity, which results in it being such a prominent presence in both the film and book.

The *leitmotif* of (post)industrialization is without doubt the combustion engine, and it is therefore appropriate that the sound of dump trucks building the dam at Aintry, juxtaposed against the engines of the vehicles transporting the city dwellers to the river, accompanies the dialogue in the opening sequence of the film. Lewis, in Deleuze and Guattari's terms, is in search of smooth space—of a space composed of an infinity of points and possible becomings between two lines (the line of "civilization" and the line of remote mountain ridges and ravines) rather than the line drawn between two prescribed points (house to work, suburbia to inner city) (see *Thousand Plateaus* 474–500). The Cahulawassee River is this smooth space but only emerges as such due to the speed and convenience of traveling in a vehicle which is itself the product of endless striations (mining, smelting, processing, manufacturing, marketing) and which must follow the line drawn between successive points on a map (inner city streets, turnpike, freeway, road, track). This is a concrete illustration of the way each remains subject to the strata even while following a molecular "calling."

Deleuze, in one of his most eloquent phrases, once remarked that "society is something that never stops slipping away" (*Foucault and the Prison* 271). One can fix an image of the social, or a subject, or a place, or a law, but the permanence of each—the apparent immutability of each—only holds good in the minds of those taken up by such efforts. "There is always something that flows or flees, that escapes the binary organizations, the resonance apparatus, and the overcoding machine: things that are attributed to a 'change in values,' the youth, women, the mad, etc." (*Thousand Plateaus* 216). Something is always being lost and created. In *Deliverance*, it is wildness that is slipping away, or more accurately, being prodded and plugged up to form a monstrous body.[6]

> *Lewis*:   You wanna talk about the vanishing wilderness?
> *Bobby*:   Lewis, why are you so anxious about this?
> *Lewis*:   Because they're building a dam across the Cahulawassee River. They're gonna flood a whole valley Bobby. That's why. Dammit, they're drowning the river, they're drownin' the river man...Just about the last wild, untamed, unpolluted, unfucked up river in the South...Don't you understand what I'm saying? They're gonna stop the river up...There ain't gonna be no more river. There's just gonna be a big dead lake...
> *Bobby*:   Well, that's progress...
> *Lewis*:   That ain't progress, that's shit...
> *Ed*:   It's a very clean way of making electric power. And those lakes up there provide a lot of people with recreation...My father-in-law has a house-boat...on Lake Bowie...
> *Lewis*:   You push a little more power into Atlanta, a little more air-conditioners for your smug little suburb and you know what's gonna

happen? We're gonna rape this whole goddamned landscape. We're gonna rape it...

Ed: That's an extreme point of view Lewis—extremist...(1 min)

There is much happening in this exchange. But the critical point is that Lewis is pleading the case as to why he and his interlocutors should seize the opportunity to place their bodies between two distinct speeds—geological time and industrial time. Lewis is lamenting the way the modern metropolis violently overcodes not just the immediate space within which it is erected, but also the most distant of bodies. For the city—that great centrifugal force—cannot tolerate wildness. Indeed the city's original and ongoing violence is inextricably bound to its hold over so-called "far flung" places— places whose remoteness, whose unknown qualities, whose dark secrets only present as such due to the city constituting itself as the privileged site *par excellence*. As Deleuze and Guattari write,

> It is not the country that progressively creates the town but the town that creates the country...The town...represents a threshold of deterritorialization, because whatever the material involved, it must be deterritorialised enough to enter the network, to submit to the polarisation, to follow the circuit of urban and road recoding. (*Thousand Plateaus* 429, 432)

As a sedentary body, it is the city that brings about a relation with the earth that is peristaltic and predatory rather than fluid and mutually agreeable. The Cahulawassee is being commanded by the city. Its diversity, its original difference, its "deep disparity" (Deleuze, *Logic of Sense* 261) stands to be reduced to a single prescribed function (that of producing electric power). But the characters in *Deliverance* still have enough of geological time on their side to experience something other than a drowned river. Critically, it is *wildness*—as distinct from nature—Lewis is searching for. In fact I would argue that it is *both* industrialization *and* the particular renderings of nature attending the construction of cities|towns (and the rise of stat(e)ic institutions) which are in the process of being put asunder by venturing to the mountains.

> Lewis: When they take another survey and rework this map...all this in here will be blue. The dam at Aintry has already been started, and when it's finished next spring the river will back up fast. This whole valley will be under water. But right now it's wild. And I *mean* wild; it looks like something up in Alaska. We really ought to go up there before the real estate people get hold of it and make it over into one of their heavens. (Dickey 3–4)

Lewis could not care less about manicured botanic gardens, local reserves, or national parks. These are, when all is said and done, heavily striated spaces with their own rules and regulations governing where one can go and how one should gaze upon or interpret the terrains within (Halsey, *Molar Ecology*). They are spaces marked by metrics (200 meters to "The Big Tree"; 500 steps to scenic lookout) and by solid things (fountains, monuments, toilet blocks, roads and paths, information and toll booths). Wildness, on the other hand, is akin to something of an altogether different force—a space where the organizational patterns and codings of Man and modern life are impossible to discern. In the lexicon of Deleuze and Guattari, wildness concerns the plane of consistency.

> The plane of consistency...is opposed to the plane of organization and development. Organization and development concern form and substance: at once the development of form and the formation of substance or a subject. But the plane of consistency knows nothing of substance and form: haecceities, which are inscribed on this plane, are precisely modes of individuation proceeding neither by form nor by the subject. The plane consists abstractly, but really, in relations of speed and slowness between unformed elements, and in compositions of corresponding intensive affects...Never unifications, never totalizations, but rather consistencies or consolidations. (*Thousand Plateaus* 507)

The plane of consistency does not reside, and nor is it formulated, as some kind of lost panacea for the ills wrought by urbano-industrialization. For how could, in any case, something be considered "lost" that is always already there? Rather, the plane of consistency (the plane of supermolecular nature) is that which serves to radically problematize the society–economy–culture–nature nexus.[7] Reaching the plane of consistency—wildness—means nothing other than to move, feel, breathe, in short, *live* in a manner which preserves as far as is possible the concept of Earth minus binary|categorical intrusions (Halsey, *Deleuze and Environmental Damage*). It is to revel in the absence of subjects and objects, of prescribed functions, of teleologies, and so forth. Lewis, and to a lesser extent Ed, are venturing to lodge themselves within this plane—to break with the plane of organization (*What Is Philosophy?* 124–126) where everything and every "proper" response is known ahead of time. And this venturing requires, of course, the complete transformation of their bodies, of their subjectivities, and of their subjection to the laws and mores of modern life. This, as Deleuze and Guattari caution, is a process fraught with danger and difficulty. There is nothing inherently liberating or positive about a line of flight. One can, as these philosophers consistently say, go too far. One can destratify to the point where nothing is able to pass over or through the body one has unwittingly become (the sodomized body (Bobby), the injured body (Lewis), the morally corrupted body (Drew), the body become killer (Ed)). To reach the plane—or to reach one moment of pure intensity on this plane (to draw one's own body without organs)—involves "a meticulous relation with the strata" (*Thousand Plateaus* 161) (with those things or abstract machines one is attempting to pull away from). This, it could be said, is the greatest folly of city dwellers who take to the mountains in search of remoteness and "adventure."

> *Bobby*:   We beat it didn't we? Did we beat that?
> *Lewis*:   You don't beat it. You don't beat this river. (24 mins)

There is no point commencing a line of flight—of letting the molecular murmurings within one's body flow unfettered—if one returns to precisely the same values and dispositions of molar life, or, worse, carries these values and dispositions to the space or place one ventures to. Bobby is the universal face of those who believe the most meaningful thing one can do once out in the "natural world" is to conquer it, subdue it, beat it—that one transforms oneself in some profound manner only when one leaves a definitive mark on the Other (a mountain, a lake, a forest, an ocean, an animal).

Law and ethics

Arguably, the key critical moment in *Deliverance*—the moment of greatest intensity—concerns the way in which the four central characters weigh up what to do with the body of the dead mountain man. This man, shot by Lewis with an arrow after he had raped Bobby and was about to defile and in all probability kill Ed, lies motionless before them on an embankment in some remote part of the Cahulawassee. The dead man's companion, meanwhile, has managed to escape into the forest and his whereabouts and intentions are unknown. This last aspect looms as a major factor influencing Lewis, Ed, Bobby and Drew as they decide what to do next. The adventure has become unstuck at this point, and urbanization—the comforts and routines of modern living—have never looked so appealing. To Lewis, of course, this is precisely the kind of situation he has been preparing for. The death is an event of manageable proportions for Lewis because he has long ago convinced himself that at such times—when there is no foundation of right or wrong to fall back on—survival and its correlates assume centre stage. Ed knows this to be Lewis's position, having heard him talk many times before about what he would do and where he would go if "the radios died."

> That's where I'd go...Right where we're going. You could make something up there...If everything wasn't dead, you could make a kind of life that wasn't out of touch with everything. Where you could hunt as you needed to, and maybe do a little light farming, and get along. You'd die early, and you'd suffer, and your children would suffer, but you'd be in touch. (Dickey 44)

The seriousness of the matter can be seen in the way Lewis responds to Ed's pithy (if somewhat logical) assertion that one could go at any time to the hills and live—that there was|is no need to await a crisis.

> It's not the same...Don't you see? It would just be eccentric. Survival depends—well, it depends on *having* to survive. The kind of life I'm talking about depends on its being the last chance. The very last of all. (Dickey 44)

This is a deceptively insightful remark. Simply, survival becomes something other than survival when it is chosen. The gravitas of survival therefore attaches to it forcing itself upon the body as a necessity. In *Deliverance*, the dead body completely transforms the spaces, sights, sounds, and rhythms of the Cahulawassee into forces which need to be outlived—survived, and more directly, reevaluated. Deleuze (*Difference and Repetition* 132, 139–141) shows that there are two kinds of signs in the world. On the one hand, there are those we recognize and which subsequently produce an array of predictable actions ensuring the routine negotiation of the world. "Recognition may be defined by the harmonious exercise of all the faculties upon a supposed same object" (133). On the other hand, there are signs which present as *encounters*, and as such are not really signs (of any known thing) at all. Encounters are characterized by "an unlimited qualitative becoming," indeed a "mad becoming" (141). These latter signs—those events which bear no "instruction label" and which do not bifurcate into recognizable blocks of action—require something new of the body. Encounters know and speak of no law. Encounters require invention and experimentation. They demand, in short, *ethical* conduct—action born from something other than the law of the land, the expectations of others, or even one's so-called conscience (which is at all times built around arborescent structures).

In this sense, the dead body is an event which unfolds in two directions—a sign to be interpreted in two distinct ways. Drew, from the very beginning, is caught up in the game of recognition. He knows what the dead body signifies—it signifies a report to police, and the formal opportunity to explain the death to authorities in a properly constituted space (court or coronial inquiry). Drew, in short, recognizes the body in terms of its connections to and implications flowing from a justifiable homicide.

> *Lewis*:   What we gonna do with him?
> *Drew*:   There's not one thing to do but take the body down to Aintry and turn it over to the highway patrol. Tell them what happened...This is a justifiable homicide if anything is...They were sexually assaulting two members of our party at gun point. Like you said there was nothin' else we could do. (48 mins)

Lewis ruminates along a very different line: "We killed a man Drew. Shot him in the back. A mountain man. A Cracker. That gives us something to consider...Shit, all these people are related. I'll be goddamned if I want to come back up here and stand trial with this man's aunt and his uncle, maybe his momma and his daddy sittin' in the jury box." Bobby and Ed are spectators at this point—weighing the rhetorical force of each viewpoint—of each manner of interpreting the signs of the body before them. The exchange continues,

> *Lewis*:   We gotta get rid of that guy.
> *Drew*:   Just how you gonna do that Lewis? Where?
> *Lewis*:   Anywhere. Everywhere. Nowhere...Did you ever look out over a lake and think about something buried underneath it? Buried underneath it. Well man that's about as buried as you can get.
> *Drew*:   Well I am telling you Lewis, I don't want any part of it.
> *Lewis*:   Well you are part of it.

And then this,

> *Drew*:   It is a matter of the law.
> *Lewis*:   The law, ha. The law. What law? Where's the law Drew? Hey?
>
> (50 mins)

In the book, there is, as Dickey points out in an interview (Anthony 109), a critical line which did not at this point appear in the film. Here, Lewis goes on to remark, "*We're the law*. What we decide is going to be the way things are" (Dickey 130, emphasis added). This exchange is one I have returned to many times in the context of inquiring after the ontology of law and right. What kind of law, if any, exists for any body within the remotest part of a river and its surrounding forests? What if doing the "lawful thing" unacceptably and unquestionably heightens the prospect of becoming a victim of judicial procedures? What if law itself becomes the major vehicle of violence in the situation one confronts?[8] In this context, it is worth noting one of Ed's earlier reflections,

> The river ran through it, but before we got back into the current other things were possible. What I thought about mainly was that I was in a place where none—or almost none—of my daily ways of living my life would work; *there was no habit I could call on*. Is this freedom? I wondered. (Dickey 93, emphasis added)

Here, the body slips into a liminal space and Ed queries whether this equates to an unqualified freedom. Deleuze and Guattari, along with Foucault, contend that freedom is only ever a matter of the kinds and degrees of constraint which present to the body. However, their term *body without organs* goes some way toward describing those moments where (one's ensuing) actions cease to be dictated by tradition, habit, prevailing opinions, or law. Ed, at the point of having to make a decision concerning the dead man, is enveloped by nothing other than intensities—by feelings which have no name, no antecedents, and no teleology or prescribed trajectory. To paraphrase Deleuze, Ed arrives at the point where he is unable to proceed faster than his own present(s) (*Difference and Repetition* 77). He is suspended (and nearly crushed) by the ethical weight of the moment. This is not, as Ed supposes, "freedom." This is not nature in the raw, at its worst, or in its most dangerous state. This is instead a concrete example of the lived reality of Nietzsche's eternal recurrence. "The eternal return is a force of affirmation, but it affirms everything of the multiple, everything of the different, everything of chance *except* what subordinates them to the One, the Same, to [N]ecessity, everything *except* the One, the Same and the Necessary" (115). This is the point at which one gets to reckon with the virtual (211), the world of the acategorical, the world of unformed things, the world minus subjects and objects. Ed, in short, stands to create or severely botch the body without organs in accordance with the decisions he takes. There is no rational path to follow; there is no "habit to call on." There is only the interminglings of bodies whose affects and trajectories are on the cusp of being transformed by desire *or* captured by the institutional traditions and constraints of the social. As Deleuze and Guattari remark, "There is only desire and the social, and nothing else" (*Anti-Oedipus* 29).

> The social machine or socius may be the body of the Earth, the body of the Despot, the body of Money...The prime function incumbent upon the socius...[is]...to codify the flows of desire, to inscribe them, to record them, to see to it that no flow exists that is not properly dammed up, channeled, regulated. (33)

Both in the book and in the film, Ed is depicted as undergoing an immense struggle concerning what to do next. He is battling the body of the law as the chief organizational and despotic machine in his life to date. He is the actualization of the phrase, *ethics precedes law*,[9] and all that this entails. As such, Ed's course of action emerges through the body and only secondarily, if at all, through the musings of the so-called mind (or what Nietzsche would call the most convenient of fictions). Phillip Goodchild has eloquently remarked on the two worlds which envelop us all.

> One may [...] distinguish between two kinds of worlds: this actual world which affects us, which our bodies contemplate from a particular perspective, and where real animals [and people] die—made up of a network of relations that only exist in a timeless present, even if it changes; and an intense, virtual world of reciprocal implications, a world that does not exist outside of its expression, a world that belongs to the ethos of a moment or a multiplicity, defining a mode of existence—a

world of phantasms that has a duration, and is experienced as past, absent, withdrawn; a world experienced as a *territory* or dwelling, a place of evaluation and desire. These two worlds are incompossible, and the difference between them produces an affect of shame. (46–47)

What leads to Ed's commitment to bury the dead man is the force of desire working its way through him. This force—the one "that belongs to the ethos of the moment"—is, in Nietzsche's terms, also known as the will to power. The will to power is that which ceaselessly *interprets* and *evaluates* within each of us (whether we acknowledge this or not) (Deleuze, *Nietzsche* 54). It is that which ultimately leads to affirmation or denial with regard to the encounters one bumps up against. Ed's affirmation—his willingness to risk his job, his liberty, his family, his whole way of life—literally arrives as a gigantic irrepressible yelp in the film. It seems to rise inexorably up from a place rarely if ever visited. Appropriately, the more rational the protestations of Drew ("you've got a family"), the more earnest his pleadings ("I'm *begging* you"), the more logical his arguments ("Think what you're doing," "Listen to reason") (Dickey 131), the further Ed moves from the line of molarity toward a truly ethical act—one which lies beyond good and evil (Nietzsche). It is a moment where one goes against the herd, against habit—where one becomes a legislator of values instead of a slave to the law of the Other. This, it might be said, is what "Nature" and its various crises need most—singular micro-political acts which engage with the acategorical so as to interrupt the molar responses of politico-corporate bodies.

Concluding remarks

At the outset of this chapter I mentioned that I wanted to explore the dynamic of descent and return in *Deliverance*. From the preceding discussion, it is clear that what descends—what gets taken to the Cahulawassee—is rationality (the Cartesian subject), law (right), and the trappings and habits spawned by urbanity (sedentary life, the rise of the *Urstaat*). What returns— what emerges—are bodies (molecular), wildness and the singular experience of an ethics. Once affirmed, Ed's determination to body through the path before him is uncompromising—fraught with danger, fraught with fear— but based nonetheless on the ethic of experimentation. Ed, like each of his companions, never pulls clear of the Cahulawassee event. It returns in all manner of ways—as nightmare, as scar, as guilt, as sin. But chiefly it returns in terms of its transformative dimensions—as a reminder of the immense gulf separating those whose lives are constrained by habit (by the logics of recognition) from those who reckon with worldly encounters.

Global warming, sustained drought, deforestation, overfishing, mass depletions of biodiversity (each of which work upon the social as so many abstract machines), all require a new understanding of the limits of the rational mind, and of the violence of urbanity and law. They require a meticulous and ongoing journeying to the body, to wildness and to singularized refrains. Presently, the abovementioned ecological crises are made to pass through the grid of recognition in order to shore up more "efficient" forms of economic growth and development. There is, in short, no active encountering. There is no sense that environmental problems might exceed orthodox ways of making sense of the world. To be sure, there are groups who know quite well what the drowning|displacement (Hurricane Katrina, New Orleans), polluting (Fly River, Papua New Guinea), crushing (mudslide, Guatemala), and malnourishment (Ethiopia) of their bodies signify. For these people are well and truly ensconced, to draw the analogy, in their own private nightmare on the banks of the Cahulawassee. The encounter has arrived and keeps arriving. But what of those (in primarily Western parts of the world) whose bodies only bump up against environmental collapse in so-called "peripheral" and (as yet) manageable ways? What of those persons|consumers who contribute most disproportionately to ecological damage but who maintain that their change in values (recycling, water-efficient shower heads, car pooling, long-life light globes) will be enough to turn the tide? I would submit that sooner, rather than later, the relative comfort of their vessels (cars, jets, offices, mansions, yachts, time-share apartments) will encounter the bank of the river and the illusion of interiority will be forced centre stage. And then, to paraphrase Lewis, everyone—regardless of profession, class, race, gender, age, religion or geography—will be forced to play the game.

Notes

1. Witness the following exchange:
    Drew: We don't really know what we're getting into...What business have I got up there in those mountains?
    Lewis: Listen,...you'll be in more danger on the four-lane going home tonight than you'd ever be on the river. Somebody might jump the divider. Who knows? (Dickey 7)
2. Ed has his own misgivings about what he may be getting himself into. One can sense the attraction–repulsion current running through his entire journey – the mark of any line of becoming.
    Griner: What the hell you want to go fuck around with that river for?
    Lewis: Because it's there.
    Griner: It's there all right. You get in there and can't get out and you're gonna wish it wasn't.
    Ed: Listen Lewis, let's go back to town and play golf. (12 mins)
    But then, once ensconced deep in the mountains around the camp fire:
    Ed: No matter what disasters may occur in other parts of the world, or what petty little problems arise in Atlanta, no one can find us up here. (31 mins)
3. I recall one particular moment where this molecular reordering struck me and put me on an entirely new path. It involved awaking one morning to see the leader of

the Australian Greens party being arrested and placed in the back of a police vehicle for protesting the logging of old-growth forest in a remote mountainous region of East Gippsland, Victoria (southeastern Australia). I had seen similar visions to this countless times before and had, for want of a better term, always *believed* passionately in the fight to halt such logging. But something on that particular morning (just over a decade ago)—something intangible, inexpressible, some acategorical force—compelled me to respond differently from all previous occasions. Within 24 hours I had managed to transport myself and two close friends to precisely the same location where the arrest took place (some 500 km to the east of Melbourne). Within another 24 hours we—along with many other protesters—were arrested. Within two years our cases had progressed all the way to the Supreme Court of Appeals where the Judge ruled that logging was being carried out illegally in contravention of the Heritage Rivers Act. All previous charges and convictions relating to the actions of almost 200 protesters were quashed and|or overturned. In late 2006, the entirety of the forest block where these and other arrests took place was declared a national park (a different kind of coding and striation which requires, in my opinion, further resistance) on account of its biological significance. If I was pushed to say why I decided to act in the way I did on precisely that occasion, and no other, I could only offer, in the words of Deleuze and Guattari, that I was unable to tolerate things as I once had—*even as I had the day or week before*. Something had changed. A threshold had been crossed from which there was no regress.

4. Dialogue and scenes from the film will be referenced in terms of the time elapsed (to the nearest whole minute) from the commencement of the feature (Boorman, *Deliverance*).

5. Lewis: I've never been insured in my life. I don't believe in insurance. There's no risk. (10 mins)

6. One must avoid here the tendency to label the visible impacts of late capitalist industrialization (skyscrapers, freeways, housing estates, hydroelectric dams) as somehow unnatural or as artifices. As Deleuze writes,

> It should be clear that the plane of immanence, the plane of Nature that distributes affects, does not make any distinction at all between things that might be called natural and the things that might be called artificial. Artifice is fully a part of Nature, since each thing, on the immanent plane of Nature, is defined by the arrangements of motions and affects into which it enters, whether these arrangements are artificial or natural. (Ethology 627)

7. A sustained engagement with Deleuze and Guattari prompts the key critical question: at what social, cultural and ecological cost do—indeed can—we presently name, divide, and regulate the plane of consistency?

8. See Derrida, *Force of Law*; McVeigh, Rush and Young, *Judgment Dwelling in Law*.

9. Emmanuel Levinas writes, "prior to all entitlement: to all tradition, all jurisprudence, all granting of privileges, awards or titles" (117).

Works cited

Anthony, Frank. "After *Deliverance*: An Exchange with James Dickey." *New England Review* 18 (1997): 108–110.

Barnett, Pamela. "James Dickey's *Deliverance*: Southern, White, Suburban Male Nightmare or Dream Come True?" *Forum for Modern Language Studies* 40 (2004): 145–159.

Boorman, John. *Deliverance*. Warner DVD, 2000.

Butterworth, Keen. "The Savage Mind: James Dickey's 'Deliverance'". *The Southern Literary Journal* 28 (1996): 69–78.

Deleuze, Gilles. *Nietzsche and Philosophy*. Trans. H. Tomlinson. New York: Columbia University Press, 1983.

——. *Spinoza: Practical Philosophy*. Trans. R. Hurley. San Francisco: City Light Books, 1988.

——. *The Logic of Sense*. Trans. M. Lester. New York: Columbia University Press, 1990.

——. *Difference and Repetition*. Trans. P. Patton. New York: Columbia University Press, 1994.

——. "Ethology: Spinoza and Us." *Incorporations*. Eds J. Crary and S. Kwinter. New York: Zone 6, Urzone, 1995, 625–633.

——. "Foucault and the Prison." *Michel Foucault: Critical Assessments, Politics, Ethics and Truth*. Ed. B. Smart. London: Routledge, 1996, 266–271.

——. *Two Regimes of Madness: Text and Interviews 1975–1995*. Trans. A. Hodges and M. Taormina. Ed. D. Lapoujade. New York: Semiotext(e), 2006.

Deleuze, Gilles and Felix Guattari. *Anti-Oedipus*. Trans. R. Hurley, M. Seem and H. R. Lane. Minneapolis: University of Minnesota Press, 1983.

——. *On the Line*. Trans. J. Johnston. New York: Semiotext(e), 1983.

——. *A Thousand Plateaus*. Trans. B. Massumi. Minneapolis: University of Minnesota Press, 1988.

——. *What Is Philosophy?* Trans. G. Burchell and H. Tomlinson. London: Verso, 1994.

Dickey, James. *Deliverance*. New York: Delta, 1970.

Derrida, Jacques. "Force of law: the mystical foundation of authority" in *Deconstruction and the Possibility of Justice*. Eds D. Carlson, D. Cornell and M. Rosenfeld. New York: Routledge, 1992, 3–66.

Goodchild, Phillip. "Deleuzean Ethics." *Theory, Culture and Society* 14 (1997): 39–50.

Guattari, Felix. *Soft Subversions*. Trans. D. Sweet and C. Wiener. Ed. S. Lotringer. New York: Semiotext(e), 1996.

Halsey, Mark. *Deleuze and Environmental Damage: Violence of the Text*. London: Ashgate, 2006.

——. "Molar Ecology: What Can the Body of an Eco-Tourist Do?" *Deleuzian Encounters: Studies in Contemporary Social Issues*. Eds P. Malins and A. Hickey-Moody. London: Palgrave Macmillan, 2007, 135–150.

Levinas, Emanuel. *Outside the Subject*. Stanford: Stanford University Press, 1993.

McVeigh, Shaun, Peter Rush, and Alison Young. "A Judgment Dwelling in Law: Violence and the Relations of Legal Thought" in *Law, Violence and the Possibility of Justice*. Ed. A. Sarat. Princeton: Princeton University Press, 2001, 101–141.

Nietzsche, Friedrich. *Beyond Good and Evil*. Trans. R. Hollingdale. London: Penguin, 1973.

14
Intensive Landscaping
Yves Abrioux

In a new introductory essay written for the 1998 edition of his influential *Social Formation and Symbolic Landscape* (1984), Denis Cosgrove acknowledges the limitations of the book's governing thesis, namely,

> that landscape constitutes a discourse through which identifiable social groups historically have framed themselves and their relations with both the land and with other human groups and that this discourse is closely related epistemically and technically to ways of seeing. (xiv)

In the comments which follow this admission, Cosgrove makes two basic points. First, he recognizes that, while substituting "social formation" for "mode of production" has the effect of avoiding economic determinism, it does so at the cost of loosening the explanatory power of what remains a Marxist approach to historiography. Second, he cites a number of increasingly influential fields of empirical and conceptual research, such as feminism, postcolonialism and psychoanalysis, as enabling the reconfiguration of the critical history of landscape through the recognition of a properly embodied viewing subject and its incorporation into landscape discourse. However, the basic causalist thrust of his argument, relating "the cultural significance of landscape" to "ways in which the land is materially appropriated and used" (1), remains substantially unaltered by the recognition of a "much less overdetermined" process, involving a more complex interplay between cultural, gendered, ethnic, and so on, consciousness and the process of social change. Indeed the notion of landscape as produced by factors which transcend it is all but definitional of standard approaches to the subject. The only real argument concerns its situation.

Landscape is sometimes taken to distinguish one physical realm from another. Often followed in geography textbooks, this approach remains operative in Trevor Rowley's recent study of *The English Landscape in the Twentieth Century*, which sets land—understood as physical terrain, the object of such sciences as ecology or geology—against land*scape* as the infrastructure imposed upon this. Cultural approaches tend to follow human geographers such as Cosgrove, in distinguishing between dimensions that are more readily identified as infra- and superstructural and also in historicizing the term itself. Cosgrove not only argues that the "material foundations" for landscape, understood as a "way of seeing," are to be sought in "the human use of the land, the relationships between society and the land;" he also states that the "idea" of landscape emerged as "a dimension of European elite consciousness" in an identifiable historical period (1–2). W. J. T. Mitchell identifies this period with European imperialism and describes how the colonial powers forced their particular pictorial conventions upon conquered territories in order to mould them into "new" Europes—socially and politically, as much as visually. Working within the wholly different conceptual framework of the philosophizing aesthete, Alain Roger opposes *paysage*, or landscape as a cultural formation, to *pays* as an unmediated unit of land, and meditates on the existence of societies devoid of a notion of landscape.

The implicit model of human behavior underlying all such conceptions is easy to discern. A recent pronouncement by a French artist makes it quite obvious. Responding to the suggestion that women artists are less boldly experimental than their male counterparts, Jean-Marc Bustamante not only proclaimed that "Yes, men need to conquer territories," in contrast to women, who are apparently happy to remain on familiar ground, but made the sexism of this remark explicit, by adding that men want "all women" and are "always in search of virgin territories"' (169; my translation). The tittering response of the woman interviewer—"I must be a man then!"— says it all. However mediated by articulations of class, genre, ethnicity, and so on, or determined by fantasies of sexual prowess, the subject confronted with land or territory is typically cast as an implicitly male actor faced with matter which ultimately veers, in intent, towards "female" passiveness. Whether understood as registering "the impact of human agency in altering the physical environment" (Cosgrove 14) or, from a more explicitly cultural viewpoint, as a "construction" or a "composition" of the world (13), landscape results from "the active engagement of a human subject [or society] with the material object" (13). Ultimately, it offers "an important element of personal *control* over the external world" (18; emphasis in the original).

The suspicion remains that this is a historically specific and ethnocentric view of landscape. Certainly, Mitchell qualifies his description of landscape as "a particular historical formation associated with European imperialism" by also defining it as a medium "found in all cultures" (7). However, this does not lead him towards an alternative definition of landscape, but rather to universalize the Western imperialist model. Thus, while celebrating the resistance of the Maoris to the invader's landscaping conventions, he characterizes their culture as "expansive" and thus as having "its own imperial ambitions" (27). If this is so, then imperialism implicitly becomes, not a historical and specifically European perversion of power, but a shared anthropological trait. It is consequently difficult to imagine how some other version of landscape, unblemished by territorial conflict, might emerge to replace the "exhausted" imperialist medium which Mitchell deems to be "no longer viable as a mode of artistic expression" (7).

It is necessary to set aside the ethnocentric triumphalism of Western imperialism, which reaches deeply into contemporary thought, in order to appreciate the existence of ecologies which regulate relations between humans and nonhumans on lines other than those of subject and object, and consequently suggest the possibility of a quite different approach to landscape. The anthropologist Philippe Descola contests the notion that all human communities enact a distinction between "what belongs to humanity and what is excluded from it" (59). He observes that there are still today any number of peoples who do not perform a "naturalization of the world" (57) and dates the preeminence of the nature|culture divide over our mode of thinking back to the birth of the modern world and industrialization—which also marked the emergence of European landscape painting (90 ff.). Descola insists that, in non-Western thought and social practice, the environment is entirely "socialized": it presents everywhere "traces of the events which have taken place" at particular locations. These may either be individual ("the fleeting signatures of biographical trajectories") or be meaningful to all the members of a community (61–62).

I have argued (in "Spectro-graphy of Gardens"), not only that such ghostly presences may be discerned in the *genius loci* which features prominently in topographical literature and was deeply involved in the inception of the English landscape garden, but furthermore that, divested of the sacred dimension of the *genius loci*, spectral traces continue to act as Deleuzian "intercessors" wherever gardens can be shown to exemplify an intensive dynamics of spacing, as opposed to a formalistic arrangement of components controlled by a stereotypical opposition of culture and nature. My purpose here is to follow up on these observations, by reformulating the articulation of landscape and territory in terms inspired by Deleuze and Guattari, which circumvent the power politics and ideological bad faith endemic to the conception of landscape as a "way of seeing," or manner of interpreting and presenting relations which transcend it. My hypothesis is that a truly operative definition of landscape can be achieved through a careful consideration of the process that Deleuze and Guattari name *territorialization* and which they associate with a simultaneous process of *de*territorialization. By forgoing any move to preemptively determine the presences which make it possible to distinguish between more or less significant sites, the secularization of Descola's traces allows gardens to emerge. It is similarly possible to detach the notion of landscape from culturally determined species of sites or ways of accommodating views, so as to associate it with a particular territorializing and deterritorializing tempo, which may similarly intervene almost at random. (De)territorialization is a matter of what Deleuze and

Guattari call *"allure,"* a term which may be translated as "pace" (*Mille Plateaux*, 382/343) so long as this conveys an aspect or quality of motion—a gait, carriage or bearing—as opposed to a mere measure of speed. Territorializing is a tempo which stabilizes in such a way as to make one feel at home, so that being-at-homeness (*"on est chez soi"*) means enjoying a calm and stable *"allure"* (383/344).

Allure is a key term in Deleuze and Guattari's definition of the *ritournelle*, as expounded in the eleventh "plateau" of *A Thousand Plateaus*: "1837—De la ritournelle," which establishes this as the principal operator of (de)territorialization. I quote the chapter heading in French, out of an ongoing dissatisfaction with the use of "refrain" to render it in English. Certainly, as a musical term, a ritornello shares something of the refrain and may be thus termed. However, several important dimensions are thus lost. There is, first of all, the sheer repetitiveness of the *"ritournelle,"* which idiomatically is not a refrain coming between verses or musical passages but a form of repetition and variation governing an entire composition. There is also the monotony which can ensue from such repetition. Both aspects are conveyed in the expression *"Toujours la même ritournelle"* ("always the same old story") and also in the use of *"ritournelle"* to describe an advertising jingle. A more eloquent translation might be "ditty."

Even if it is not etymologically an onomatopoeia,[1] "ditty" phonetically integrates the repetitiveness of a *ritournelle*. More significantly, through the minimal differentiation between its voiced "d" and unvoiced "t," it incorporates a further feature which is central to Deleuzian thinking: the notion of difference through repetition. Were this to be absent, we should have merely a catchy song or a catchphrase—what Deleuze (*Différence et Répétition* 14) calls a *"rengaine,"* rather than a *"ritournelle."* A *"rengaine"* is mere repetition; it transforms into "circular simplicity" something which has a quite different musical quality—*"ce qui est d'une autre musique"* (*Logique du sens* 305).[2] It nevertheless remains possible to sense the risk of difference-in-repetition collapsing into the mere repetition of the same, in the fact that *"Toujours la même ritournelle"* can equally be phrased *"Toujours la même rengaine"*: the expressions are synonymous. It should further be noted that, in its second set of definitions of the world "ditty," the Oxford English Dictionary includes "a short, simple song; often used of the songs of birds, or applied depreciatively," the reference to birdsong being supported by a quotation from Cowper's "Poplar Field" (1800) which highlights its link with landscape or territory: "The blackbird has fled...And the scene...Resounds with his sweet-flowing ditty no more."

The initial examples of the *ritournelle* provided by Deleuze and Guattari (*Mille Plateaux*, 382–384/343–345) adduce the features listed by the *OED*. There is the simplicity of a child singing under its breath, either to reassure itself in the dark or to help it concentrate on its homework, but also the superficially depreciative conduct of a woman (*sic*) singing to herself or listening to the radio as she goes about the housework and, finally, birdsong as way of staking out a territory. Two points need to be made here. The first is that, however modest, each of these examples—including that of bird-song—is of an *artistic* practice. Art is not, for Deleuze and Guattari a matter of high or low. They can thus protest against the idea that *art brut* is either "pathological" or "primitive" (389/349) while citing Wagner, Messiaen and *lieder*, among others, and reviewing the traditions of classicism, romanticism and modernism. Moreover, they describe art as an ethological phenomenon, rather than an anthropological one: birdsong, among other forms of animal behavior, shows that "art is not the privilege of human beings" (389/349). My second point follows on from this ethological dimension. It is that each of the examples just given is an instance of an artistic *practice*. Jumping, skipping and singing are not just aspects of behavior. A child may skip or jump in the dark. However, the song which provokes or accompanies this behavior is itself an action. Better still, it is an event: *"Il se peut que l'enfant saute en même temps qu'il chante [...] mais c'est déjà la chanson qui est elle-même un saut"*/"Perhaps the child *skips* as he sings [...] But the song itself is already a skip: it jumps [...]" (382/343). The song is a "jump," in the sense that its performance makes a difference.

A ditty provokes a jump into being-at-homeness: the difference it makes is that it invents a homely mode of habitation. It follows from this that what Ian Buchanan and Gregg Lambert (6) describe as the "problem of habitation"—the fact that space has come to be regarded as "uninhabitable" or "unhomely" (3)—cannot, contrary to what they claim, be reduced to "a problem of *recognition* under the regime of representation," which is to say a discomfort provoked by the fact that "the modern subject no longer recognizes the space in which it is located" (6). In contrast, say, to Marc Augé (cited by Buchanan and Lambert 4) and contrary to what certain descriptions of postwar Europe in *Cinéma 2* may seem to imply, Deleuze is not primarily a symptomatologist of our sense of no longer being at home in the world. In consequence, however sympathetically one may view Buchanan's Deleuze-inspired identification of postmodern space with "new type of [frictionless] social space" that has "lightened our step on earth" (28) and his mobilization of the notion of deterritorialization as a corrective for the fact that "our accounts of space do not yet reflect an awareness" of the hypermobility of the postmodern subject (26), it remains that the stimulation of recognition or awareness is far from being the primary concern of Deleuze's thought—or of Guattari's. In its territorializing function, the *ritournelle* or ditty no more functions "under the regime of representation" than economics as the *nomos* of *oikos* or ecology as its *logos* function as representations of home. Indeed, this is the reason why the definition of the ditty in *A Thousand Plateaus*, as simultaneously a territorializing and a deterritorializing force, promises to provide a conceptual framework for a consideration of landscape stripped of its cultural and|or ideological way-of-looking slant.

A *ritournelle* has three aspects or phases which must be understood together. In phase one, it is a jump out of chaos, considered as a jumble of affects and percepts such as those that confront a child out alone in the dark. The child seeks shelter or direction in a "little song" which is not a representation of these threatening forces but an attempt to combine them into a specific tempo, which can be speeded up or slowed down as required and is perpetually in danger of breaking apart. The ditty constructs a fragile stabilizing centre (*Mille Plateaux* 382/343). This is not yet a *territory* but rather, as will be seen later, a *medium*. Alternatively, instead of being immersed in chaos and having perpetually to find the tempo which will momentarily provoke a centering, one may actually already be at home. However, home "does not pre-exist" (ibid.). It emerges in or as a limited space—a circle around a stabilized centre—in phase two of the *ritournelle*, now described as a "territorial assemblage" (383/344). How exactly a *ritournelle* assumes this territorializing function requires some detailed explanation.

The formation and stabilization of territory require what Deleuze and Guattari call a "sound barrier"—"*un mur du son*." I have modified the published translation, to allow for the fact that the French expression alludes to a Mach number as much as to a Phil Spector "wall of sound" or "*mur de son*." The suggestion is that the "wall" is produced by a change of "pace" which provokes a "jump": the black hole of chaos "*a donc sauté d'état*"—has thus jumped from one state to another (383/344). Furthermore, it will be seen that a *ritournelle* necessarily involves more than one dimension. Sound nevertheless remains a typical component of a territorial barrier, which the authors describe as "a wall with some sonic bricks in it" (382/343), as when radios and television sets blare out from households. However, while loud music functions as a defensive wall of sound which angers neighbors when it invades their own territories, the ditty is never purely defensive. There is always more to territoriality than this. The jump which constitutes a territory is, once again, a qualitative one.

The discernible cosmological analogy is of Earth emerging from Chaos. The circumscribed earth as home—or again, home as a parcel of earth—is a configuration of "interior forces" which "resist" chaos and may even borrow something from it "across the filter or sieve of the space that has been drawn [or traced out: *tracé*]" (382/343). When a housewife sings to herself or listens to the radio, sound participates in "marshal[ling] the antichaos forces of her work"; similarly, a child hums in order to "gather into itself the forces"[3] which it needs to get its homework done. In a word, the forces interiorized *as* earth by a territory are forces of creation: "*les forces intérieures de la création*" (382/343). Consequently, in a play on "*terre*" (earth, estate, property, region, territory, land), it is said that a *ritournelle* "always carries earth [*de la terre*] with it; it has as its concomitant, an estate [*une terre*], even if it is a spiritual one" (384/344; tr. mod.).

As a (re)territorializing phase in continuous interchange with its two other aspects, the *ritournelle* thus never simply effects the attribution of "home value" to a new object, so that—merely repeating the same old story (*toujours la même rengaine*)—this may act as "a compensation and substitute for the home that has been lost" (Buchanan 30). (Re)territorialization always provokes a reorganization of forces as earth. Creativity is not only ethological; it is cosmological. One gets from cosmos to home and|or society and (as we shall see) on to world, via animal behavior and by way of a succession of jumps. As a principle of invention which stretches back to chaos and creates a world, art does not function under the regime of representation. It always acts so as to make a difference.

Certainly, as Abode (*"la Demeure"*)—that is, as what abides as a particular distribution of forces—earth has an originary or native feel to it: the *ritournelle* has "an essential relation to a Natal, a Native" (384/344). However, failing to get the speed, rhythm or harmony right would once again provoke catastrophe and "bring back the forces of chaos, destroying both creator and creation" (382/343). Another way of leaving home nevertheless beckons. For, if the forces of chaos belong to phase one of the *ritournelle* and terrestrial forces to phase two, to phase three belong cosmic forces.

The pressure of the territorialized "interior forces of creation" opens a crack in the circle of earth. One may thus venture from home "on the thread of a little tune [*une chansonnette*]" or by risking an improvisation (383/344; tr. mod.). This does not imply sliding back into chaos, as when the little song of a frightened child breaks down, since the organizing force of the circle provides thrust. Opening out on to a future "as a function of the working forces it shelters," home as territory has in effect created "another region," which Deleuze and Guattari name "the World" (383/344). If phase one corresponds to fixation or infra-assemblage, phase two to organization or intra-assemblage, phase three, as inter-assemblage, corresponds to passage through an assemblage or indeed into flight, which is not a passage to an identifiable or neighboring territory, but towards something else entirely (384/345). This is deterritorialization. Once again, however, each of the three phases inheres in the other two: chaos, earth and World are forces or tempos confronting one another and converging within the *ritournelle* as, simultaneously, territorialization and deterritorialization.

A number of steps remain to be taken in order to properly appreciate the complex dynamics involved here and its significance for the study of landscape. Deleuze and Guattari's starting point is constituted by the notions of *milieu* and *rhythm*. When a frightened child reassures itself by skipping and humming, it wards off a chaotic jumble of sensations by harnessing energies into a repetitive ditty, so that affects and percepts which went off haphazardly in all directions are better directed. The components of phase one of the ditty are multi-"directional" and the child finds shelter *in* orientation—that is, by constituting a bloc of space-time through the "periodic repetition" which gives its component forces a certain consistency (384/345). Such a "vibratory" entity is a milieu. Although Deleuze and Guattari do not make the point explicitly, the terrified child's ditty is a milieu. Acting within this milieu, the child reassures itself.

It is, however, possible to imagine one scared child meeting another in the dark and the differing and internally varying tempos of their ditties locking into step; or to observe a child's own scared movements locking into harmony with its singing, in a skip and a jump; or again, to imagine its song and its step adjusting to the waving of the branches in the trees or the blinking of a light, and so on. Quoting Bachelard, Deleuze and Guattari declare such convergences to be essential to establishing a tempo: "the link between truly active moments (rhythm) is always effected on a different plane from the one upon which the action is carried out" (385/346). The child's action has the effect of reassuring itself. Rhythmical linking makes this possible but it also does more: it orchestrates milieus. Throughout this stage of the argument, Deleuze and Guattari's examples are taken from living beings and their milieus, whether exterior (materials), interior (elements and substances), intermediary (membranes) or annexed (energy sources; actions-perceptions). Citing Jacob von Uexküll, they suggest that, milieus being ditties, the "productive repetition" or rhythm of a milieu (386/346) constitute a *motif*. Counterpointed motifs form *melodies*, which in turn come together in further contrapuntal relations to compose the "music" of nature, understood as a proliferation of nested functionalities. Deleuze and Guattari insist that there is always an element of chance in this emergent music, since it does not result from the imposition of meter (that is, measure), but through resonating inequalities and heterogeneities, whenever the periodic repetition which constitutes one milieu interacts with another in a context of dynamical criticality.

I am now approaching the properly operative definition of landscape I have been looking for. Because Deleuze and Guattari's argument does not establish agency independently of a milieu but makes it coemerge with milieus whose resonance enable its stabilization, it does not make of these an object of relations of power, either for or between subjects. Nevertheless, landscape is something other than the symphony of nature. There remains a further step to take. Rhythmically emerging milieus do not of themselves constitute territories. Deleuze and Guattari explicitly state that a territory is not a milieu. Neither is it an additional, containing milieu—a milieu of milieus. Nor is it the rhythm or passage between milieus. As was suggested above, with reference to the transition between phases one and two of the ditty, territory is an "act." It affects milieus and rhythms, "territorializes" them (386/346–347). There is said to be territory when milieu components cease to be directional and become dimensional, as in the protective sonic wall without which a home-like space would not emerge. This happens when these components cease to be purely functional and become *expressive*. Territory requires that rhythm and melody, in the sense just described, become "matters of expression," which is to say qualities—color, odor, sound, silhouette, and so on (388/348)—that enjoy the "temporal constancy" and "spatial range" necessary for them to become territorializing marks (387/347). Any such mark is a "signature." Territory thus has "two notable effects," since the reorganizing of functions just described goes hand in hand with the regrouping of forces through the "attribution of all diffuse forces to the earth" (394–395/353–354; tr. mod.), which has already been discussed.

A territory *has* such and such qualities, which define or sign it. A signature functions expressively as a "poster" or a "placard": it is hazarded as the "constituting mark of a domain," rather than composing the "constituted mark" of the subject which coemerges with it (389/349). Once again, the defensive action of a subject or agent within a milieu—not, this time, the child singing more emphatically in order to harness its fear but, say, someone turning up the volume because, for whatever reason, real or imagined, they feel the neighbors are getting too close—does not exhaust the matter. Expressive qualities are "auto-objective:" they "find an objective existence in the territory they trace out" (390/349; tr. mod.). This is what allows a signature not to be merely a poster or a placard but to become what Deleuze and Guattari call a *style*.

The combined definition of expression and style constitutes the final important element in Deleuze and Guattari's account of the complex territorializing phase of the *ritournelle*. Just as resonating or contrapuntal milieus formed motifs and composed melodies out of the forces of chaos, so does signature as "style" compose territorial motifs, described as the rhythmic "face" of a territory, and territorial counterpoint, which is its melodic "landscape" (390–391/350–351). Deleuze and Guattari refer rhythmic faces—or characters—to the interior milieu of impulses which, it will be recalled, are the creative forces of earth. Melodic landscapes are referred to the exterior milieu of circumstances, in which motifs are caught up in contrapuntal relations. There is style when "expressive qualities or matters of expression enter shifting relations with one another that 'express' the relations of the territory they trace to the interior milieu of impulses and exterior milieu of circumstances" (390/349–350; tr. mod.).

Understood in the terms defined by Deleuze and Guattari, the expressive dimension of style allows a transformation in the definition of landscape. It will be recalled that territory is "dimensional." Nevertheless, even "in extension," it does not merely relate to the possessive behavior of its inhabitants, since it also "separates the interior forces of the earth from the exterior forces of chaos." Deleuze and Guattari add that this separation does not hold "in 'intension' ": in the "dimension of depth," the two types of force "clasp and are wed in a battle whose only criterion and stakes is the earth" (395/354). This territorializ*ing* struggle is to be distinguished from imperialistic or territorializ*ed* battles for control over a portion of land, on which landscape is indexed, so long as it is considered as an ideological formation.

As Deleuze and Guattari say, in a manner which indirectly invokes the notion of an unpredictably located *genius loci*, "There is always a place, a tree or a grove, in the territory where all the forces come together in a hand-to-hand combat [*un corps-à-corps*] of energies"; furthermore, "Earth is this close embrace [*ce corps-à-corps*]' (395/354; tr. mod.). Even as it unfolds the clasp in which these forces are held, it folds them together again otherwise, in a creative act which makes a difference.

There are at least two reasons why it is important not to adopt a constrained notion of territory in *ex*tension, nor to confuse this with territory in "*in*tension," if one wishes to develop a fully operational concept of landscape. In the first place, it is reductive to focus exclusively—or indeed primarily—on territorial conflict. It would, for example, be helpful (and indeed urgent) to investigate just how Maori culture, say, is involved in a territorializing struggle effort to hold together the configuration of forces which emerges from the specific invention that is its territory, before assuming, as Mitchell too readily does, that, in resisting an imperialistic invader, it simply manifests its own imperial expansionism. Deleuze and Guattari deplore the "dangerous political overtones" of assuming aggressiveness to be the basis of territory, when one should rather investigate the particular tempo ("*allure*") of behavior or "impulse" which corresponds to species-specific violence, as resulting from the reorganization and creation of forces brought about by territorialization (388/348).[4] In the second place, the struggle "in intension" with the "exterior" forces of chaos is not a purely defensive one. Even if the difference in context supports diverging interpretations, what the published translation first calls "hand-to-hand" combat in a spot where all forces come together is the selfsame "*corps-à-corps*" which then becomes the "embrace" of these forces. Consequently, the "intense" centre is "simultaneously inside the territory and outside." There is therefore something misleading in the supposedly originary dimension of earth as source (the "*natal*"), since this always incorporates a dimension of deterritorialization. Home is, in an important sense, unknown:

> Within itself or outside itself [*En lui ou hors de lui*], the territory refers back to an intense center [*renvoie à un centre intense*] which is like the unknown homeland, terrestrial source of all forces, friendly or hostile, where the entire issue is settled [*où tout se décide*]. (395/344; tr. mod.)

Any such decision or "settlement" will be a complex tempo, folding territorialization and deterritorialization into one another. Thus, even as territory emerges as an individuated modulation of motif and counterpoint, these coevolve as the twin poles of intensive landscaping, or style. Rhythms are amplified, diminish and vary; melody develops unpredictably.

Going beyond the literal thrust of Deleuze and Guattari's text, I assimilate landscaping to style in order to highlight the *geo*philosophical dimension of their agenda. This is formulated more forcefully than ever in the fourth chapter of *What is Philosophy?*, which starts by declaring it impossible to acquire a proper understanding of thought so long as one considers this in terms of subject and object, before going on to state that thinking gets done "in the relationship of territory [*territoire*] and earth [*terre*]" (82/85; tr. mod.). The latter "constantly carries out a movement of deterritorialization on the spot, by which it goes beyond any territory: it is deterritorializing and deterritorialized" but at the same time "restores territories" (82/85–86). Allowing these references to stand as testimony to the fact that my subject is far from being incidental to Deleuze and Guattari's thought, I return in conclusion to questions of landscape, in the more ordinarily recognizable sense of the term, but always bearing in mind Deleuze's insistence on the fact that acts of creation concern "ideas" understood as potentials embedded within a particular mode of expression and inseparable from it ("Qu'est-ce que l'acte de création?" 291). An idea, in this sense, is never general; it is always technical. What, then, is the potential of the idea in landscape?

While Deleuze and Guattari insist in *A Thousand Plateaus* on the decisive nature of the discovery of rhythmic character and melodic landscape in allowing art to move beyond the stage of the placard, they also acknowledge that this may not be art's "last word" (393/352). This is essentially a historical statement applying to Liszt in relation to Wagner. However, it is more generally applicable to the historiographical issues raised by what I have called style as intensive landscaping (or *vice versa*). Indeed, I have repeatedly insisted on the creative unpredictability of the "jumps" which characterize the *ritournelle* on a cosmological and ethological scale, as well as in what is more readily recognized as artistic practice, up to and including the autonomous development of matters of expression, which is, properly, a question of becoming.

The eleventh plateau of *A Thousand Plateaus* formulates a series of observations relating artistic style to the ideas intimately connected with landscape. These are presented chronologically, the authors respectively articulating Classicism with milieus, Romanticism with territory, earth and the people, and Modern art with the cosmos. Quite apart from the fact that Deleuze's periodizations of style raise problems which are philosophical as much as historical,[5] there is no room to detail Deleuze and Guattari's arguments on these matters here. Let me more succinctly return to the point made by Mitchell, when he declares landscape an imperialist medium to be exhausted and denies its continued validity as a mode of artistic expression, while nevertheless hoping for its return under a different aspect. In a not dissimilar vein, Cosgrove's 1984 *Social Formation and Symbolic Landscape* situates landscape's period of cultural ascendancy between 1400 and 1900, in relation with the ascendancy of Europe. However, this position is revised in the new introduction written a decade and a half later, when Cosgrove insists that he was mistaken not to recognize that Romantic nationalism found "intense artistic expression" through landscape in the period he previously imagined to coincide with the decline of the medium, and furthermore acknowledges that, in addition to the persistence of a recognizable landscape aesthetics throughout the twentieth century, Modernist aesthetics repeatedly appealed to the "spiritual power" of landscape (xxi; xxiii). This does not significantly affect Cosgrove's acclaim for the work of Richard Long in the concluding pages of the original book, where he relates the fact that Long refers to his work as "land art" rather than "landscape" to his practice of "[e]ntering the land rather than seeing it" and hails this innovation as "perhaps a sign of hope for our future cultural relations with land" (270). The question remains whether art is properly concerned with "giv[ing] force to our own inalienable experience of home, of life, of the rhythms of diurnal and seasonal life" (270) or whether, in terms of what Deleuze and Guattari have taught us, it can participate in the repeated inventions of "home" as a territory which is "itself" a "place of passage" already "in the process of passing into something else" (*Mille Plateaux* 397/356).

One might begin answering this question by noting that significant examples of recent land art do indeed confirm Deleuze and Guattari's suggestion that art now concerns itself with cosmic forces. This is particularly the case with American land artists working in direct confrontation with the elements, frequently in barren, uninhabited areas. However, when one thinks of such works as Michael Heizer's *Double Negative* (1969–1970), Robert Smithson's *Spiral Jetty* (1970), Robert Morris's *Observatory* (1971), Walter de Maria's *Lightning Field* (1977) or James Turrell's *Roden Crater* (1982),[6] one cannot help observing that they operate on a grand scale. Smithson aside,[7] they provide little or no evidence of Deleuze and Guattari's notion of a *ritournelle* which "becomes at once molecular and cosmic" (*Mille Plateaux* 423/378; my translation: the sentence is missing from the published translation). That is to say, they do not grapple with or embrace "densities" and "intensities" (423/378) as forces of creation or style, but rather measure human against cosmic power in an essentially territorialized act of defiance and|or homage which, remaining firmly within its milieu, provokes no rhythmic variation or contrapuntal development. Witness the obsession of these works with scale, their careful orientation in space and their equally minute positioning of the viewer in relation to the elements. Once again, the exception is the variable mobility inherent to the viewer of *Spiral Jetty* and the way Smithson's film shows the jetty's involuted silhouette|trajectory to function as a rhythmical cell within a montage in which it is intensively counterpointed across many scales and milieus.

If, alternatively, one takes up Cosgrove's suggestion and looks at the work of Richard Long, one is left with the conviction that his signature forms—typically, straight lines or circles which, once again, are characteristically inscribed in deserted or desolate locations, in confrontation with the elements and other forces outwith society, but sometimes invoke "originary" cultures, such as Ancient and Roman Britain, or the Incas—are precisely that: signatures. Following the demise of straightforwardly representational landscape as a significant form in high Modernist art, the interest of Long's geometrical interventions lies in their showing the signature to have a territorializing function, as opposed to acting as the "constituted mark of a subject." However, Deleuze and Guattari might have described almost any of Long's works as "a silent painting on a signboard," devoid of the "properly rhythmic character" necessary for the self-development of a melodic landscape (*Mille Plateaux* 393/352). Consequently, the sheer repetitiveness of the artist's "signature" finishes by functioning as an operator of institutional reterritorialization, as his works achieve instant recognition value in the milieu of the art book or the gallery. The counterexample to Long (in spite of the fact that the two artists are regularly associated) would be the more complex rhythmical and melodic work of Hamish Fulton. However, rather than closing on the relevance of Deleuze and Guattari's thought to a necessary revaluation of land art, which would seek to identify lines of becoming in specific styles of work, as opposed to discerning nuances in a body of production that is still regarded as essentially homogenous, I wish to evoke briefly a modest landscaping project which demonstrates the creative power of intensive landscaping, which is to say, landscaping which is expressive in the terms developed by Deleuze and Guattari.

A few years ago, gardeners in the employment of the town of Fontenay-sous-Bois in the outskirts of Paris created, on a traffic island near a roundabout housing a monumental modernistic sculpture, a miniature Dutch-style garden, barely a couple of meters long, complete with gate, bridge and windmill. Everything about this garden appeared kitsch and|or incoherent: its picture-postcard iconology inserted at the foot of an imposing piece of modern art, the garish colors of its flowers, its inaccessibility, the discrepancies of scale among its built features.[8] However, these very qualities engaged the passer-by in such a way as to make it impossible to visually accommodate the garden in a stable manner, either within itself or in relation to its environment. The eye was perpetually impelled to attribute variable distance and size to any given detail, built or grown, according to how it was gauged in relation to any other feature, in terms of relative size or chromatic intensity. As a result, perception went haywire. The garden appeared to expand and contract uncontrollably, while the monumental sculpture momentarily receded into the distance and|or mimicked the road signs in the foreground, before reassuming its rightful place and dignity. In a word, height and depth became intensive; perspective ceased to function as a straightforward scale of measurement and acquired variable density.

The garden was a temporary piece of municipal landscaping, created in the spirit of a "city in bloom." Its "Dutch" aspect might legitimately have seemed to reterritorialize its decorative energies in a bucolic fiction, in accordance with a city directive that its gardeners should evoke places from around the world (there was also a "Chinese" garden, and so on). However, the context suggested otherwise. The neighboring sculpture was an imposing piece of official art, placed there by the local authority and resented by many of those who lived nearby or used the shopping mall across the road. The way the brightly colored wings of the garden's miniature windmill gently echoed its gigantic, rust-colored silhouette suggested that the homely exoticism of the scene also participated in the embrace of forces that made of this particular garden in this particular environment a freely evolving contrapuntal composition, incorporating not only percepts—color, silhouette, depth, and so on, in constantly varying intensities—but also sociopolitical affects—large and small, high and low, gaudy and austere, near and distant, etc., as expressive of status.

There is no suggestion that the municipal gardeners of Fontenay-sous-Bois were engaged in a deliberate critique of a piece of public artwork which many locals admittedly dislike, or even in a protest against the fact that their employers do not allow them to landscape the roundabout on which the sculpture stands, on the grounds that the responsibility is too great for them. The garden did not stage a power struggle over some physical or institutional territory. Rather, through the exercise of the gardeners' practical knowledge, objectively existing perceptive and the affective qualities were made to resonate unpredictably, thereby liberating the creative potential of the idea of landscape across milieus and holding out the promise that the forces that had thus begun to fluctuate wildly might jump into another state. This was a garden that sought to invent relations, rather than assert ideological or cultural control. It was, in the spirit of Deleuze and Guattari, landscaping as style, as the promise of a social spacing yet to come, that would readjust articulations of hierarchy and the distribution of the public and the private sphere for a people to come.[9]

Notes

1. It derives from the Latin *"dictare,"* to dictate, by way of the Old French *"dité"* or *"ditté,"* originally *"ditié."*
2. For a brief review of the *ritournelle* in Deleuze, see also Sasso and Villani, 2003.
3. The published translation renders Deleuze and Guattari's *"recueillir en soi les forces"* as "to summon the strength."
4. In many ways, the entire focus of their "Capitalism and Schizophrenia" project involves just such an analysis of capitalism.
5. For a brief discussion of this issue in relation to painting, see Abrioux, "Diagramme, histoire, devenir."
6. All these works are illustrated in Tieberghien, *Land Art*.
7. See his *Spiral Jetty Film* (1970) and his artworks and writings which use entropy as a paradoxical creative force.

8. For a more detailed description, see Abrioux, "Spectro-graphy of Gardens."
9. For the people, in relation to the *ritournelle* and Romantic art, see *Mille Plateaux*, 419/375.

Works cited

Abrioux, Yves. "Diagramme, histoire, devenir. Ou, quand Deleuze fait de l'histoire de l'art." *TLE* 22 (2004): 165–176.

——. "Spectro-graphy of Gardens." in *Le Spectre des jardins/The Specter of Gardens*. Ed. Yves Abrioux. Saint Rémy-lès-Chevreuse: Fondation de Coubertin, 2007, 39–83.

Buchanan, Ian. "Space in the Age of Non-Place." in *Deleuze and Space*. Eds Ian Buchanan and Greg Lambert. Edinburgh: Edinburgh University Press, 2005, 16–35.

Buchanan, Ian, and G. Lambert (Eds). *Deleuze and Space*. Edinburgh: Edinburgh University Press, 2005.

Bustamante, Jean-Marie. *Bustamante*. Paris: Flammarion, 2005.

Cosgrove, Denis E. *Social Formation and Symbolic Landscape*. Madison: University of Wisconsin Press, 1998.

Deleuze, Gilles. *Différence et Répétition*. Paris: P.U.F., 1968.

——. *Logique du sens*. Paris: Éditions de Minuit, 1969.

——.*Cinéma 2—L'image-temps*. Paris: Éditions de Minuit, 1985.

——."Qu'est-ce que l'acte de création?" Lecture given at the Paris Film School FEMIS; transcript in *Deux Régimes de fous. Textes et entretiens 1975–1995*. Paris: Éditions de Minuit, 1987, 291–302.

Deleuze, Gilles, and Félix Guattari. *Mille Plateaux*. Paris: Éditions de Minuit, 1980. *A Thousand Plateaus*. Trans. Brian Massumi. London and New York: Continuum, 2004.

——. *Qu'est-ce que la Philosophie?* Paris: Éditions de Minuit, 1991. *What is Philosophy?* Trans. Hugh Tomlinson and Graham Burchell. London: Verso, 1994.

Descola, Philippe. *Par-delà Nature et culture*. Paris: Éditions Gallimard, 2005.

Long, Richard. *Walking in Circles*. London: The South Bank Centre, 1991.

Mitchell, W. J. T. "Imperial Landscape" in *Landscape and Power*. Ed. W. J. T. Mitchell. Chicago: Chicago UP, 1994/2002, 5–34.

Roger, Alain. *Court Traité du paysage*. Paris: Éditions Gallimard, 1997.

Rowley, Trevor. *The English Landscape in the Twentieth Century*. London: Continuum, 2005.

Tiberghien, Gilles A. *Land Art*. New York: Princeton Architectural Press, 1995.

Sasso, Robert and Arnaud Villani (Eds). *Le vocabulaire de Gilles Deleuze*. Paris: Vrin, 2003.

15
Art for Animals

Matthew Fuller

If art is genuine it is creative revolution regardless of who looks at it
(Laszlo Moholy-Nagy 87)

A crowd of apes and monkeys sit clustered upon a box gawping and grinning and staring at a canvas. They've seen nothing like it; or they are bored by it; or they raise their arms in delight at the general hullabaloo. They are of a number of sorts, baboons, gibbons and others; all, however, have the painting as the primary focus of their attention or reaction. What is on the canvas is hidden from view; all we see is the gilded side of a carved frame. Gabriel von Max's turn of the century comedy in oils, *The Jury of Apes* (1889) points at the trade of the art critic, utter monkey business, but also at the viewer of art, a mug, an enthusiast, or, in the stare of an ape turned to address the viewer through half-closed lids, a rare specimen in itself. For apes to look at a canvas makes the pretensions of those who look with a mind to judge also minds to be judged, or at least, to be sniggered at.

Pliny the Elder's *Natural History*, a book which places painting and sculpture amongst an inventory of animals, plants, and minerals, gives us another story along these lines (see Book XXXV, Section XXXVI: 309). In a competition between two painters in *trompe l'oeil* technique, Zeuxis and Parrhasius face off in front of a crowd. The first artist pulls away the curtain protecting his work to reveal the most perfectly rendered bowl of fruit, so lucidly real in fact that a flock of birds immediately descends upon it and starts to peck away the paint. Impressed, Parrhasius stirs, but does not move. He simply stands and watches. The annoyed Zeuxis demands that he remove the curtain from his canvas. The second artist does indeed reveal his painting, but by stating that he has no curtain to remove, that it is a painting of a curtain. This painting has deceived the eyes of an artist, not a mere bird. Parrhasius wins the competition and perhaps brought to a temporary close a current in art which is only just reemerging, art for animals.

Art for animals is art with animals intended as its key users or audience. Art for animals is not therefore art that uses animals as a substrate or a carrier, nor as an object of contemplation or use[1] (needless to say, given these criteria, it does not fall into the category of transgenic art, with its all too frequent tendency to animal abuse and naive sensationalist celebration of genetic engineering). It is not art that, like *The Jury of Apes*, depicts animals for human viewers, or that incorporates animals into living tableaux, but work that makes a direct address to the perceptual world of one or more nonhuman animal species. There are only a very small number of works that make such an address. This essay will make a brief survey of them and then go on to discuss their implications. Where it differs from Pliny's tale is in that it works, not on the level of successful imitation, of setting up perception as a means by which one is duped, but in rendering perceptual dynamics as both somewhat more unresolved and more powerful.

A further important category of work that does not usefully fall into this current consists of objects such as dog-kennels by celebrity architects (such as Frank Gehry[2]) or housings for birds. Some work in zoo design, notably for Carl Hagenbeck by Johannes Baader, and the aviary in London Zoo by Cedric Price, does attempt to engage with animals' behaviors, in a way that Berthold Lubetkin's famous double spiral-ramped penguin pool at the latter zoo does not.[3] Thomas Schütte installed a work originally entitled *Hotel For Birds* on a plinth in London's war monument congested Trafalgar Square. Made of brightly colored layers of perspex, this is a sculpture in the style of an architectural maquette designed to catch light, and to act as a "public space" for urban rock doves displaced by a cleansing policy established by a different branch of the body commissioning the work. (Indeed on installation the work was re-named Model for a Hotel.) Whilst being of interest, it is primarily a "housing." David Nash, an artist who works with the materiality of wood, and whose aim is for the work to integrate into natural processes, has made shaped blocks of oak for use in a small copse, by sheep who gather there to escape the rain. They use the blocks for "shelter, safety and scratching" (see Sutton).[4] More recently, the sociology artist Jeremy Deller is using the device of an architectural competition to produce a design for a *Bat House* for the Wetlands Centre in South London (see Deller). Whilst these are interesting projects, they largely address animals in terms of ergonomics, making spaces that physically "fit" them.

At the same time, because many animals experience and shape a locale by literally inhabiting it, there is no absolute distinction between what is proposed here as art for animals and work that produces scenarios that animals live in, work on, and complete, or render definitively unfinished. Equally, other projects that involve moving animals from one context to another, as in the case of Hans Haacke's *Ten Turtles Set Free* (1970), or sorting systems for animals, as in Robert Morris' *A Method for Sorting Cows* (1967), are assumed to engage some aspects germane to this project, such as the categorical systems, including property, to which animals are assigned, but fall outside the scope of this essay (see Morris and Haacke).[5] Equally, durational performances of coexistence with animals are related but sit to the side of the present text.[6]

Other areas which would possibly suggest further development, but which are outside of the present discussion, include the production of visual material by animals (famously including paintings by chimpanzees or elephants). Other perhaps more promising research includes findings that indicate pigeons' capacity to distinguish between styles of picture making —that is, for example Shigeru Watanabe's research that showed pigeons could learn to distinguish between works by Monet and Picasso and, subsequently, that they were able to carry over this capacity for distinction to categorically related art by Cézanne and Braque (see Watanabe et al.).

A weakness of some of the main streams of cultural theory over the past decades is that, in its emphasis on the constructive aspects of culture, biological questions are neglected or considered reactionary. At the same time, a thread of biologically based research, functioning largely by an unsophisticated positivism, makes any chance of a dialogue between disciplines and styles of research difficult. There is a certain laboriousness in getting through the clunky formulations that are dredged up by instruments incapable of finding anything but what is expected and that are proudly displayed as having "explained" culture. Certain currents in contemporary biology have made an attempt to perform a "land-grab" on culture, to suggest that biology provides a baseline level of explanation for all forms of behavior. Often these are characterized as being simplistically "Darwinian" in motivation, with characteristics of culture identified as mere epiphenomena. It is not necessary to get locked into simply refuting the shrillest voices or those advocating the most absolute reductionism as an *a priori*. However, this kind of argument has not come solely in the form of a landgrab on culture, nor has it come only from scientists. A "recall to biology" has been a ruse often played by those in the domain of art discourse who attempt to enforce a "shared symbolic order" of the kind once supposedly provided by religion (see, for example, Peter Fuller). I would suggest that much of this work is a betrayal of the subtlety and speculative nature of the current of thought set in play by Darwin.

Much of such work prefaces its findings by a complaint. In this scenario, biological approaches to culture are refused out of hand because of a conformist consortium of Marxists, poststructuralists, feminists, queers, and others who bunker culture off from questions of innateness or predilection. When Marx has written about species being, Foucault on biopolitics, Cixous on *ecriture feminine*, and there is a plethora of more recent research and art emphasizing corporeality, it is unfortunately mistaken to describe those primarily concerned with culture as somehow assuming that they entirely surpass biology. Ellen Dissenyake suggests that art is a refusal to "grow up," a prolongation of the sense of exploring the world for the first time, of maintaining sensual delight in novel growth and experience, the capacity to escape from a subordinate role (see Dissenyake). Perhaps certain participants in science too are undergoing such a thrill in their discovery of culture, and their entry into culture as a previously taboo domain. If so, this is entirely to be welcomed, but perhaps they should calm down just a little. At least, in a society such as ours, for scientists to borrow the Cultural Studies ruse of presenting one's arguments as the knowledge of the oppressed at least has the virtue of being amusing.

Art for animals intends to address the ecology of capacities for perceptions, sensation, thought and reflexivity of animals. The capacity for art is part of the rather mobile boundary line that performs the task of annihilating the animal in human and in demarcating the human from animality. The purpose of this text is not so much to legislate upon the placing of this line, but rather to suggest that the sensual and cultural capacities of various kinds of being, whether ordered into species or not, can be explored, and to follow a few ways in which this has been done. Paul Perry has installed a small robotic device to spray bobcat urine high up a tree to stimulate an imaginary of pheromone responses. Natalie Jeremijenko makes a robotic goose, the aim of which is to set up interactions with a small group of geese; in a number of other projects she sets up devices for interspecies communication. Louis Bec attempts to set up a dialogue between two speciated parts of the same genus of fish. Anthony Hall also works on communications and perceptual reflexivity with weakly electric fish. Marcus Coates stages a series of actions with animal materials and behaviors with interaction with other species as the prime goal. Some of this work is rightfully absurdist, whimsical, self-trivializing. But all of it moves towards setting up actual, multi-scalar and imaginal relations with animals that involve a testing of shared and distinct capacities of perception.

Deleuze and Guattari, following von Uexküll, Kafka and Maturana and Varela amongst others, have placed animal subjectivity at the core of their reinvigoration of thought. In this, they provide some dynamic formulations of conceptual personae as animal-beings and of animals as engaged in reciprocal relations of life shaped by color, growth and habitat formation. In their book *What is Philosophy* art and nature are described as being alike because they combine an interplay between House and Universe, the homely and the strange, and the specific articulation of the possible with the infinite plane of composition. "Art for Animals" takes up such work for the category of art.

In engaging animal cultures and sensoria, these projects also make art step outside itself, and make us imagine a nature in which nature itself must be imagined, sensed and thought through. At a time when human practices are rendering the earth definitively *unheimlich* for an increasing number of species, abandoning the human as the sole user or producer of art is one perverse step towards doing so. More widely, a core process of Guattari's writing, one which it amplifies in that of Deleuze, is the project of understanding ecology at multiple scales, from the social, to the medial, technical and esthetic, to that of subjectification. This text draws upon such processes to develop the question of animal–human subjectivation as a cultural and inventive process. Within a web of interconnected capacities and materials a set of processes and instances, set-ups, ruses, devices, work to establish what Rosi Braidotti has called "affirmative interrelations" (209) between not simply a fixed set of innate behaviors and predilections but the capacities for becoming that might exist between different forms of life and esthetic dynamics.

It is not the intention here to suggest that there is a necessary continuum between human and animal; a continuum is a figure that implies fixed ends and a neat metric running between them. Rather, what is suggested in this initial sketch of a possible field is a myriadic ecology of perceptual–cognitive sets, some of which may overlap or share functions and capacities. As the primatologist Frans de Waal notes in his reflections on culture, "[o]ne cannot expect predators to react the same as prey, solitary animals the same as social ones, vision-oriented animals the same as those relying on sonar, and so on" (55). Equally, we cannot expect sensual experience to stay the same amongst members of what is logged as the same species. Humans, for instance, have domesticated themselves since the advent of agriculture, with, at the genetic scale, changes in composition equivalent in the degree of change to that found to be involved in the transition from wild corn to domestic corn today. In certain populations such changes manifest in the ability to digest foods associated with a sedentary mode of life (such as the developed ability to digest lactose linked with the unfortunate tendency to drink cow's milk). At a sensory level, rather than a genetic one, our habituations tend towards similarly substantial changes: one recent study, for instance, suggests that it is possible, with a little retraining, for humans to acquire an equivalent capacity of smell to that of dogs (see Geddes). Regardless of whether this is desirable or not, or whether it might also suggest the need for an uptake of the scenting and smelling habits of dogs, art for animals does send a tingle along the edges of what we take for granted as our current capacities. It suggests that we search out and test the discontinuities and overlaps between our sensual and intelligent capacities and those of others. What would it be like, for instance, to be able to see just the very edge of ultraviolet in the iridescence of a petal or on the wing of a butterfly? How would such a change in sensual capacity re-order us, make life bulge? Is there a market for drugs that temporarily reconfigure nervous and perceptual systems to those of other species?

Gilles Deleuze laughingly describes the sensorial world of the spider: a juicy fly can be placed in front of it; it doesn't care. All it wants to feel are a few small twitches on the far reaches of its web. Just a few details, a muttering in the background, that's what is appetizing. This, says Deleuze, is the same sense of the world as the narrator of Proust's *Recherche*. Deleuze himself mobilizes various nonhuman sensoria, ticks, lobsters, dogs, lice, bees, wolves, bowerbirds, flies, the horse–knight assemblage. Such creatures become ethological devices to overstep what can be sensed, thought or said. They are paths of becoming, gravitational lodes of traction which pull the human out of its skin, and pull the singular animal into the multiplicity of packs, of evolution and of ecology.

There are a number of ways and particular domains in which such becoming can be seen to occur, at the scale of brains, that of bodily elements and organization, and that of means and kinds of communication, amongst other things. Paul Rozin, for instance, catalogues a number of ways in which human cultural processes and evolutionarily accrued predispositions are interwoven in the case of food (see Rozin). What such work reveals is that the bodies of individuals in evolutionary conditions are means by which forms of life scan for potential adaptions; they are also means by which ecosystems arrange themselves, and the platforms for cultures to articulate and be experienced, revised and produced. They are in turn worked on and produced by cultures. Ecologies emerge in a multi-scalar way. What Deleuze and Guattari argue for is that an understanding of the virtual be added as a specific scale within ecologies, as a dimension of relationality that exists at every scale within such a system, and a diagonal which connects them.

Evolution by natural selection is often characterized as a process of the survival of the most fit. Fitness is a relative and distinctly processual term. A whale is fit for its habitat, but, as the current representative of a mammalian lineage that reentered the water, it is also the result of massive and quite possibly awkward adaptational change (see Zimmer). It cannot be understood as being perfectly fit, but as the ongoing result of many interlocking morphogenetic, material and adaptive capacities that may involve substantial shifts in the use or function of bodily elements. This given, it is useful to consider the question of the virtual in relation to the way in which bodies, entities that can be regarded as their components (such as genes or organs), their aggregates, and those of their products, such as cultures, explore, adapt to, make adaptations of, coevolve with and form ecologies.

It is a commonplace that organs, behaviors or other entities in ecologies can change or add functions over time. Julian Huxley, in his early work of ethology, notes that the behavior of grebes in courtship includes adaptations and appropriations of movements, such as dives, that might have primarily developed as feeding movements but which are repurposed as displays of fitness and of courtship interest. These are elaborately linked and synchronized in a distinctive and beautiful set of behaviors (see Huxley). In a further dislocation of signaling into mimicry across species, when showing aggression meerkats raise and curve their long tails over their backs. In this, they are thought to be mimicking the posture of their enemy and food source, scorpions. North American chickadees (red-breasted nuthatches) are able to distinguish between the alarm calls of black capped chickadees, according to whether the species being alerted of is likely to predate them, so the signaling of information crosses between species.[7] Signs given for one purpose are used for another. Such chains of dislocation are potentially end-less; the mouth, originally used for biting and eating, over time gains additional functions such as speech and, in humans and a few other primates, sexual activity. Chains of dislocation constitute a form of primary experimentation of the capacities and materials of bodies and of life. They may occur across all scales of a body or at those of individuals or populations.

Aside from adaptions and accumulations of function and behavior, coevolutionary assemblages, such as the wasp–orchid reciprocation machine described by Deleuze and Guattari, set up consistencies across scales and discrete objects or organisms, by means of which each probes the virtuality of the other, but also interacts more generally, as an assemblage, with wider formations and compositional dynamics. Thus an entity or a process might be imagined to occur in the liver of one being, be sensed as creepy sizzle by the automatic fight or flight responses of another, stimulate pheromone exchange between two members of different species, determine the use of grammatical tense in an essay by a specimen of another, but exist as much more than these. There is no teleology in such occurrences, but rather a drift of reciprocal relays established more or less directly by potentially thousands of interacting and diverging entities.

The question of the exploration of virtuality within an ecology is also carried out at an experiential scale in play. The kinds of play associated with different species are equally heterogeneous. The field of comparative psychology is developing understanding of multiple forms of consciousness: mirror recognition (a test of self-awareness); theory of mind; tool use; emotions and empathy; the capacity to imitate; the capacity to think about thought, metacognition; language; reflection recognition, and other capacities which in turn become affordances for entities, capacities and dynamics, which almost weekly produce experimental results widening recognition the domain of intelligence, and the distribution of skills and aptitudes once thought exclusive to *homo sapiens*. In his landmark survey of play in a multitude of species, Gordon Burghardt states that "[p]lay with objects is behaviour in which an animal investigates not just their nature ... but what he or she can do with them" (386). This would also suggest that play acts as a context in which animals not only probe potential affordances amongst their conspecifics and the things that surround them, but also count themselves amongst the things that, at multiple scales, are being so probed. Play behaviors can also be autotelic, independent of adaptiveness or function, or producing a reserve of "anticipatory adaption"; as such it is at once something that is absolutely live, but also a gateway into the virtual, the plethora of forces and possibilities that interact to produce the actual.

In Deleuze and Guattari's account of ecology as melody (*What is Philosophy?* 184–185), affordances become counterpoints, relays between one set of compositional dynamics, such as the bumblebee and the snapdragon, that trip, not simply in tight coevolutionary couples, but out, from *oikos*, home, the root word of ecology, to the cosmos. Extending this cosmological dimension, if we concur that "a work is always the creation of a new space time" (Deleuze 289), art for animals also allows us a way of thinking through the processes of intersubjectivation that we experience in ecology, a move that chimes with Guattari's critique of the "pure intentional transparency" (*Three Ecologies* 37) of phenomenology. Guattari calls instead for a means of recognition of components of subjectification which meet each other by means of transits that are relatively autonomous from one another (36). The cosmos figured here is one that moves towards openness. The works considered below as art for animals can be thought of as specific articulations of such a process of opening.

Paul Perry—*Predator Mark*

In his work on the literature of wilderness, Gary Snyder suggests that

> [o]ther orders of being have their own literatures. Narrative in the deer world is a track of scents that is passed on from deer to deer with an art of interpretation which is instinctive. A literature of blood-stains, a bit of piss, a whiff of estrus, a hit of rut, a scrape on a sapling and long gone (112).

In encounter with changes in the use of land, these literatures find themselves recomposed. Urban foxes in London, for instance, are notorious for their habit of shitting on children's toys left outside overnight in gardens and yards. Their territory-marking habits have been displaced and appear as cunning acts of deposition.

Paul Perry's 1995 installation *Predator Mark* is a subtle reordering of such a literature of scents. The work consists of a device made up of an electronic timer, a compressed gas spray mechanism and a flask of bobcat urine. This mechanism was installed high on a tree in a wooded estate, Landgoed Wolfslaar, in Breda in the east of The Netherlands. Bobcats are native to North America and Mexico. Their scents are thus not part of the vocabulary of ecology of the area.

Bobcat urine is, however, commercially available in North America, along with that of other local predators such as wolves. Its commodification, and provision for credit units over the internet, allows its dislocation from territory. Once bought by the user it is judiciously sprinkled to deter certain animals from crossing into the space that the scent suggests is inhabited as territory by another. Other scents, such as the urine of doe deer in heat, are used as lures by hunters, in this case to draw deer away from trails into the line of sight of hunters. The urine of both predator and prey animals, like other animals products available for retail, spells out a new kind of literature, one of commodification, of humans gaining the capacities of cunning shitters, and the grisly promise of meat on a stick.[8] Whether, like mosquito repellent, these products have anything more than fetish value for men investing in quality time alone with nature remains questionable.

In *Predator Mark*, introducing the scent of any animal, predator or not, is imagined to shift the register of references to presence within the place. It suggests an openness to the possible that resingularizes experience as an event in which the dimensions of relationality surging through it require recognition. This is a speculative literature of piss, involving floods, drips and sprays of matter, energy and signs, and the intelligences they invoke to sense and comprehend them.

Whilst one form of experiment is to set things out, to wait and see what gathers or grows in the manner of Duchamp's early artificial life work, *Elevage de Poussiere* (Breeding Ground of Dust).[9] Perry did not set out to observe whether there were any differences in behavior associated with the installation of this work, as would be characteristic of a scientific experiment proper, in which one variable only is isolated and probed for the conditions of its variation. Indeed, it is not even clear whether the species most drawn to the scent-marking activity of art was even aware of the work's existence. This gratuity of the work, that it addresses itself primarily to animals, those who read no press releases, and its operation in a way that is imperceptible, indeed, by its height from the ground and position deep within a wood, almost impossible to experience, distinguishes it from an entity operating within the normal dynamics of art systems. If, to make one comparison, conceptual art made the move towards experiencing the materiality and multiply structurating forces of ideas and language, such work suggests a means for such conceptuality in multiple species and across many means of sensing, acting in and interpreting the world.

Natalie Jeremijenko—*OOZ*

Natalie Jeremijenko is engaged in an ongoing series of works called *OOZ*,[10] which test human–animal cohabituation of city spaces and set up novel kinds of instruments and infrastructure for urban and feral animals. *OOZ*, as a series of works and ongoing revisions of projects, establishes situations for animal and human interaction in contexts in which, unlike that of a zoo, the animals are free to leave. The *OOZ* series has involved work adopting the housing paradigm, such as an installation on the roof of the Postmasters Gallery in New York in 2006.[11] Whilst this was largely to do with providing amenities such as houses, perches, a supply of fresh water and the growth of plants with medicinal function, there were also two other key directions to this work. One included anthropomorphic architectural organizations of space, such as a "shopping mall," and architectural work offering ironic recognition for the benefit of human viewers, such as components testing the mechanical understanding of what is normal for animal provision by applying architectural notions of "luxury" to fittings and spaces. There is an air of the flea circus about aspects of this project, dinky versions of high-end contemporary architectural concerns and urban systems. To achieve these, the project involved commissioning elements from a number of architectural studios, perhaps inevitably leading to a tendency towards calling-card architecture. Such elements might perhaps work as lures, sparkly things that attract attention and draw humans towards them. Perhaps anthropocentrism can work as an interpretative layer for one species, whose cognition is partly organized by glamour, without ruining the primary emphasis on addressing the perceptual and experiential capacities of another. More importantly, the project tests the notion of what the feral condition implies; might there be an outgrowth of provision from urban systems in order to provide more edges, and habitats for displaced and incoming nonhuman inhabitants of cities? Such provision might entail the imagination of multiscalar "green corridors," micro-to-macro-scale affordances built on into and through cities for ameliorating, or even improving on, the kinds of ecological condition they erase, build into or establish.

A common thread between the different components of the OOZ series is that of experimental forms of communication. The Postmasters installation, titled *OOZ (for the birds)* included a "concert hall" space for pigeon calls. Whilst this functioned as something of an architectural in-joke, being a miniaturely scaled version of *Casa de Musica*, the Office for Metropolitan Architecture's 2005 concert hall in Porto, it allowed for the amplification of voices and calls. In other work, *Comm. Technology,* (2006) Jeremijenko has set up novel devices for pigeons to amplify their vocalizations (see *OOZ* website). A series of perches to be attached to buildings consists of a hollow plastic horn fitted with a small microphone and speaker. The noises made by the pigeon whilst using the perch are powered up to address the street. Jeremijenko's wager is that the pigeons will recognize this, and note the changes in reaction of humans using the street, including possible food sharing, and begin to favor the use of the perch. Unlike Perry's *Predator Mark*, therefore, there is a sense in which the use of the work is monitored and evaluated, even if only informally. This is in part because Jeremijenko's work sites itself very much in dialogue with design, and the critical design discourse also involving Anthony Dunne, Beatriz da Costa,[12] Phoebe Sengers[13] and others. Here, design without a direct client or a customer and with animals as its users enters a modality that is enormously suggestive.

An early component of the OOZ project was *Robotic Geese* (2005 onwards), one unit of which, in an installation with the Bureau of Inverse Technology, *Romancing the Geese*, was placed in a small stretch of water next to the De Verbeelding art centre in Flevoland. The goose, a basic plastic decoy body with added features including motorized legs, an articulated neck, a head-mounted camera, microphone and speaker, was remotely controlled from a seat which allowed a visitor to view the eye view of the robot, to steer it and to "make utterances" through it.[14] The idea is to stage interactions with a small population of Greylag and feral domestic geese that inhabit the area. In the projected full iteration of the work, each speech interaction will trigger the recording of short bursts of audio-visual information to a database. Once it becomes public, items on the database can be correlated so that users can gradually, through standard collaborative filtering algorithms, aggregate opinions on the semantic content of the utterances of the non-robot geese.

Communication amongst humans is increasingly configured as a means of the delivery of order words and the management of the distribution of micro-compulsions to respond, advise, participate, collaborate and to organize attention. Against this figure of the regime of responsiveness, to think about communication outside the boundary of a species sets up a number of possibilities. Perhaps *OOZ* allows us to imagine a form of taxonomy in which speciation was marked not by the matter of which animal could engage in effective genetic transfer with another, but on the basis of those which engage in semiotic (memetic) relays.

Marcus Coates—Out of Season, Sparrowhawk Bait, and Dawn Chorus

Marcus Coates has embarked upon a body of work which maps out a certain set of figurations of interactions with animals, with birds in particular. Only a few pieces of his work fall into the art-for-animals current and are early, perhaps more minor, more throwaway or institutionally indetermined than the larger-scale projects he has more recently embarked upon. They may indeed be pointing towards something that he, with his continued interest in "animal becoming," will return to. Before we address these, some of the other works are also worth mentioning. In a second work entitled *Dawn Chorus* (2007)[15] high-quality field recordings of bird songs are slowed down 16 times until they reach a pitch easily matched by a human throat. The resulting sounds are played to volunteers who learn to repeat them. These enactments are videoed, and then played back as a projection. It seems that, at least in terms of their reenaction, only the relative size of the vocal apparatus distinguishes the calls of the birds and humans.

In *Journey to the Lower World* (2003),[16] Coates uses a persona suggested by brief training in the rituals of Siberian Shamans. He performs a ritual for residents of a soon-to-be-demolished tower block in Liverpool, wearing the skin of a deer, mimicking the work of a shaman, apparently communing with a number of bird spirits and in so doing bringing back a vision of hope for the bemused ladies and gentlemen attending his ritual. The latter work is interesting because it knows that it is weak but makes use of this. The action is awkward, based on a relatively shabby, slightly embarrassing, day of training with the kind of guru who acquires their flock through post-cards in health shop windows, and carried out by a denizen of the upper world. Nevertheless this specimen of the contemporary European, gawkily decked out in the culled, shameful trappings of authenticity, as compromised as it knows it is, attempts to get something going. There is an earnestness achieved through a reflexive mimicry, of ritual, and of animal calls, especially Coates's constant attention to those of birds, that carries through into his work fitting more precisely into the art-for-animals current. Mimicry is a means to set up ruses, initiatives that skirt the edge of multi-directional fraud in which the everyday and ideas of the wild, the primitive and capacities of sensual perception that overlap between species can be mobilized. Here mimicry unfolds both as play and as learning; in bird calls with their worlds of call and refrain, or their remobilization of surrounding sounds; and in contemporary art and its constant reversioning of appropriation, pastiche, copy, plagiarism, found materials, how to deal with and configure what exists, what repeats, in relation to the creation of the new. These are vectors in the generation of what Coates calls "animal becoming" but, partially overlapping, they also shift each other.

During a series of short live works in the Grizedale Forest, Coates set up three interactions with local bird populations. They share some of the "do it and see (or imagine) what happens" approach of Perry's *Predator Mark*. The experiment is done for its experiential value rather than the extraction of unequivocal data. In *Sparrowhawk Bait* (1999), Coates makes himself the target for a predator. The corpses of a Blackbird, a Blue Tit, a Mistle Thrush, a Grey Wagtail and a Green Finch are tied to his hair. He runs through the forest, with the anticipation that a local Sparrowhawk will be attracted by and pounce on the momentarily reanimated bodies. In *Dawn Chorus* (2001) a shaven-headed male actor enters an area of young pines and shouts football chants, fan-versus-fan abuse in good spittle-flinging style. Taking place in a deciduous wood, *Out of Season* (2000), another short video, documents the same kind of performance, with another actor and the addition of a Chelsea shirt. Aside from its relay and remediation as a video, the primary audience is the birds whose territorial and mating calls normally fill the spaces. In the work concerned with mimicry and imitation, whether of the shaman or of birds, making these chants and calls, listening out for any response, Coates has to link himself as an apprentice to the song domain of the birds, the processes of learning and training of listening and responding, which they establish. Taking the football chants to the forest does not only set out an idea of how human communications may often be so similar in their territoriality to those of birds. It shows too how demented and dreamy the possibility of talking to the animals really is, but also makes us wonder whether it could ever really be anything more than an unreturnable "fuck you."

Louis Bec—*Stimutalogues,* and Anthony Hall—*Enki*

Louis Bec describes himself as a Zoosystémicien, a sole participant of this discipline working with an extended conception of artificial life, an abstraction of life in more general terms, and some developed ideas as to how to proliferate interrelations between technologies of information and different biological manifestations of signification and intelligence. His work tends towards a science fiction in practice and Bec is an adept at the time-accredited techniques of neologism, fabulation, mind-boggling and acronym usage. His manifesto text "Squids, elements of technozoosemiotics" strives for a moment in which hyperbole and a series of programmatic and poetic statements achieve a density of semantic condensation sufficient to bring a world to life.

Aside from a number of projects developing interactive animated versions of artificial life projects, Bec has worked with various species of fish which use electrical pulses released by special electric organs located in certain parts (varying across species, generally transmission towards the tail, reception in foveal regions at the head) of their bodies. According to a document describing the research program, this series, the Stimutalogues project, includes:

Logognathe Artefact (interactive customizable loop of communication between the living, artifact and interactive agent)
Logomorphogenesis (modeling by dynamic morphogenesis of information exchanges between three *Gnathonemus Petersii*)
Ichthyophonie|PanGea (setting up a communication device allowing exchanges between Mormyridées in Brazil and Gymnarchidées in Africa, trying to connect two continents which are gradually being separated by the tectonic plates) (Bec "Arapuca").

These fish are nocturnal; as well as having good hearing, they use their electric organs over short ranges to signal mating readiness or aggression, to locate food and to navigate in the dark water. Research by the sensory ecologist Gerhard von der Emde suggests that their complex sensory system is capable of using the way in which an object resists or stores mild electrical currents to determine its shape, and are able to categorize what they find. The movement of the fish, and the tail bending required for ordinary motion, allow the process of electric organ discharge to effectively "triangulate" objects.

Anthony Hall, is leader of a related project called *Enki* (2006), which also uses a number of species of weakly electric fish, including Black Ghost Knife fish (a species which breeds quite comfortably in captivity). The technique is to place them in a tank containing sensors which pick up the electrical signaling of the fish. The signals are then converted into waves which are played at a seated user by means of sound and flickering LEDs. A lead travels from the arm of the user carrying electrical pulses from the human body to an electrode in the water in which the fish swims.[17]

As with the *Logognathe Artefact* and *Logomorphogenesis* proposals, the fish are placed in conditions in which, compared to their native habitat, they are sensorially and behaviorally deprived. Elephantnose fish (*Gnathonemus Petersii*) do not breed in captivity, and will therefore, in every case of their use as a component in such projects, have been captured from the wild, from areas, Nigeria and Brazil, already subject to significant pillaging for materials. In terms of the development of species-specific art, the question of how markets in animals and animal products intersect with the organization of art, and with the global distribution of habitats and organisms, is essential to recognize. By comparison with the emphasis on the capacity for animals to come and go in *OOZ* projects, most of the work done with elephantnose fish has substantial problems in terms of its ethical composition. The one clear exception to this is a version of the *Ichthyophonie/PanGea* project which will be discussed last.[18]

In versions of the *Enki* project which also involve a human subject, it is not clear whether, from the perspective of the fish due to their modeling in the system that receives them, and their mediation by layers of devices, it might not be simpler to replace them, or indeed the human user, with an entity in software equally capable of providing aleatory stimulus to the mechanism. The latter is the approach of Bec's *Logognathe Artefact*.

Underneath the generalizations about possible therapeutic implications and pastel fractals of one early iteration of the *Enki* project website it becomes clear that certain aspects of the project are potentially quite welcomely dark. Gregory Bateson, in work discussed by Guattari in *The Three Ecologies*, suggests that decisions and learning may be made by systems "immanent in the large biological system—the ecosystem" (466) or "at the scale of total evolutionary structure" (466) that are analogous to or developing qualities characteristic of mind. Such minds, systems of learning, occur between interacting elements; they are not isolatable to one single entity bounded by a membrane, but arise from cybernetically describable relays of entities bound at such a scale. One spin on the *Enki* project is that what we might be seeing here is the production of a mind or mentality, a mind that is at once fish and human but not reducible to either. That the fish part at least (when *petersii* are used), in its refusal to breed, is displaying classic signs of confinement stress suggests significant questions about the ethico-esthetic dimensions of art for animals involving captive life. Extreme doubt must be applied to any project that involves confinement, and especially confinement with such negative consequences. Here the question of the conjunctive form ethico-esthetics proposed by Guattari is useful to draw upon. The *Three Ecologies* emphasizes processes of subjectification that are artistic in style and inspiration, in imaginal power, rather than being quasi-scientific. Ethics does not consist of the completion of a series of tick boxes of an approvals committee. More fundamentally, to make of the fish an instrument, even one whose cognitive and communicational processes "complete" the work, is to curse it. Art for animals proposes instead that animals have a necessarily ontological world-making dimension. As such an ethico-aesthetic approach disrupts the normal great chain of thought that starts with ontology, proceeds through epistemology and ends with the mere implementation details of ethics and aesthetics. It suggests that each moment of each scalar state is riven through with such figurations and modes, without any gaining an *a priori* superiority or precedence to the others. Electronic art is trivial and boring when it simply confirms the interrelation between sensors and responses. Art using animals is trivial and abusive when it locks animals into devices that deplete their involvement in and creation of the world rather than supplementing it.

This given, the last listed of Louis Bec's projects in this series is particularly interesting to attend to. *Ichthyophonie\PanGea* is an attempt to develop a communication network between two families of fish using electric signaling, location finding and, more fully, echoperception. These two families, the Mormyrids in located in South America and Gymnarchids in West and Central Africa, originally sharing an early common ancestor, were split apart into different phylogenetic branches by the movement of continental plates as they broke from the early supercontinent, PanGea (or Panagea). As yet unrealized, the plan involves setting a network of sensors\actuators in the habitats of these fish which are to be connected to each other via internet. This would allow the communicatory behaviors of these fish, at least those transferable by such means, to enter into some kind of sense of colocation with the possibility for sensorial interplay; perhaps evoking and probing remnants of shared signaling; or perhaps simply adding a small sizzle of now meaningless noise to a particular patch of water. Perhaps too, it is something else, a paradox: something that tickles the fishes' curiosity, changes the economy of their attention, dislocating their access to the virtual.

In this respect, *Enki* also establishes some interesting possibilities for further development. Electroperception in electric fish has some very special qualities. Electric waves move in curved rather than straight lines, and the reflections produced typically become larger the further they are from the object—so this is something rather different from the capacity for orientation via sonic echolocation or by vision. These fish can also produce concepts of the objects in the sense of abstract categories that are transferable across entities they may encounter. In other iterations of the project, Anthony Hall set up a context in which no human was attached. The fish's signal was picked up by one or more electrodes, typically placed in the corner of their familiar tank. This signal was then fed back to the fish in a different corner of the tank. Because the fish perceive the world in waves, the effect of this can be imagined as being something similar to pushing a limb towards a mirror only to have it "reflect" via a wall behind you, an experience Hall recounts as provoking much curiosity in the fish. When two weakly electric fish of either of these families meet they go through a process of modulating the individual frequency of the current they give off in order that each can maintain their own signal or refrain. Interestingly, the signals produced by the fish in this context do not carry this "handshake," suggesting that they recognize themselves in this substantially distorted context, one which they spend time in exploring.

"Je weet nooit hoe een koe een haas vangt"[19]

One way in which art for animals might progress is along the lines suggested by biosemiotics or zoomusicology.[20] Biosemiotics is concerned with the transmission of information as part of living processes, expanding the domain of signaling from that of DNA, to molecules, the interoperation of body parts and systems to the function of organisms and out into other scales of ecologies. Coupled with this, it is a field which develops an idea of a more generalized domain of semiosis, such as communication, subterfuge, courtship and ludic enjoyment configured at the level of the organism or, as with Bateson's ecology of mind, in interactions between organisms. Of importance here too is a notion of aesthetics, of the configuration of beauty. This is something that has been present in a certain way in biology from Darwin's work on sexual selection, and threads through to sociobiological accounts of beauty configured as attractiveness. Amongst other creatures, Deleuze and Guattari draw upon the stagemaker bird, whose pergola is an example both of an extended phenotype and an exuberant courtship display. It is usually taken to be a highly nuanced example of aesthetic judgment involving dimensions that are spatial, coloristic, to do with the freshness of materials and their intercomposition. For them, this constant act of the compilation, sorting and arrangement of materials epitomizes an enactment of territory as rhythm within the melody of ecology.

In many accounts of a possible animal aesthetics there is a dance performed around the threshold of functionality or expressivity configured as being demarcated as that which is gratuitous. This dance may pass through various subthresholds according to whether expressivity corresponds to a given stack of drives and needs, to evoke curiosity, to learn, to mate, to eat, to dominate, to play. Where this dance gets stuck is to read these as purely obligatory functions or, in a bipolar switch, as being utterly "free"—without interrelation with other compositional forces or constraints. This is part of the terms of their composition, but the dance around their thresholds might also usefully recognize the dance within each of these scales themselves. For instance, in a dance within the scale of play as play comes the dance of the mimicry of mimicry, one which opens out onto all other scales. Such a dance between gratuitousness and functionality needs to be recognized within the context of the general economy, Bataille's substantial contribution to the intellectual work of ecology in which all, drives included, are ultimately gratuitous. As such it is a liberation and a curse which can only be remedied, or modulated, by being entered into with adequately vivid forms of life. Any point in this stack, or others not named or yet to be invented, may tip this dance into a new rhythm. Each element of this stack, whether operating as drive, function or play, may become more dislocated or increase its capacity of dislocation for a moment yet to come. Equally, in this dance between scalar function and cosmological gratuitousness, elements may exist across many assemblages functioning in different terms in each, as anchors, blocks, voids or torrents. It is taking part in this movement, doubling it by means of reflexivity, in this case not simply the reflexivity of a single mind or within the scalar boundary of a compositional entity, but its multiplication by an ecology of sensoria, that art for animals emerges.

Whether it is paint, wood, chrome, text, scent, move, sound, leaf, art works with and through materials that are direct to hand, to thought or to experience, but which also anticipate their coming into composition, their recomposition, with, or by means of, other elements; art may require work from primary natural forces in order to become complete. Think of Edward Munch's habit of leaving his oil-painted canvases out in the rain for weeks in order that they might be worked upon by it. It may be suspected that something of the same happens in the philosophy of Deleuze and Guattari, something which both brings it closer in practice to art and allows it to produce itself as a receptive domain in which ecologies of texts, histories and ideas occur, spawn and leave their traces. This is philosophy which leaves itself out in too many weathers. In doing so, they form new relays with ecologies.

Before they too become mulch, those who advocate purity of the discipline now have their turn to rain upon this work, so go the almost inevitable recalls to reason. But this is philosophy. With 2000 years' worth of beard to avoid tripping over it is almost compelled to immobility. This, disciplinary automatism masked up as a holy stillness, allows it to position itself as a metadiscourse towards which all other fields, not simply philosophers, must measure their orbit and meet their judges. Art is in a certain way equally ambitious; it will admit no limits, but only in so far as it provides a means by which, in a deeply amateur way, by means of the art methodology of unreadiness, it comes into composition with other techniques of working. Whilst other discursive frameworks cannot by these means become mastered, they can always be used. Whether this capacity really does extend to the sensual, semiotic and world-making capacities of animals is something too that needs to be left outside, to see what happens.

Notes

1. Notable examples would be Jannis Kounellis's installation, *Horses*, Rome, 1969, in which a dozen horses were stabled in the Galleria L'Attico, setting up a situation in which the physical presence, movement, smell and palpability of the horses went straight to matter conjugated by the multiple kinds of expectation and viewing accentuated in art systems. Paolo Pivi's work follows somewhat in this trajectory but with an emphasis on exoticism and absurdist conjuncture, an alligator covered

in whipped cream, zebras transported to a snowy landscape, a leopard prowling amongst plastic replica cappuccino cups.

2. See http://arts.guardian.co.uk/news/story/0,1718642,00.html. Last accessed January 14, 2008.

3. The development of such architectural work in the London Zoo was at the initiative of Julian Huxley, then secretary of the Zoological Society. Lubetkin also worked later at Dudley Zoo, which, almost in reverse of *OOZ (for the birds)* provided a miniature example of modern town planning. For an analysis of the development of the architecture of London Zoo, see (Steiner). The Penguin Pool was eventually abandoned after about 70 years of occupation, with the penguins being moved to a more "organic" site with various kinds of surface and housings. It remains standing as a grade one listed building, but, as of this writing, remains unused.

4. The Sheep Spaces sculptures were made in 1993 as part of the TICKON Project, Langeland, Denmark. The same exhibition also included an oversize thatch beehive by Jan Norman.

5. Haacke's intervention consisted of buying ten turtles and releasing them into the wild. The methods of the Animal Liberation Front have by and large improved on such approaches.

6. See, for instance, Jospeh Beuys, *I Like America and America Likes Me* (1974), a durational performance in which a room was shared with a coyote. Bonnie Sherk's *Public Lunch* (1971) was held at the Lion House in San Francisco Zoo, during which the artist would introduce herself to the Lions' enclosure during feeding times.

7. See "Why Chickadee Calls Spook Other Birds." *New Scientist* (March 24, 2007), and also http://students.washington.edu/ctemple2/chickadee.html, last accessed January 14, 2008, and (Templeton).

8. See http://www.hotdoe.com/, last accessed January 14, 2008.

9. This work is a photograph by Man Ray of the reverse side of Duchamp's *The Bride Stripped Bare by Her Bachelors, Even* (1915–1923). It shows the glass in 1920 after having accumulated a landscape of dust.

10. See http://www.nyu.edu/projects/xdesign/ooz/, last accessed January 14, 2008.

11. OOZ, Inc. (for the birds) Infrastructure and facilities for high-density bird cohabitation on the roof of Postmasters Gallery, installed at the Postmaster's Gallery, September 7–October 7, 2006.

12. See the *PigeonBlog* project in which tame pigeons are fitted with environmental pollution data-gathering equipment, http://www.pigeonblog.mapyourcity.net/, last accessed January 14, 2008.

13. Compare the Culturally Embedded Computing research group at Cornell University, http://cemcom.infosci.cornell.edu/, last accessed January 14, 2008.

14. Whilst it might be imagined that the robot is clunky relative to a goose, a number of parallel experiments in animal behavior, including birds, suggest that devices of this sort can be extremely useful in establishing communication. For a survey of such work, see (Young).

15. *Dawn Chorus* was first shown at the Baltic in Gateshead in February 2007. It takes part in a thread of work in contemporary art involving animal imitation such as Lucy Gunning's video of people imitating horses, *The Horse Impressionists*, 1994.

16. Documented in (Coates 2005).

17. One aspect of the project which is not covered here is that Hall works informally with an acupuncturist to apply galvanic skin response sensors to places on the human body with the suggestion that the fish might respond to different currents

from the human subject. Additionally, the kind of electrode used is important; carbon electrodes give a soft profile, metal ones a very hard edge, quite distinct from anything they might encounter in the wild.

18. In a sense, this distinction recapitulates the difference between lab-based cognitive psychology work with animals and ethology's insistence on observation of animals in their habitats.

19. Trans.: "You'll never know how a cow catches a hare" (Dutch proverb).

20. A summary of possible divulgations of aesthetics by means of this approach is given in (Martinelli).

Works cited

Barker, Steve. *The Postmodern Animal*. London: Reaktion Books, 2000.

Bateson, Gregory. *Steps to an Ecology of Mind*. Chicago: University of Chicago Press, 2000.

Bataille, George. *The Accursed Share. Vol. 1*. Trans Robert Hurley. New York: Zone Books, 1991.

Bec, Louis. "Squids, elements of technozoosemiotics, a lesson in fabulatory epistemology of the scientific institute for paranatural research" in *Technomorphica*. Eds Joke Brouwer, Carla Hoekendijk. Rotterdam: V2_organisatie, Rotterdam, 1997, 279–311.

——. Explications sur le fonctionnement matériel et logiciel des stimutalogues. Exemple avec le gnathonemus petersii. Not dated.

——. "Arapuca." *Alterne: Creation and Technology Proposals*, EU IST proposal no. 39575, July 2003. See Alterne, 'ARAPUCA' in *Alternate Realities in Networked Environments*, http://www.alterne.info/node/42. Last accessed January 14, 2008.

Braidotti, Rosi. *Transpositions*. Cambridge: Polity, 2006.

Bruner, Jerome, Alison Jolly and Kathy Sylva. *Play, its Role in Development and Education*. London: Penguin, 1976.

Burghardt, Gordon M. *The Genesis of Animal Play. Testing the Limits*. Cambridge: MIT Press, 2005.

Coates, Marcus. *Journey to the Lower World*. Ed. Alec Finlay. London: Platform projects, Morning Star, Film London, 2005.

——. *Marcus Coates*. Ambleside: Grizedale Books, 2001.

Coe, Peter, and Malcolm Reading. *Lubetkin and Tecton, Architecture and Social Commitment*. London: The Arts Council of Great Britain, 1981.

Deleuze, Gilles. "The Brain is the Screen" in *Two Regimes of Madness. Texts and Interviews 1975–1995*. Ed. David Lapoujade. Trans. Ames Hodges and Mark Taormina. New York: Semiotext(e), 2006, 282–291.

Deleuze, Gilles, and Félix Guattari. *What is Philosophy?* Trans. Graham Burchell and Hugh Tomlinson. London: Verso, 1994.

Deller, Jeremy. *The Bat House Project*. 2006 onwards. http://www.bathouseproject.org/. Last accessed January 14, 2008.

De Waal, Frans. *The Ape and the Sushi Master. Cultural Reflections of a Primatologist*. New York: Basic Books, 2001.

Dissenyake, Ellen. *What is Art For?* Seattle: University of Washington Press, 1988.

Dunne, Anthony. *Hertzian Tales. Electronic Products, Aesthetic Eexperience and Critical Design*. London: Royal College of Art Computer Related Design Research, 1999.

Ellenbroek, Frans. *The Biological Evolution of the Arts*. Tilberg: Natuurmuseum Brabant, 2006.

Fuller, Peter. *The Naked Artist. Art and Biology*. London: Readers and Writers, 1983.

Fuller, Matthew. "Towards an Ecology of Media Ecology." *x med a, experimental media arts*. Brussels: Okno/Foam/Nadine, 2006.

——. *Media Ecologies. Materialist Energies in Art and Technoculture*. Cambridge: MIT Press, 2005.

Geddes, Linda. "Unleash your inner bloodhound—start sniffing." *New Scientist* (December 17, 2006), http://www.newscientist.com/channel/being-human/dn10810-unleash-your-inner-bloodhound--start-sniffing.html. Last accessed January 14, 2008.

Geissmann, Thomas. "Gibbon songs and human music from an evolutionary perspective" in *The Origins of Music*. Eds N. Wallin, , B. Merker and S. Brown. Cambridge: MIT Press, 2000, 103–123.

Grandin, Temple, and Catherine Johnson. *Animals in Translation. The woman Who Thinks Like a Cow*. London: Bloomsbury, 2005.

Guattari, Félix. *The Three Ecologies*. Trans. Ian Pindar and Paul Sutton. London: Athlone, 2000.

——. *Cartographies Schizoanalytique*. Paris: Editions Galilée, 1989.

Hall, Anthony. *Human to Fish Interface Project—Image Gallery*. September 2006, http://www.variableg.org.uk/enki_project.htm. Last accessed January 14, 2008.

Haraway, Donna. *Companion Species Manifesto. Dogs, People, and Significant Otherness*. Chicago: Prickly Pear Press, 2003.

Haacke, Hans. *Ten Turtles Set Free*. St. Paul-de-Vence, France, July 20, 1970.

Herbert, Martin. "Tales of the Unexpected." *Frieze* 106 (April 2007): 106–113.

Huxley, Julian. *The Courtship Habits of the Great Crested Grebe*. (1st ed. 1914.) London: Jonathan Cape, 1968.

Lippincott, Louise and Andreas Blühm. *Fierce Friends, Artists and Animals 1750–1900*. London: Merrell, 2005.

Martinelli, Dario. *Liars, Players, Artists. A Zöösemiotic Approach To Aesthetics*. Online at http://www.zoosemiotics.helsinki.fi/, last accessed January 14, 2008.

Matilsky, Barbara C. *Fragile Ecologies, Contemporary Artists' Interpretations and Solutions*. New York: The Queens Museum of Art, 1992.

Morris, Robert. "A Method for Sorting Cows." in *Information*. Ed. Kynaston McShine. New York: Museum of Modern Art, 1970.

Nagel, Thomas. "What Is It Like to Be a Bat?" *The Philosophical Review* 83:4 (October 1974): 435–450.

Pliny. *Natural History. Books 33–35*. Trans. H. Rackham. Cambridge: Loeb Classical Library, Harvard University Press, 2003.

Rozin, Paul. "About 17 (+/−2) Potential Principles about Links between the Innate Mind and Culture. Preadaptations, Predispositions, Preferences, Pathways and Domains" in *The Innate Mind. Vol.2: Culture and Cognition*. Eds Peter Carruthers, Stephen Laurence, Stephen Stich. Oxford: Oxford University Press, 2006, 39–60.

Sutton, Gertrud Købke. "David Nash: The Language of Wood." *Art and Design* 36 (1994): 28–73.

Steiner, Hadas A. "For the Birds." *Grey Room* 13 (2003): 6–31.

Templeton, C. N. and E. Greene. "Bilingual birds and eavesdropping: Nuthatches respond to subtle variations in 'chick-a-dee' alarm calls." *Proceedings of the National Academy of Sciences* DOI: 10.1073/pnas.0605183104.

Uexküll, Jacob von. "A Stroll Through the Worlds of Animals and Men." Trans. Claire H. Schiller. *German Essays on Science in the 20th Century*. Ed. Wolfgang Schirmacher. New York: Continuum, 1996, 171–178.

von der Emde, Gerhard. "Non-visual environmental imaging and object detection through active electrolocation in weakly electric fish." *Journal of Comparative Physiology* A 192 (2006): 601–612.

Watanabe, Shigeru, with Junko Sakamoto and Masumi Wakita. "Pigeons' Discrimination of Paintings by Monet and Picasso." *Journal of the Experimental Analysis of Behaviour* 63 (1995): 65–74.

Wolfe, Cary. *Animal Rites. American Culture, the Discourse of Species and Posthumanist Theory.* Chicago: University of Chicago Press, 2003.

Young, Emma. "Undercover Robots Lift Lid on Animal Body Language." *New Scientist* (January 6, 2007): 22–23.

Zimmer, Carl. At the Water's Edge. Macroevolution and the Transformation of Life. New York: Free Press, 1998.

Name Index

Subject Index

Printed and bound by CPI Group (UK) Ltd, Croydon, CR0 4YY